UNDERSTANDING LASERS

IEEE PRESS Understanding Science & Technology Series

The IEEE PRESS Understanding Series treats important topics in science and technology in a simple and easy-to-understand manner. Designed expressly for the non-specialist engineer, scientist, or technician as well as the technologically curious—each volume stresses practical information over mathematical theorems and complicated derivations.

Other books in the series include:

Understanding the Nervous System
An Engineering Perspective
by Sid Deutsch, Visiting Professor, University of South Florida, Tampa
Alice Deutsch, President, Bioscreen, Inc., New York

1993 Softcover 408 pp ISBN 0-87942-296-3

Understanding Telecommunications and Lightwave Systems
An Entry-Level Guide
by John G. Nellist, Consultants, Sarita Enterprises Ltd.

1992 Softcover 200 pp ISBN 0-7803-0418-7

Tele-Visionaries: ***The People Behind the Invention of Television***
by Richard C. Webb

2005 184 pp ISBN 978-0471-71156

UNDERSTANDING LASERS

An Entry-Level Guide

Third Edition

JEFF HECHT

IEEE Press Understanding Science & Technology Series

IEEE Lasers and Electro-Optics Society, *Sponsor*

IEEE Press

A JOHN WILEY & SONS, INC., PUBLICATION

IEEE Press
445 Hoes Lane
Piscataway, NJ 08855

Published by John Wiley & Sons, Inc., Hoboken, New Jersey
Published simultaneously in Canada.

Library of Congress Cataloging-in-Publication Data is available.

ISBN 978-0470-08890-6

10 9 8 7 6

To the memory of family, friends, and members of the laser community who have passed away since the last edition: My parents, George T. and Laura Hecht; Howard Rausch, who taught me how to write about lasers; laser pioneers Theodore Maiman, Gordon Gould, and Arthur Schawlow; and my friend and editor Heather Messenger.

CONTENTS

PREFACE

"For Credible Lasers, See Inside."

THE LASER IS LESS THAN three years younger than the space age. Just days after the Soviet Union launched Sputnik I on October 4, 1957, Charles Townes and Gordon Gould had two crucial discussions at Columbia University about the idea that would become the laser. As the United States and Soviets launched the space race, Townes and Gould went their separate ways and started their own race to make the laser. On May 16, 1960, Theodore Maiman crossed the laser finish line, demonstrating the world's first laser at Hughes Research Laboratories in California.

Bright, coherent, and tightly focused, laser beams were a new kind of light that excited the imagination. Science fiction writers turned their fictional ray guns into lasers with a stroke of the pen. Science writers inhaled deeply of the technological optimism of the early 1960s and wrote breathless predictions about the future of "the incredible laser." An article in the November 11, 1962 issue of the Sunday newspaper supplement *This Week* revealed U.S. Army schemes for equipping soldiers with a "death-ray gun . . . small enough to be carried or worn as a side-arm." It quoted Air Force Chief of Staff Curtis E. LeMay predicting that ground-based lasers could zap incoming missiles at the speed of light.

The reality was something else. A bemused Arthur Schawlow, who had worked with Townes on the laser, posted a copy of "The Incredible Laser" on his door at Stanford University, along with a note that read, "For credible lasers, see inside." Irnee D'Haenens, who had helped Maiman make the first laser, called the laser "a solution looking for a problem," a joke that summed up the real situation. The infant laser had tremendous potential, but it had to grow up first.

D'Haenens's joke lasted many years. So did the popular misconception that lasers were science-fictional weapons. If you told your neighbors you worked with lasers in the 1970s, they inevitably thought you were building death rays. That began to change as supermarkets installed laser scanners to automate checkout in the early 1980s. Then lasers began playing music on Compact Discs. Laser printers, laser pointers, CD-ROMs, and DVD players followed. Laser surgery became common, particularly to treat eye disease. Surveyors, farmers, and construction workers used lasers to draw straight lines in their work. Lasers marked serial numbers on products, drilled holes in baby-bottle nipples, and did a thousand obscure tasks in industry. Lasers transmitted billions of bits per second through optical fibers, becoming the backbone of the global telecommunications network and the Internet.

The incredible laser has become credible, a global business with annual sales in the billions of dollars. Lasers have spread throughout science, medicine and industry. Lasers are essential components in home electronics, buried inside today's CD and DVD players, and vital to tomorrow's high-definition disk systems. It's a rare household that doesn't own at least one laser. Yet lasers have not become merely routine; they still play vital roles in Nobel-grade scientific research.

This book will tell you about these real-world lasers. To borrow Schawlow's line, "For credible lasers, see inside." It will tell you how lasers work, what they do, and how they are used. It is arranged somewhat like a textbook, but you can read it on your own to learn about the field Each chapter starts by saying what it will cover, ends by reviewing key points, and is followed by a short multiple-choice quiz.

We start with a broad overview of lasers. The second chapter reviews key concepts of physics and optics that are essential to understand lasers. You should review this even if you have a background in physics, especially to check basic optical concepts and terms. The third and fourth chapters describe what makes a laser work and how lasers operate. The fifth chapter describes the optical accessories used with lasers. Try to master each of these chapters before going on to the next.

The sixth through tenth chapters describe various types of lasers. Chapter 6 gives an overview of laser types and configurations, and explains such critical concepts as the difference between laser oscillation and amplification, the importance of laser gain, and tunable lasers. Chapter 7 describes the workings of gas lasers and important types such as the helium–neon and carbon-dioxide lasers. Chapter 8 covers solid-state and fiber lasers, including neodymium lasers and fiber lasers and amplifiers. Chapter 9 covers the hot area of semiconductor diode lasers, including the important new blue diode lasers. Chapter 10 describes other types of lasers, including tunable dye lasers, extreme ultraviolet sources, free-electron lasers, and efforts to develop silicon lasers.

The final three chapters cover laser applications, divided into three groups. Chapter 11 describes low-power applications, including communications, measurement, and optical data storage. Chapter 12 covers high-power applications, including surgery, industrial materials processing, and laser weapons. Chapter 13 focuses on research and emerging developments in areas including spectroscopy, slow light, laser cooling, and extremely precise measurements. The appendices, glossary, and index are included to help make this book a useful reference.

To keep this book to a reasonable length, we concentrate on lasers and their workings. We cover optics and laser applications only in brief, but after reading this book you may want to study them in more detail.

I met my first laser in college and have been writing about laser technology since 1974. I have found it fascinating, and I hope you will, too.

JEFF HECHT

Auburndale, Massachusetts
January 2008

CHAPTER 1

INTRODUCTION AND OVERVIEW

ABOUT THIS CHAPTER

This chapter will introduce you to lasers. It will give you a basic idea of their use, their operation, and their important properties. This basic understanding will serve as a foundation for the more detailed descriptions of lasers and their operation in later chapters.

1.1 THE IDEA OF THE LASER

Optics was a sleepy backwater of physics when Theodore Maiman demonstrated the first laser in 1960. His announcement made headlines, and for many years afterward, lasers were novelties that attracted attention. Today, lasers are commonplace in developed countries. Thanks in large part to the laser, optics has become a dynamic field, expanding far beyond the binoculars, cameras, and spectacles that were the main products of the optical industry half a century ago.

We take lasers almost for granted today, as just another wonder of our technological age along with satellites and electronic chips. Most of us think of lasers as cylindrical devices that emit pencil-thin beams of red or green light, and shine bright spots on the wall. The first kind of laser to come to your mind is likely to be the pen-like laser pointers you can buy for $10 or less at an electronics or stationary store.

But lasers come in many other sizes, shapes, and forms. Most of them are tiny semiconductor chips that we never see because

Understanding Lasers: An Entry-Level Guide, Third Edition. By Jeff Hecht

they are hidden inside electronic equipment such as CD players, CD-ROM drives, and DVD, or Blu-Ray players. Others are tubes filled with gas that emit laser light. Some are boxes the size of a filing cabinet or a refrigerator that emit powerful beams to cut or drill holes in metal or plastic. The largest lasers fill the interior of a building and generate pulses of light that for a fleeting billionth or trillionth of a second can deliver more power than the whole U.S. electric power grid. Laser output may not be visible; many lasers emit at infrared or ultraviolet wavelengths invisible to the human eye.

What makes them all lasers is that they generate light in the same way, by a process called "light amplification by the stimulated emission of radiation." The word "LASER" is an acronym for that phrase. It is the process of amplifying stimulated emission that makes laser light special. The sun, light bulbs, flames, and other light sources emit light in a different way, spontaneously. That leads to important differences between laser light and other kinds of light, which we will explain later.

Most of us also are familiar with fictional weapons that resemble lasers and sometimes are called lasers. The deadly heat rays used by the Martian invaders of Earth in *The War of the Worlds* seem uncannily like lasers, emitting beams of invisible infrared light. Yet H. G. Wells wrote the book in 1896, long before anyone had thought of stimulated emission or lasers. Wells just imagined a searchlight beam that could burn rather than illuminate.

Pulp science fiction writers soon churned out tales of ray guns or death rays, which fired deadly beams of light or other (often undefined) forms of radiation. The writers may have heard rumors that legendary inventor Nikola Tesla and a handful of other scientists were working on death rays in the 1920s and 1930s, but there was no real science behind their weapons. They were just futuristic props to avoid arming 25th century heroes with six-shooters. But thanks to those stories, when the laser was invented the public thought of it as a "death ray," much to the annoyance of the people working with real lasers.

It is true that military researchers are trying to develop laser weapons. That is not new; it has been going on since the 1960s and so far has consumed many billions of dollars to shoot down a few targets. As you will learn in Section 12.8, laser weapons are big, and they try to destroy targets by focusing a lot of light energy on them. In short, it is not easy to make lasers into weapons.

This book is about real lasers, so we will start by looking at the fundamental concepts behind real-world laser technology, briefly explaining what they are and how they developed.

1.2 WHAT IS A LASER?

You have already seen than the word "laser" is shorthand for the phrase "light amplification by the stimulated emission of radiation." Each part of that phrase has a special meaning, so we will look at it piece by piece, starting from the end.

Radiation means *electromagnetic radiation,* a massless form of energy that travels at the speed of light. It comes in various forms, including visible light, infrared, ultraviolet, radio waves, microwaves, and X-rays. Light and other forms of electromagnetic radiation behave like both waves and particles (called *photons*). You will learn more about the details in Chapter 2.

Stimulated emission tells us that laser light is produced in a special way. Ordinarily, atoms or molecules spontaneously emit energy in the form of light or other types of electromagnetic radiation. The sun, flames, and fluorescent lamps all release energy by emitting light spontaneously. However, in certain cases atoms and molecules can be stimulated to emit that extra energy as light. This process is called stimulated emission, and you will learn more about it in Chapter 3.

Amplification means increasing the amount of light. In stimulated emission, an input wave stimulates an atom or molecule to release energy as a second wave, which is perfectly matched to the input wave. The stimulated wave, in turn, can stimulate other atoms or molecules to emit duplicate waves, causing further amplification. It may be easier to think of stimulated emission as one light photon stimulating an atom or molecule to emit an identical photon, which in turn can stimulate the emission of another identical photon. In both cases the result is amplification, producing more light.

Light describes the type of electromagnetic radiation produced. In practice, that means not just light visible to the human eye, but also electromagnetic radiation that our eyes cannot see because it is either longer in wavelength (infrared) or shorter in wavelength (ultraviolet.)

It took a long time to put the pieces of the idea together. Albert Einstein first suggested the possibility of stimulated emission in a paper published in 1917. Although stimulated emission was first observed in the 1920s, physicists long thought that spontaneous emission was much more likely, so stimulated emission would always be much weaker. The first hints that stimulated emission could be stronger came in radio experiments shortly after World War II, but the key experiment came in the 1950s.

Charles H. Townes, then at Columbia University, conceived of a way to build up stimulated emission at microwave frequencies in 1951. His idea was to isolate ammonia molecules with extra energy, then stimulate them to emit their extra energy at a particular microwave frequency as they were passing through a cavity that reflected the microwave frequency emitted by the ammonia molecules. He called his device a "maser," an acronym for microwave amplification by the stimulated emission of radiation.

It took until 1954 for Townes and his graduate student James Gordon to make the maser work. It could serve either as an amplifier or an oscillator. Some ammonia molecules spontaneously emitted microwaves at a frequency of 24 gigahertz, and that spontaneous emission could stimulate other excited ammonia molecules to emit at the same frequency, building up a signal that oscillated on its own. Alternatively, an external 24-GHz signal could stimulate the ammonia molecules to emit at 24 GHz, amplifying the signal.

In principle, the maser process could be extended to other types of electromagnetic radiation if the right materials could be found. The next logical step was to optical wavelengths, and a number of people thought seriously about the possibility. However Townes was the first to start serious research in 1957. In the course of gathering information, he talked with Gordon Gould, a Columbia graduate student who was using an important new idea called optical pumping in his doctoral research project. Townes thought he could use optical pumping to excite atoms in a laser, and the laser idea intrigued Gould.

Townes went on to enlist the help of his brother-in-law, Arthur Schawlow, who knew more about optics, to work out how to amplify stimulated emission of light. Meanwhile Gould quietly tackled the problem with a pile of reference books on his kitchen table. They essentially independently solved the same physics problem, and both proposed building cylindrical laser resonators with mirrors on opposite ends so the light would bounce back and forth between the mirrors while it was being amplified. Gould set out to patent the his ideas; Townes and Schawlow published their proposal in a scientific journal, *Physical Review Letters.* Their work launched a race to build a laser, which I chronicled in *Beam: The Race to Make the Laser* (Oxford University Press 2005).

Townes shared in the 1964 Nobel Prize in physics for his pioneering work on "the maser/laser principle," and after a long series of legal battles, Gould earned tens of millions of dollars from his patent claims. However, the winner of the laser race was Theodore Maiman, who on May 16, 1960 produced laser pulses from a fingertip-sized crystal of synthetic ruby at Hughes Research Laboratories in Malibu, California. Figure 1-1 shows Maiman and

Figure 1-1. Theodore Maiman and Irnee J. D'Haenens with a replica of the world's first laser, which they made at Hughes Research Laboratories in 1960. (Reprinted from Hughes Research Laboratories, courtesy of AIP Neils Bohr Library.)

his assistant Irnee D'Haenens holding a replica of his elegant little device, the world's first laser.

The ruby laser was in many ways typical of the many other types of lasers that followed it. Energy from an external source—in this case, a bright flash of light from a photographic flash lamp—excited chromium atoms in a ruby cylinder. Some excited chromium atoms spontaneously emitted light, and that light stimulated other excited chromium atoms to release their excess energy as an identical light wave. Silver film mirrors coated onto the ends of the ruby rod formed a resonant cavity, so light bounced back and forth between them, stimulating more emission from chromium atoms and amplifying the red light to build up a beam. The laser beam emerged through a hole in one of the silver coatings on the ends of the rod. The laser light was at a single wavelength—694 nanometers (1 nm = 10^{-9} meter) at the red end of the visible spectrum. The light waves were coherent, all aligned with each other and marching along in step.

The lasers that followed generally shared key properties of the ruby laser, generating coherent beams of monochromatic light.

LASER OSCILLATION

Stimulated emission amplifies light in a laser, but the laser itself is an oscillator. So why, you may wonder, does the word "laser" come from "light *amplification* by the stimulated emission of radiation"? There's an interesting bit of history behind that.

Charles Townes created the word "maser" as an acronym for microwave amplification by the stimulated emission of radiation. When he began thinking of an optical version of the maser, he called it an optical maser. When Gordon Gould sat down to tackle the same problem, he wrote "laser" at the top of his notes, inventing the acronym for light amplification. As the competition between Townes and Gould became intense, each side pushed its own term.

Arthur Schawlow was a jovial soul, and at one conference pointed out that because the laser was actually an oscillator, it should be described as "light oscillation by the stimulated emission of radiation," making the laser a "loser." Everybody laughed, but the word laser proved a winner.

Maiman's ruby laser was pulsed; many others generated continuous beams. Some generated stimulated emission from longer, thinner rods of other crystals. Others stimulated emission from gases inside a tube with mirrors at its two ends. The most common lasers today are tiny chips of semiconductor compounds such as gallium arsenide. But some lasers occupy entire rooms in buildings, and the most powerful lasers—like the U.S. Air Force's Airborne Laser—occupy whole buildings or aircraft.

Lasers operate at wavelengths from the infrared all the way to soft X-rays. They can generate modest powers far below one watt, steady powers of thousands of watts, or concentrate light into pulses lasting less than a billionth of a second. Figure 1-2 shows commercial gas, semiconductor, and solid-state lasers designed for a variety of applications.

1.3 LASER MATERIALS AND TYPES

Maiman won the laser race because he had studied the optical properties of ruby, and carefully designed his laser to take advantage of them. Matching the laser design to the material properties was critical. Most materials won't work as lasers under most conditions. What is needed is a material containing atoms or molecules that can be excited into a state ready to be stimulated to emit light energy.

Ruby worked because it contains chromium atoms, which absorb energy as visible light, then eventually release much of that energy as a photon of red light. Maiman found that if he slipped a ruby rod inside a coiled flash lamp, the bright flash would excite most of the chromium atoms, leaving them ready to emit red light. If one chromium atom spontaneously emitted a red photon, that photon could stimulate another chromium atom to emit a second photon; and both of those photons could stimulate more emission, eventually producing a cascade of red light in a laser pulse. Figure 1-3 shows the basic idea.

Ruby is an important example of a *solid-state laser.* In these lasers, light from an external source, such as a flash lamp, excites atoms distributed within a solid. The solid must be transparent at the wavelength of the pump light so it can excite the atoms that produce the stimulated emission. In ruby, the transparent material is sapphire (aluminum oxide or Al_2O_3) and the light emitting

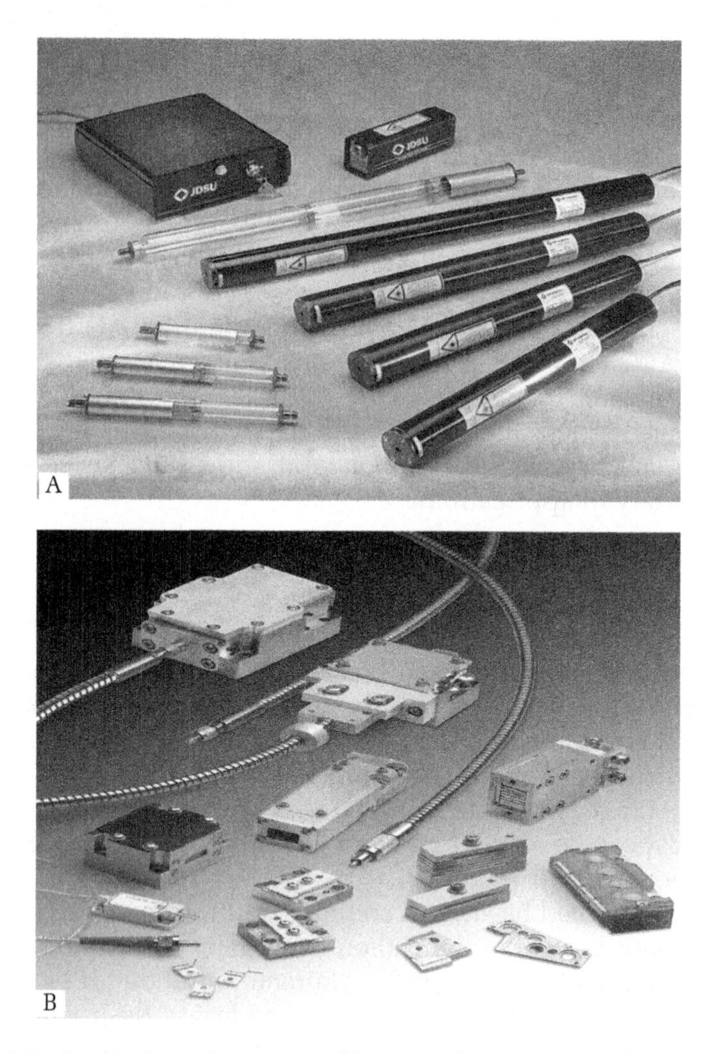

Figure 1-2. A sampling of commercial lasers. (A) A sampling of milliwatt-class helium–neon lasers, including both packaged heads and bare tubes (courtesy JDS Uniphase). (B) A sampling of commercial diode lasers packaged for various types of applications (courtesy of Spectra-Physics, a division of Newport Corporation).

atoms are chromium. Another common choice is adding a rare-earth element called neodymium to transparent materials such as the crystal yttrium–aluminum garnet (YAG) or certain types of glass. Two other rare-earth elements, erbium and ytterbium, can be added to glass that is drawn into optical fibers, to produce solid-state fiber lasers.

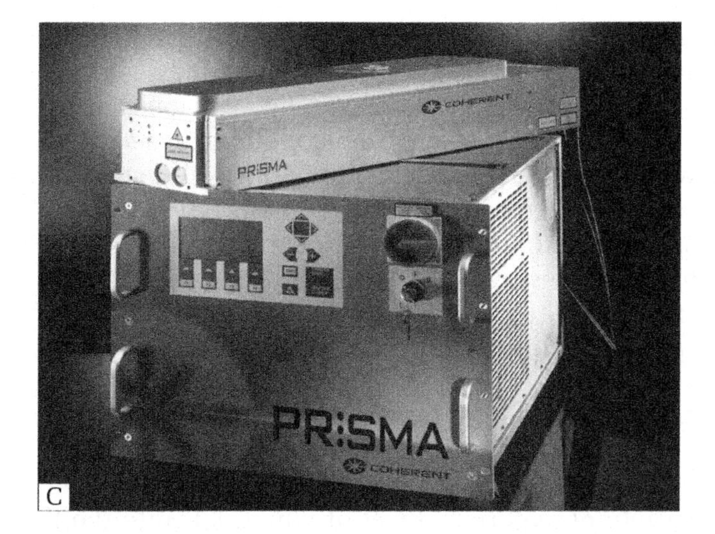

Figure 1-2. *Continued* (C) A diode-pumped, solid-state neodymium laser producing an average power of several watts for materials-working applications (courtesy of Coherent Inc.).

In all cases, light from an external lamp or laser passes through the transparent host material to excite the light-emitting atoms. This process is called optical pumping. Described in Chapter 8, these lasers can generate pulses or continuous beams.

A second broad class of lasers are *gas lasers*, covered in Chapter 7, in which a light-emitting vapor is confined inside a hollow tube

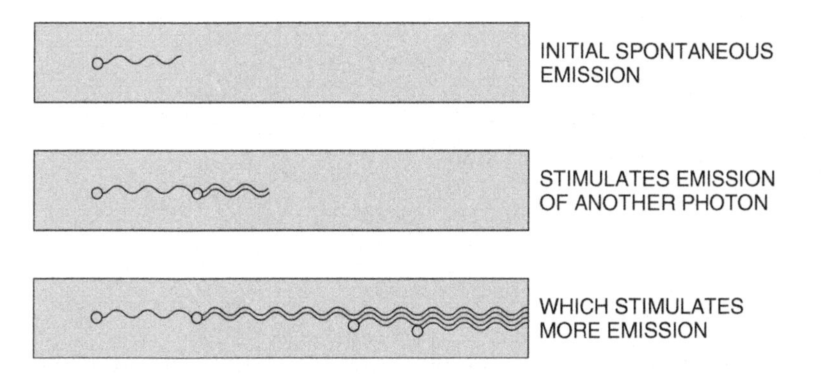

Figure 1-3. A single spontaneously emitted photon triggers stimulated emission from excited atoms, building up a cascade of stimulated emission. In ruby, the excited atoms are chromium.

with mirrors on the ends. Passing an electric discharge through the gas excites the gas atoms to states in which they can generate stimulated emission. Many gas lasers emit continuous beams.

A third broad class are *semiconductor lasers,* which in the laser world are considered distinct from solid-state lasers. Standard semiconductor lasers are actually *diode lasers* or *laser diodes,* in which current flows through a semiconductor chip in a way that generates light at a junction between two zones of different composition, as you will learn in Chapter 9. Semiconductor lasers are tiny, cheap, and efficient, and have become by far the most common types of laser. Every CD or DVD player contains at least one semiconductor laser, and hundreds of millions are produced each year.

Laboratory researchers have demonstrated laser action in a wide variety of materials, most of which fall into the three broad categories of gas, solid-state, or semiconductor. A few other types with limited applications are described in Chapter 10, including organic dye lasers, extreme ultraviolet lasers, and free-electron lasers.

1.4 OPTICAL PROPERTIES OF LASER LIGHT

Lasers can be fascinating devices in themselves, but their practical importance comes from the unusual properties of the light in a laser beam. These properties are crucial for applications of lasers ranging from cutting sheets of plastic or metal to making extremely precise and sensitive measurements in scientific research. The most important of these optical properties are:

- Wavelength(s)
- Beam power
- Variation of beam power with time (e.g., pulse duration)
- Beam divergence and size
- Coherence
- Efficiency

1.4.1 Wavelength(s)

The wavelength emitted by a laser depends both on the laser material and the design of the laser cavity. The range of possible wavelengths depends on the material, with the optical design of the laser selecting which wavelengths can be emitted. Table 1-1 lists

Table 1-1. Wavelengths of some important lasers

Type	Wavelength Range
Fluorine (F_2) excimer	157 nm
Argon–fluoride excimer	193 nm
Argon-ion (ultraviolet)	229–264 nm
Krypton–fluoride excimer	248 nm
Neodymium solid-state (4th harmonic)	266 nm
Argon-ion (ultraviolet)	275–303 nm
Xenon–chloride excimer	308 nm
Helium–cadmium (ultraviolet)	325 nm
Argon-ion (ultraviolet)	330–360 nm
Nitrogen gas (N_2)	337 nm
Xenon–fluoride excimer	351 nm
Neodymium solid-state (3rd harmonic)	355 nm
GaN/InGaN family diodes	375–440 nm
Organic dye (in solution)	320–1000 nm (tunable)
Helium–cadmium (blue)	442 nm
Argon-ion (visible)	454–515 nm
Krypton-ion (visible)	472–800 nm
Neodymium solid-state (doubled)	532 nm
Helium–neon (green)	543.5 nm
Helium–neon (yellow)	594 nm
Helium–neon (orange)	612 nm
Helium–neon (red)	632.8 nm
AlGaInP/GaAs family diodes	620–680 nm
$Ga_{0.5}In_{0.5}P$/GaAs family diodes	670–680 nm
Krypton (strongest line)	647 nm
Ruby	694 nm
Titanium–sapphire	675–1100 nm (tunable)
Alexandrite	701–826 nm (tunable)
GaAlAs family diodes	750–905 nm
InGaAs family diodes	915–1050 nm
Ytterbium-fiber	1030–1100 nm
Neodymium–YLF	1057 nm
Neodymium–YAG or Nd–YVO_4 (primary)	1064 nm
InGaAsP family diodes	1100–1650 nm
Chemical oxygen–iodine	1315 nm
Neodymium solid-state (secondary)	1320 nm
Erbium–glass or fiber	1530–1570 nm
Hydrogen fluoride chemical	2600–3000 nm
Quantum cascade (semiconductor)	3000 nm–50 μm
Deuterium fluoride chemical	3500–4500 nm
Carbon monoxide	5000–6000 nm
Carbon dioxide	9000–11000 nm

important types of lasers and their typical output wavelengths in nanometers.

Most lasers are called "monochromatic" (single-colored) and actually emit at a narrow range of wavelengths at any one time. However, some can be operated across a range of wavelengths, as shown in Table 1-1. Exactly what that range means depends on the type of laser. In some cases, the specific composition of the laser material or the design of the optics limit the laser to a narrow range of wavelengths. In other cases, the laser can emit across much or all of the range with suitable optics. You will learn more about the specifics of these lasers in Chapters 7–10.

1.4.2 Beam Power

Beam power measures the amount of energy a laser beam delivers per unit time. It is measured in watts and defined by the formula

$$\text{Power} = \frac{\Delta\ \text{energy}}{\Delta\ \text{time}} \tag{1-1}$$

One watt of power equals one joule (of energy) per second.

The powers of steady laser beams range from less than a milliwatt (0.001 watt) to kilowatts (thousands of watts), but instantaneous power can be much higher in brief intervals. No single laser emits across that entire range of power, and many types of lasers cannot be scaled beyond milliwatt or watt levels.

Note that the beam power means the total power in the whole beam, not the power per unit area, which is also important. One attraction of the laser beam is that it concentrates light energy onto a small area, and that suitable optics can also focus the light to a very small spot, producing high powers per unit area.

1.4.3 Time Variation of Output Power

Lasers may generate pulses or continuous beams, but some types can only produce pulsed beams. Pulses vary in length, typically from milliseconds (10^{-3} second) to femtoseconds (10^{-15} second). Typically, pulses are repeated at a steady rate, ranging from once a minute to billions of times a second; the repetition rate is often given in the units of hertz, or pulses per second.

The instantaneous power is the rate of energy emission, which rises and falls during a pulse. Short pulses concentrate energy in time, so the peak power in a pulse can be very high although the total amount of energy remains modest. For example, if a pulse delivers 10 millijoules of energy in 10 femtoseconds, average power during the pulse is 10^{12} watts, or 1 terawatt.

Some lasers inherently operate in a pulsed mode. Others are modulated externally or internally to generate pulses of desired characteristics. Time variation of laser output is very important both for communicating information and for controlling how the laser beam interacts with materials.

1.4.4 Beam Divergence and Size

A laser beam in dusty air looks as thin as a string or a pencil line, but beyond a meter or so from the laser, the beam actually is spreading at a very small angle. This spreading is called *divergence,* and is shown in Figure 1-4. Beam divergence depends both on the type of laser and on the external optics. Semiconductor lasers naturally have a high divergence, but external optics can focus their output into a pencil-like beam.

Typically, laser beam divergence is measured in milliradians, or thousandths of a radian, a unit equal to 0.057 degree. As long as the beam divergence is small, this is quite handy, because you can estimate the radius of a laser spot at a distance D from the laser by multiplying the distance by the divergence in radians. Thus, a 2-milliradian beam spreads to a 0.2 meter spot at a distance of 100 meters. This high directionality of the laser beam is important in many applications.

1.4.5 Coherence

Light waves are *coherent* if they are all in phase with one another. Figure 1-5 compares coherent and incoherent light waves. The

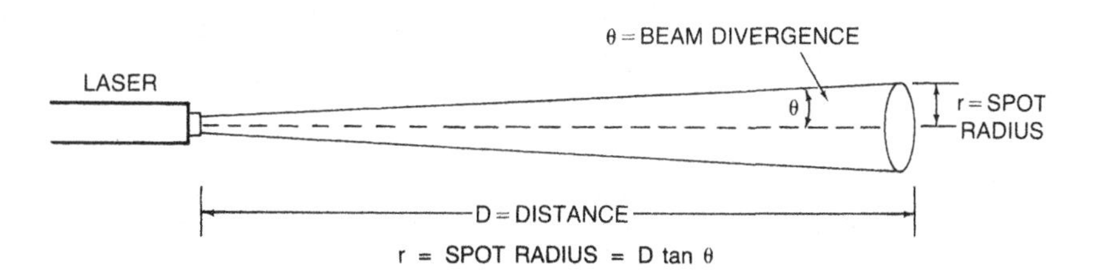

Figure 1-4. Calculating the size of a laser spot from the beam divergence.

peaks and valleys of coherent light waves (top of Figure 1-5) are all lined up with each other. The peaks and valleys of incoherent light waves (bottom of Figure 1-5) do not line up. Stimulated emission is in phase with the light that stimulates it, so laser light is coherent. The sun, light bulbs, flames and other sources that generate spontaneous emission are incoherent.

The coherence of laser light is related to the narrow range of wavelengths emitted. The more monochromatic the light, the more coherent it is, and the longer the distance over which the light waves will remain in phase. Monochromatic light need not be coherent, but light that is not monochromatic cannot stay coherent over a long distance. Lasers are the only light sources that can readily generate light that is coherent over relatively long distances.

1.4.6 Efficiency

How efficiently lasers can convert input energy into output light varies widely. The least efficient types may convert only 0.001% of the input light into laser energy, but the most efficient types can convert well over half the input power into light. Efficiency is particularly important in high-power applications, because all the

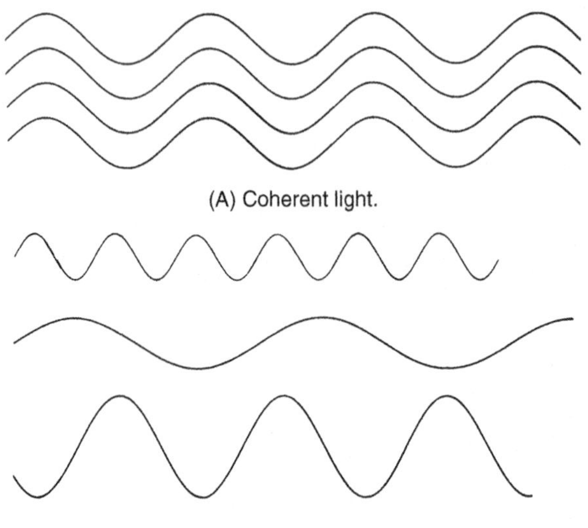

Figure 1-5. Coherent (A) and incoherent (B) light.

input energy that does not emerge in the laser beam winds up as heat that must be dissipated from the laser. Generating a kilowatt of laser power requires five kilowatts of input if the laser is 20% efficient, but 1000 kilowatts is needed if the laser is only 0.1% efficient.

1.5 HOW LASERS ARE USED

Scientists and engineers began playing with lasers almost as soon as they could lay their hands on one after the laser was invented. They fired lasers at just about everything that could not run away. They shot so many holes in razor blades that for a while laser power was informally measured in "gillettes." Yet few practical applications emerged quickly, and for a while the laser seemed to be, as Irnee D'Haenens told Ted Maiman soon after they made the first one, "a solution looking for a problem."

We are long past that stage. Lasers have become standard tools in industry and research. They align construction equipment, transmit voice and data around the globe, and perform exquisitely sensitive measurements that have earned a fair number of Nobel Prizes. Table 1-2 lists a sampling of laser applications, and Chapters 11–13 cover those applications in more detail.

Lasers are used in diverse ways. The final three chapters divide laser applications into three broad categories.

Chapter 11 covers low-power applications. One broad family of such applications uses lasers as sources of highly controlled light for transmitting and processing information, such as reading or writing data or transmitting signals. For example, the laser in a CD player is focused onto a tiny spot as the CD spins beneath it, and the player reads data recorded on the disk by observing how the light is reflected. The coherence of lasers makes it possible to create and display three-dimensional holographic images.

Another broad category of low-power applications are in measurement. Laser beams travel straight through the air, so they can draw straight lines to assist in alignment of walls or pipes during construction projects. Precision techniques can take advantage of the coherence of lasers to measure distances to within a fraction of the wavelength of light. Laser radars can measure the velocity of objects, including speeding cars for police speed traps. Scanning laser systems can record three-dimensional profiles of the surfaces of objects.

Table 1-2. A Sampling of laser applications

Information handling
Fiber-optic communications
Laser printers for computer output
Playing DVD or Blu-Ray video
Playing CD audio
Reading and writing computer data on CDs and DVDs
Reading printed bar codes for store checkout and inventory control

Measurement and Inspection
Detecting flaws in aircraft tires
Exciting fluorescence from various materials
Illuminating cells for biomedical measurements
Measuring concentrations of chemicals or pollutants
Measuring small distances very precisely
Measuring the range to distant objects
Measuring velocity
Projecting straight lines for construction alignment and irrigation
Studies of atomic and molecular physics

Medicine and Dentistry
Bleaching of port-wine stain birthmarks and certain tattoos
Clearing vision complications after cataract surgery
Dentistry
Refractive surgery to correct vision
Reattaching detached retinas
Shattering of stones in the kidney and pancreas
Treatment of diabetic retinopathy to forestall blindness
Surgery on tissue rich in blood vessels

Materials Working
Cutting, drilling, and welding plastics, metals, and other materials
Cutting cloth
Cutting titanium sheets
Drilling materials from diamonds to baby-bottle nipples
Engraving wood
Heat-treating surfaces
Marking identification codes
Semiconductor device manufacture

Military
Range-finding to targets
Simulating effects of nuclear weapons
Target designation for bombs and missiles
War games and battle simulation
Antisatellite weapons
Antisensor and antipersonnel weapons
Antimissile weapons

Table 1-2. *Continued*

Other applications
Basic research
Controlling chemical reactions
Displays
Holography
Laser light shows
Laser pointers
Three-dimensional profiling and modeling

Chapter 12 covers high-power applications, in which a laser beam delivers energy that alters the material it hits. Lasers deliver small bursts of energy to mark painted metal surfaces; the laser vaporizes the paint, exposing the shiny metal. More powerful lasers can drill holes through materials ranging from baby-bottle nipples to sheets of titanium. The laser beam does not bend soft materials like latex nipples, and does not grow dull like a drill bit.

Laser surgery works in the same way. Pulses from an ultraviolet laser can vaporize tissue from the surface of the eye, precisely removing just the right amount to correct vision defects. By selecting the right laser wavelength, surgeons can bleach dark birthmarks or tattoos.

The ultimate in high-power lasers are high-energy laser weapons. You can think of them as performing materials-working on unfriendly objects. A laser weapon might blind the sensor that guides a missile, causing it to go astray. Or a higher-energy laser might heat or punch holes in the fuel tank of a missile so it explodes before reaching its targets.

Chapter 13 covers laser applications in scientific research. Laser techniques can slow atoms to a virtual crawl and probe their energy states with exquisite precision. Laser beams can manipulate tiny objects, from bacteria to single atoms. These laser applications have led to several Nobel Prizes.

1.6 WHAT HAVE WE LEARNED?

- Most lasers are tiny semiconductor chips inside of electronic equipment.
- Lasers have become so commonplace you can buy a laser pointer as a toy for $10 or less.

- LASER is an acronym for "light amplification by the stimulated emission of radiation."
- Many people initially thought of lasers as science-fiction "death rays."
- Stimulated emission of light by excited atoms generates laser radiation.
- Charles Townes conceived of the amplification of stimulated emission for microwaves.
- Theodore Maiman demonstrated the first laser using a ruby rod pumped by a photographic flashlamp.
- Successful operation of a laser requires both an optical resonator and a suitable material.
- The three main classes of lasers are gas, semiconductor, and solid-state lasers. Note that solid-state differs from semiconductor in the laser world.
- Lasers can emit a very narrow range of wavelengths.
- Laser light is concentrated in a beam, which generally is tightly focused.
- Laser light is coherent.
- Low-power laser applications include measurement and information processing.
- High-power laser applications modify materials for tasks including surgery, machining, and weapons.
- Lasers can make precision measurements for scientific research.

WHAT'S NEXT?

The first step in understanding lasers is to learn the basic principles of physics and optics that are involved in laser operation. Chapter 2 introduces the essential physical concepts. Some of this material may be familiar if you have been exposed to physics before, but you should review it because later chapters assume that you understand it.

QUIZ FOR CHAPTER 1

1. The word laser originated as
   a. A military codeword for a top-secret project
   b. A trade name

c. An acronym for Light Amplification by the Stimulated Emission of Radiation
d. The German word for light emitter

2. The first laser was made by
   a. Charles Townes
   b. Theodore Maiman
   c. Gordon Gould
   d. Arthur Schawlow
   e. H. G. Wells
3. Most lasers today are
   a. Semiconductor devices used inside electronic equipment
   b. High-power weapons used in ballistic missile defense
   c. Gas-filled tubes emitting red light
   d. Ruby rods powered by flash lamps
   e. Ruby rods powered by LEDs
4. Laser light is generated by
   a. Spontaneous emission
   b. Gravity
   c. Stimulated emission
   d. Microwaves
   e. Mirrors
5. What emits light in a ruby laser?
   a. Aluminum atoms
   b. Sapphire atoms
   c. Oxygen atoms
   d. Chromium atoms
   e. Mirrors on the ends of the rod
6. Why is a semiconductor laser sometimes called a diode laser?
   a. Because the first diode lasers had to be installed in vacuum tubes so the semiconductor would not evaporate.
   b. Because the semiconductor is electrically a diode, powered by current flowing through it.
   c. Because it is powered by light from an external light-emitting diode.
   d. Because it's an acronym for "damn idiotic optical device exploded," which is what happened to the first one.
7. Which is not considered a low-power laser application?
   a. Laser illumination of a dark birthmark called a port-wine stain to bleach it
   b. Playing a CD or DVD
   c. A computer laser printer

d. A laser system used by surveyors for measurement
e. A laser radar used by police to spot speeders

8. Why might a laser be attractive for use as a weapon to hit a target a long distance away?
   a. It's easy to make lasers very powerful.
   b. Lasers look neat and the enemy would run away if they saw them.
   c. Laser beams travel at the speed of light, so they would be easier to aim at a moving target than a missile that took a long time to reach it.
   d. You can buy laser pointers for $10 each.
9. Stimulated emission generates light waves that are in phase with each other. This makes them
   a. A beam
   b. Coherent
   c. Pulsed
   d. Span a range of wavelengths
10. How many lasers do you own? There's no single "right" answer, but it's fun to take a mental inventory. Don't forget that some devices contain multiple lasers, such as combination DVD/CD players.

CHAPTER 2

PHYSICAL BASICS

ABOUT THIS CHAPTER

Lasers evolved from concepts of modern physics that emerged early in the twentieth century. To understand lasers, you need to understand the basic concepts that underlie laser physics, including light, atomic energy levels, quantum mechanics, and optics. This chapter starts with the nature of light, then moves on to how light is generated, the interactions of light and matter, and some fundamentals of optics, to give you the background you need to understand lasers themselves.

2.1 ELECTROMAGNETIC WAVES AND PHOTONS

Early physicists debated long and loud over the nature of light. Isaac Newton held that light was made up of tiny particles. Christian Huygens believed light was made up of waves, vibrating up and down, perpendicular to the direction in which the light travels. Newton's theory came first, but Huygens' theory explained early experiments better, so for a long time it was assumed to be right.

Today, we know that both theories are partially right. Much of the time, light behaves like a wave. Light is called an *electromagnetic* wave because it consists of electric and magnetic fields perpendicular to each other, as shown in Figure 2-1. Because the electric and magnetic fields oscillate perpendicular to the direction in which the waves travel, they are called *transverse waves.*

At other times, light behaves like massless particles called *photons.* Photons are not exactly what Newton envisioned, but the

Understanding Lasers: An Entry-Level Guide, Third Edition. By Jeff Hecht

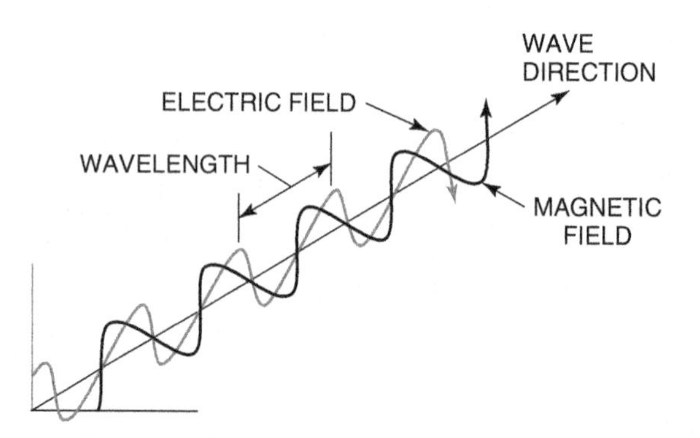

Figure 2-1. Structure of an electromagnetic wave.

particle view is also critical to understanding light. We will move back and forth between the wave and particle views of light, but you should remember that they are just different ways to look at the same thing.

2.1.1 Waves and Wavelength

The nature of an electromagnetic wave is measured by its *wavelength,* the distance between successive peaks, illustrated in Figure 2-1, and denoted by the Greek letter lambda (λ).

Wave properties also depend on their *frequency,* the number of wave peaks passing a point in a second, denoted by the Greek letter nu (ν). Frequency is measured in oscillation cycles (from peak to valley and back) per second, a unit called hertz, after Heinrich Hertz, the 19th-century discoverer of radio waves. Multiplying the length of the wave by the number of waves per second gives the wave's velocity, in units of distance per second:

$$\text{Wavelength} \times \text{Frequency} = \text{Velocity} \quad (2\text{-}1)$$

The velocity of electromagnetic waves is a universal constant, denoted by c, the speed of light in vacuum, which is close to 300,000 kilometers per second or 186,000 miles per second. (The exact value is 2.99792458×10^8 meters/second, listed in Appendix B, but for practical purposes remember the round numbers.) This relationship makes wavelength inversely proportional to frequency—the shorter the wavelength, the higher the frequency. Using the standard symbols for wavelength, frequency, and the speed of light, we can write simple formulas to convert between wavelength and frequency (assuming, for the moment, that we are working in a vacuum):

$$\lambda = c/\nu \tag{2-2}$$

and

$$\nu = c/\lambda \tag{2-3}$$

We can also view light as made up of particles called *photons,* which are chunks or quanta of energy. Photons can travel at the speed of light because they are massless. The energy of a photon E equals the frequency of the wave multiplied by a constant h called Planck's constant, after the German physicist Max Planck, who worked out the formula. Planck's constant equals 6.63×10^{-34} joule-second, so photon energy is

$$E = h\nu = 6.63 \times 10^{-34}\ \nu \tag{2-4}$$

Because frequency ν is measured in waves per second, the result gives photon energy in joules when Planck's constant is given in joule-second units.

Combining this formula with the relationship between wavelength and frequency shows that wavelength times photon energy equals Planck's constant multiplied by the speed of light c, or 1.99 $\times 10^{-25}$ joule-meter:

$$\lambda E = hc = 1.99 \times 10^{-25} \text{ joule-meter} \tag{2-5}$$

As you work with light and other electromagnetic waves, you will come to remember these conversions. For now, the most important lesson is that the nature of a light wave can be measured in three ways: as photon energy, wavelength, or frequency. For example, light with 1-micrometer wavelength has a frequency of 3×10^{14} Hz, and photon energy of 2×10^{-19} J.

Physicists often measure photon energy in electron volts, the energy an electron acquires by moving through a one-volt potential. One electron volt equals 1.6022×10^{-19} joule, so a photon energy of 2×10^{-19} joule equals 1.24 electron volts, a more convenient unit of measurement.

It is easy to make mistakes in converting units, so it helps to remember these simple rules of thumb:

- The higher the frequency, the shorter the wavelength.
- The higher the frequency, the larger the photon energy.
- The shorter the wavelength, the larger the photon energy.

Electromagnetic waves are often called *electromagnetic radiation* because objects emit or *radiate* them into space. If the word "radiation" sounds distressingly like something that comes from a leaky nuclear reactor, it is because the vital difference between nuclear radiation and other types is too often lost. The electromagnetic spectrum spans a broad range of photon energies, and only the most energetic photons—X-rays and gamma rays, which have the highest frequencies and shortest wavelengths—are dangerous.

2.1.2 The Electromagnetic Spectrum

We usually think of the spectrum as the colors that spread out in a rainbow, or appear when we pass sunlight through a prism. Colors are the way our eyes and brains sense the differences in wavelength across the visible spectrum, from the short violet waves to the long red waves. However, our eyes can see only a narrow slice of the entire range of wavelengths in the *electromagnetic spectrum.* Table 2-1 lists the components of that spectrum, ranging from extremely low-frequency waves many miles long to gamma rays a trillionth of a meter long.

Table 2-1 lists all frequency and wavelength values in the same units to simplify comparison. However, in practice the val-

Table 2-1. Wavelengths and frequencies of electromagnetic radiation

Name	Wavelengths (m)	Frequencies (Hz)
Gamma rays*	under 3×10^{-11}	over 10^{20}
X-rays*	3×10^{-11} to 10^{-8}	3×10^{16} to 10^{20}
Ultraviolet light*	10^{-8} to 4×10^{-7}	7.5×10^{14} to 3×10^{16}
Visible light	4×10^{-7} to 7×10^{-7}	4.2×10^{14} to 7.5×10^{14}
Infrared light	7×10^{-7} to 10^{-3}	3×10^{11} to 4.2×10^{14}
Microwaves	10^{-3} to 0.3	10^{9} to 3×10^{11}
Radio waves	0.3 to 30,000	10^{4} to 10^{9}
Low-frequency waves*	over 30,000	under 10,000

*Denotes no widely accepted standard boundaries.

ues of frequency, wavelength, and other quantities such as time and power are expressed in metric units with the standard prefixes shown in Table 2-2. You probably already know some of these prefixes, but you are likely to discover others as you explore the world of lasers. Virtually everything optical is measured in metric units, and we will follow that practice in this book. (Some older books give visible wavelengths in Ångstroms, Å, a unit equal to 10^{-10} meter or 0.1 nanometer, which is not a standard metric unit.)

The different parts of the electromagnetic spectrum were discovered at different times, and the divisions between them remain somewhat arbitrary and hazy. Instruments are needed to detect infrared and ultraviolet light beyond the visible spectrum. Radio waves, X-rays, and gamma rays all were discovered independently, and only later did physicists realize that all those waves were part of the same family of electromagnetic waves. Many dividing lines between types remain informal.

Even the limits of visibility to the human eye are not rigidly standardized. The reason is that the eye's sensitivity to light drops gradually, as shown in Figure 2-2. The limits of visibility often are defined as 400 to 700 nanometers, but the eye can detect slightly shorter ultraviolet wavelengths, and responds very weakly to longer infrared wavelengths. You will find some books that define the visible range as extending from 380 nm to as long as 780 nm.

The terms "light" and "optical" are not limited to the part of the electromagnetic spectrum sensed by the human eye. In prac-

Table 2-2. Prefixes used for metric units

Prefix	Abbreviation	Meaning	Number
exa	E	quintillion	10^{18}
peta	P	quadrillion	10^{15}
tera	T	trillion	10^{12}
giga	G	billion	10^{9}
mega	M	million	10^{6}
kilo	k	thousand	1,000
deci	d	tenth	0.1
centi	c	hundredth	0.01
milli	m	thousandth	0.001
micro	μ	millionth	10^{-6}
nano	n	billionth	10^{-9}
pico	p	trillionth	10^{-12}
femto	f	quadrillionth	10^{-15}
atto	a	quintillionth	10^{-18}

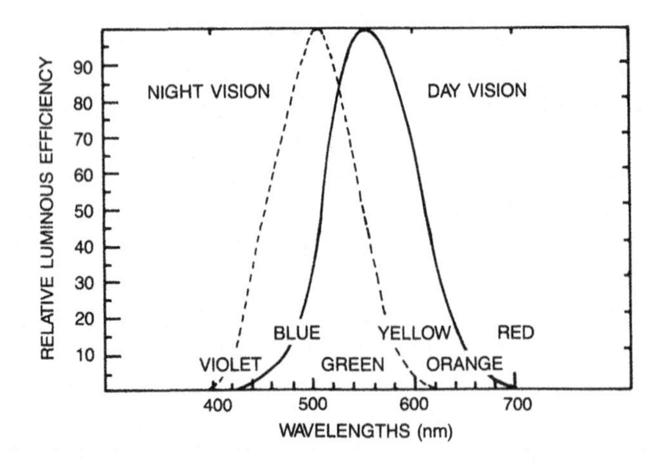

Figure 2-2. Sensitivity of the human eye to different wavelengths.

tice, light or the optical band includes nearby parts of the infrared and ultraviolet because these parts of the spectrum behave much like visible light, except that our eyes cannot see them. Thus, the same optics used for visible light can focus near-infrared light and, in practice, "optical" wavelengths are those for which glass optics work well.

The electromagnetic spectrum is a continuum of waves from extremely low to extremely high frequencies. The divisions that we make depend largely on how waves in different parts of the spectrum interact with matter. For example, the eyes of humans and all other terrestrial creatures evolved to see the sunlight transmitted by our atmosphere. The atmosphere becomes opaque at wavelengths shorter than about 200 nm, which is called the "vacuum ultraviolet" because you need a vacuum to transmit it. We will talk more about light interactions with matter in Section 2.3.

Lasers normally operate in the infrared, visible, and ultraviolet parts of the spectrum. Masers, the microwave counterparts of lasers, operate at microwave frequencies, which are lower than the frequencies of infrared waves, but microwave wavelengths are longer than those in the infrared. A few lasers have operated at the long-wavelength edge of the X-ray spectrum.

2.1.3 Interference and Waves

Light has a dual personality, sometimes behaving as a wave and sometimes as a series of particles. The best illustration of the wave nature of light is the interference of two waves, revealed by illuminating two parallel slits with a single bright light. As shown in Figure 2-3, light passing through the slits forms a pattern of light and dark bands.

We can understand that pattern by thinking of the light passing through each slit as waves spreading from the slit that eventually overlap with the waves spreading from another slit, so the amplitudes of the two waves add together at the screen in the back. Assume that exactly the same amount of light passes through each slit, so the waves emerging from slit each have the same amplitude, and consider how the amplitudes of the two waves add together.

Start at the midpoint on the screen, halfway between the two slits; moving closer to one slit takes you further from the other. Suppose the midpoint is exactly 20 wavelengths of light from each slit. In that case, the waves from both slits strike the screen at the peak of their cycle, as shown at the top of Figure 2-4, so the two waves add and we see a bright spot. This phenomenon is called *constructive interference.*

As we move away from that point, the waves slip slightly out of phase, and do not add together exactly. Eventually, we reach a point where the waves from the closer slit travel 20.25 wave-

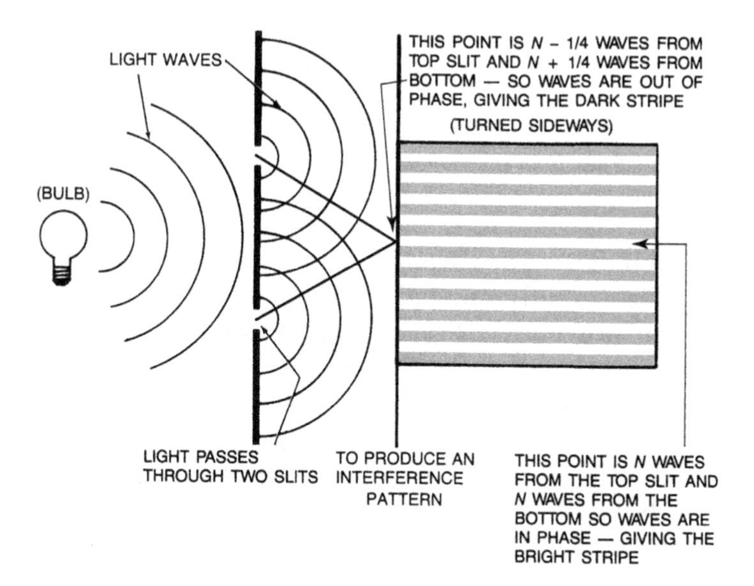

Figure 2-3. Bright light illuminating two slits causes interference.

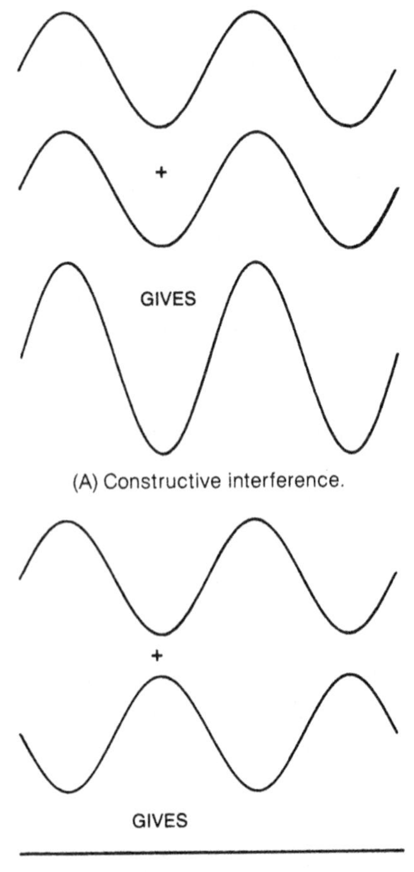

Figure 2-4. Addition and subtraction of light-wave amplitudes causes interference. (A) Constructive interference. (B) Destructive interference.

lengths, and the waves from the more distant slit travel 19.75 wavelengths—exactly one-half wavelength apart. That means the peak of one wave arrives at the same time as the trough of the other wave, as shown at the bottom of Figure 2-4. At this point, the waves are 180 degrees out of phase and cancel each other out precisely in *destructive interference,* producing a dark spot.

Interference is a general effect, and occurs whenever waves combine. If the two waves are precisely in phase, they add together without loss. If two equal-amplitude waves are exactly one-half wavelength out of phase, they cancel each other out at that point. In either case, the waves keep on going unless something blocks them, so they can interfere with other waves at other points in space.

No light energy is really lost in destructive interference. If nothing blocks the waves, they just keep on going as if they had never encountered each other, like ripples spreading across a pond. If something does block the light so you see a pattern of light and dark zones, the total amount of energy across the pattern remains the same, but interference changes its distribution.

Interference is a very important property of light, and it is particularly important for lasers because their coherent light remains in phase over relatively long distances.

2.1.4 Light As Photons

The photon side of light's dual personality shows most clearly when light interacts with matter.

One of the most famous examples is called the *photoelectric effect,* which causes certain metals to emit electrons if light strikes them in a vacuum. Initial experiments were puzzling because they showed a peculiar dependence on the wavelength. The metal did not emit electrons when illuminated by long wavelengths, even if the light was turned up very bright. But at wavelengths shorter than a threshold that was different for each metal, the metal emitted a number of electrons that depended on the light intensity. That is, doubling the illumination doubled the number of electrons released.

This did not make sense if light was purely a wave; adding more energy at a long wavelength eventually should build up enough energy to free an electron. But in 1905, Albert Einstein explained that the threshold occurred because light energy was bundled as photons, and the photon had to have enough energy to free the electron. Once you had a wavelength short enough to get a photon with enough energy to free the electron, adding more light could free more electrons. Einstein's explanation of the photoelectric effect helped lay the groundwork for quantum mechanics, and it eventually earned him the Nobel Prize in Physics.

2.2 QUANTUM AND CLASSICAL PHYSICS

So far, we have mostly been considering light, but laser physics is not just about light; it is about how we organize matter to produce a beam of coherent light. That means we need to look at quantum mechanics and the atom.

The laser is a quantum mechanical device because its operation depends on the quantum mechanical properties of matter. The classical physics described by Isaac Newton assumed that energy could vary continuously, like an absolutely smooth liquid. In contrast, quantum mechanics tells us that energy comes in discrete chunks called quanta, so everything in the universe can have only discrete amounts of energy. In classical physics, energy can change continuously; in quantum mechanics, it changes in steps. The wave picture of light is classical; the photon picture is quantum mechanical.

The central issue for the laser is the quantization of energy levels within atoms. We will start with atomic energy levels, then look at how that leads to the atomic physics behind the laser.

2.2.1 Energy Levels

To start, let us look at the simplest atom, hydrogen, in which one electron circles a nucleus that contains one proton. The hydrogen atom looks like a very simple solar system, with a single planet (the electron) orbiting a star (the proton). The force holding the atom together is not gravity but the electrical attraction between the positive charge of the proton and the negative charge of the electron.

In a real planetary system, the planet could orbit at any distance from the star. However, quantum mechanics allows the electron in the hydrogen atom to occupy only certain orbits, shown in Figure 2-5. We show the orbits as circles for simplicity, but we cannot see exactly what the orbits look like. Their nominal sizes depend on a "wavelength" assigned to the electron because matter, too, sometimes can act like a wave. The innermost orbit has a circumference of length equal to one wavelength, the next orbit a circumference of two wavelengths, and so on.

If we added energy to a planet, it would speed up and move further from its star. The electron in a hydrogen atom also can move to more distant orbits when it absorbs extra energy. However, unlike the planet, the electron can occupy only certain orbits, usually called energy levels, which are plotted at the side of Figure 2-5, with labels indicating the corresponding orbits. The atom is in its lowest possible energy level—the ground state—when the

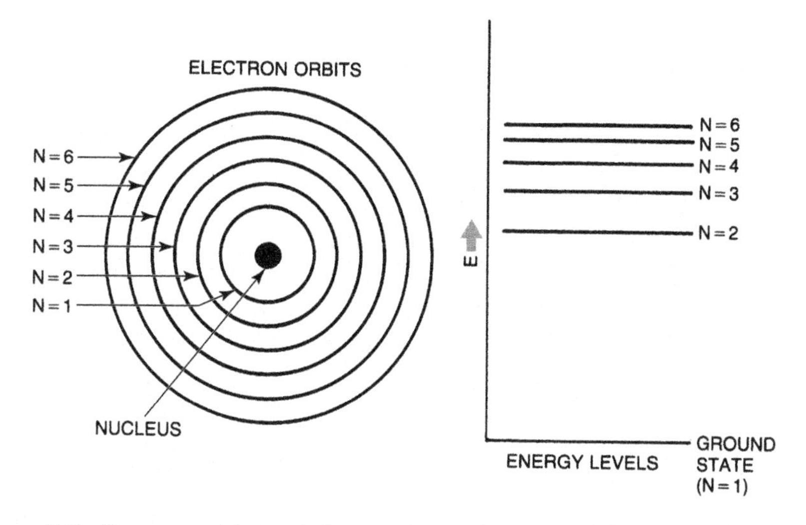

Figure 2-5. Electron orbits and the corresponding energy levels of the hydrogen atom.

electron is in the innermost orbit, closest to the proton. (The electron cannot fall onto the proton.)

The spacing between energy levels in the hydrogen atom decreases as the energy levels become higher above the ground state and, eventually, the energy difference becomes vanishingly small. If the electron gets enough energy, it escapes from the atom altogether, a process called ionization.

If we define the energy of the ionized hydrogen atom as zero, we can write the energy of the unionized hydrogen atom E as a negative number using the simple formula

$$E = -R/n^2 \tag{2-6}$$

where R is a constant (2.179×10^{-18} joule) and n is the quantum number of the orbit (counting outward, with one being the innermost level).

Hydrogen is the simplest atom in the universe, so physicists use it as the standard model to explain quantum mechanics and atomic energy levels. Naturally, things get considerably more complicated in atoms with more electrons. Electron energy levels become far more complex, with each electron occupying an energy state that is specified by multiple quantum numbers, which physicists interpret as identifying quantities including the shell, the subshell, position in a shell, and spin. You do not need to know all the details to understand the basics of laser physics, so we will not cover them.

An important rule called the Pauli exclusion principle says that each electron in an atom must occupy a unique quantum state, with a unique set of quantum numbers. Shells and subshells usually are identified by a number and letter, such as 1s or 2p. The number is the primary quantum number. The letter identifies subshells formed under complex quantum-mechanical rules. Each shell can contain two or more electrons, but each electron must have a unique set of quantum numbers. In the simplest case of two-electron shells, the two electrons spin in opposite directions.

If atoms are in their ground state, electrons fill in energy levels from the lowest-energy or innermost shell until the number of electrons equals the number of protons in the nucleus. The primary quantum number of the last ground-state electron corresponds to the row the element occupies on the periodic table. Table 2-3 shows how electronic energy levels fill up for elements with increasing numbers of electrons, for elements up through strontium (atomic number 38).

These energy levels are the same ones that play a crucial role in chemistry, determining the chemical behavior of the elements and the structure of the periodic table. The elements react with others to fill their outermost electron shell. Atoms with filled outer shells, such as the rare gases argon, neon, helium, krypton, and xenon, tend not to react with other atoms. On the other hand,

Table 2-3. Electronic energy levels or orbitals, by increasing energy, up to element 38 (strontium)

1s								
1s	2s							
1s	2s	2p						
1s	2s	2p	3s					
1s	2s	2p	3s	3p				
1s	2s	2p	3s	3p	4s			
1s	2s	2p	3s	3p	4s	3d		
1s	2s	2p	3s	3p	4s	3d	4p	
1s	2s	2p	3s	3p	4s	3d	4p	5s
Numbers of electrons in shell								
2	2	6	2	6	2	10	6	2

atoms such as sodium, with a single outer electron, or chlorine, with an outer shell missing just one electron, are highly reactive because they need little energy to lose or gain electrons so that they have a full outer shell.

The energy levels that electrons occupy in atoms are simple to describe, but there also are many more types of energy levels. Molecules, like atoms, have electronic energy levels, which increase in complexity with the number of electrons and atoms in the molecule. Molecules also have energy levels that depend on vibrations of atoms within them and on the rotation of the entire molecule. All these energy levels are quantized.

2.2.2 Transitions and Spectral Lines

The transitions that atoms and molecules make between energy levels are crucial to laser physics as well as many other things. To understand how they work, let us start again with the hydrogen atom.

The electron needs an extra increment of energy to move from the ground state to a higher energy level. Conversely, the electron must release energy to drop from a higher level to a lower one. Electrons can do this by absorbing or emitting electromagnetic energy in the form of photons. The amount of energy absorbed or emitted equals the difference in energy between the two states of the electron. (They can also transfer energy in other ways, such as by interacting with other atoms, but for now we will focus on light absorption and emission.)

To move one step up the energy-level ladder from the ground state to the first excited level, an electron must absorb a photon with energy equal to the difference between the two levels, called the transition energy. To drop back down to the ground state, the electron must emit a photon with the transition energy. The same is true for all transitions between any two energy levels—the electron must absorb (or emit) a photon with energy equal to the energy difference between the initial and the final states.

For the simple case of hydrogen, this means that the atom's single electron can absorb or emit light only at certain wavelengths, corresponding to transition energies. Figure 2-6 shows wavelengths corresponding to transitions between the ground state of the hydrogen atom and higher energy levels. The longest wavelength, 121.6 nm, corresponds to the 10.15-electronvolt tran-

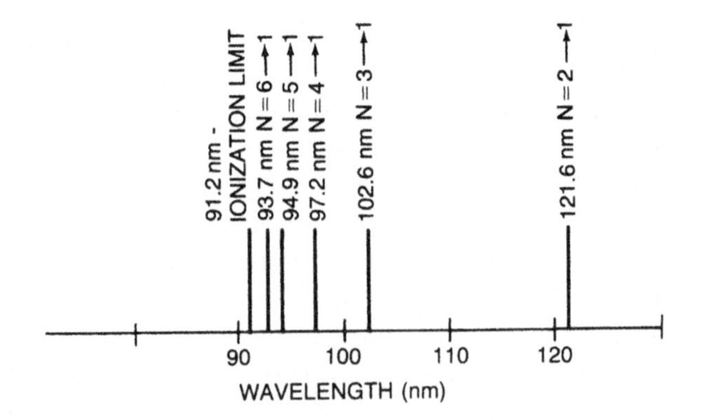

Figure 2-6. Lyman lines of hydrogen, showing energy levels involved.

sition energy between the ground state and first excited level. The next-longest wavelength, 102.6 nm, is the transition between the ground state and the second excited state. The wavelengths of transitions from the ground state to higher energy levels are even shorter, but they eventually reach a limit at 91.2 nm, where the photon has enough energy to remove the electron from the atom, ionizing it.

The series of wavelengths in Figure 2-6 are called the Lyman lines, after American physicist Theodore Lyman who discovered them in the laboratory. There are similar series of transitions at longer wavelengths starting from the first, second, and subsequent excited levels. The wavelengths (λ) of all these transitions are given by the formula

$$\frac{1}{\lambda} = R\left(\frac{1}{m^2} - \frac{1}{n^2}\right) \qquad (2\text{-}7)$$

where m is the quantum number of the lower energy level, n the quantum number of the higher level (the numbers start from 1 for the ground state), and R is the Rydberg constant, 109,678 cm^{-1}. (The cm^{-1} stands for inverse centimeters, and is the number of waves per centimeter; that scale is another way of measuring frequency.)

Hydrogen has such neatly ordered energy levels because it has only a single electron. Add more electrons, and the energy-level structure becomes much more complicated. Electrons interact with each other and with the nucleus, creating additional energy levels, and more transitions, but imposing some restrictions on what transitions are possible. This creates a multitude of possible spectral lines for gas atoms.

Each transition has its own probability, which depends on the population of the two states and their quantum characteristics. Certain types of quantum transitions are far more likely than others, so each transition has its own characteristic lifetime, a factor that enters into laser operation.

Complex interactions between adjacent atoms in liquids or solids blur energy levels so that simple absorption and emission lines are not visible. That creates profound differences between gas and solid-state physics, and between gas and solid-state lasers.

2.2.3 Types of Transitions

Transitions of electrons in the hydrogen atom are good examples because they are simple to describe and understand, but they are only one of several types of energy-level transitions in atoms and molecules. Table 2-4 lists some important examples and their wavelength ranges.

Transitions like those of the hydrogen atom are called *electronic transitions* because they involve electrons moving between two electronic energy levels. Specifically, they are transitions within the outer shell of electrons, the valence shell involved in chemical reactions. Hydrogen only has one populated electron shell, so all its electrons are valence electrons, but heavier elements also

Table 2-4. Representative transitions and their wavelengths

Transition type	Wavelengths	Spectral range
Nuclear	0.0005–0.1 nm	Gamma ray
Inner-shell electronic (heavy element)	0.01–100 nm	X-ray
Inner-shell electronic (Cd^{+20})	13.2 nm	Soft X-ray
Electronic (Lyman-alpha in H)	121.6 nm	Ultraviolet
Electronic (argon-ion laser)	488 nm	Visible, green
Electronic (H, levels 2–3)	656 nm	Visible, red
Electronic (neodymium laser)	1064 nm	Near infrared
Vibrational (HF laser)	2700 nm	Infrared
Vibrational (CO_2 laser)	10600 nm	Infrared
Electronic Rydberg (H, levels 18–19)	0.288 mm	Far infrared
Rotational transitions	0.1–10 mm	Far infrared to microwave
Electronic Rydberg (H, levels 109–110)	6 cm	Microwave
Hyperfine transitions (Interstellar H gas)	21 cm	Microwave

have electrons in inner shells, which do not take place in chemical reactions, Outer-shell transitions span a range of wavelengths from about 100 nm in the ultraviolet to a couple of micrometers in the near infrared. Electronic transitions can occur in molecules as well as atoms, although most electronic laser transitions are in atoms.

Electronic transitions can have significantly higher energies if they involve electron shells that lie inside the valence shells of heavier atoms. Inner-shell electrons are tightly bound to a highly charged nucleus, so it takes a lot of energy to remove them from the inner shell, and a large amount of energy is released when electrons drop back into those shells. Laser action at wavelengths down to about 10 nm has been produced by blasting atoms with intense pulses of visible or infrared light that strip many electrons from the atoms, leaving a highly charged ion. Electrons that rapidly drop into the inner shells emit short-wavelength photons. For example, blasting 20 electrons from cadmium atoms produces Cd^{+20}, which produces laser pulses at 13.2 nm when an electron falls back into the exposed inner shell.

In principle, transitions between energy levels in the nuclei of atoms could release gamma rays, photons with even more energy than X rays. However, we do not know how to make a gamma-ray laser.

On the other hand, very high-lying electronic energy levels (say, levels 18 and 19 of hydrogen) are spaced so closely that transitions between them involve very little energy, with wavelengths in the far infrared, microwave, or even radio-frequency range. These are called Rydberg transitions. They are not likely under normal conditions because very few atoms have electrons in such high energy levels, which are barely bound to the atom. Interestingly, some interstellar hydrogen clouds have been seen to emit light as the result of Rydberg transitions.

Molecules have two other sets of quantized energy levels with lower energy than electronic transitions. Transitions involving vibrations of atoms in the molecule typically have wavelengths of a few to tens of micrometers. Those involving rotation of the entire molecule typically correspond to wavelengths of at least 100 micrometers. Laser action can occur in both vibrational and rotational transitions.

Transitions in two or more types of energy levels can occur at once. For example, a molecule can undergo a vibrational and a rotational transition simultaneously, with the resulting wavelength close to that of the more energetic vibrational transition. Many infrared lasers emit families of closely spaced wavelengths on such vibrational–rotational transitions. Molecules that make vibrational transitions generally make rotational transitions at the same time, which can spread the combined transition over a range of wavelengths.

Transition energies or frequencies add together in a straightforward manner:

$$E_{1+2} = E_1 + E_2 \tag{2-8}$$

where E_{1+2} is the combined transition energy, and E_1 and E_2 are the energies of the separate transition. The same rule holds if you substitute frequency ν for energy. However, you need a different rule to get the wavelengths of combined transitions:

$$\frac{1}{\lambda_{1+2}} = \frac{1}{\lambda_1} + \frac{1}{\lambda_2} \tag{2-9}$$

2.2.4 Absorption

For an atom to make a transition to a higher energy level, it must absorb a photon of the right energy to raise it to that higher energy level. You may think it unlikely that a photon of exactly the right energy to make a particular transition would happen to wander by just at the time an atom is in exactly the right state to absorb it. In fact, it is not that improbable because the world is more complicated than we have admitted.

The first complication is that atoms tend to be in the lowest possible energy state they can occupy, especially at cold temperatures. If you have a cloud of hydrogen atoms floating in intergalactic space, they are likely to be cold and sitting around in the ground state. That means they are likely to absorb light on the Lyman lines of hydrogen shown in Figure 2-6. Sure enough, if you take spectra of the light from distant galaxies, you can see dark lines where interstellar hydrogen atoms absorbed the light. But you are less likely to see dark lines at wavelengths at which hydrogen would have to be in a much higher energy level to absorb the light, because there are few hydrogen atoms in that state. In short, the population of particular states differs in important ways, something you will learn more about shortly.

The second complication is that atoms around us rarely are truly isolated. Atoms and molecules in the air are moving constantly, bumping into each other, and transferring energy. Atoms in solids and liquids are constantly in contact with their neighbors, transferring energy back and forth. An atom in a solid does not have to wait for a photon with exactly the right energy to match a transition; if the energy is close, the atom can transfer some to or from a neighbor to match the transition energy. As a result, absorption is not sharply defined when matter is packed together.

2.2.5 Stimulated and Spontaneous Emission

Light emission seems simpler than absorption. An atom in a high-energy state can make a transition to a lower state whenever it "wants to" simply by emitting a photon of the right energy spontaneously. Spontaneous emission, shown in Figure 2-7A, occurs all by itself, and produces the light we see from the sun, stars, television sets, candles, and light bulbs. An atom or molecule in an excited state releases energy and drops to a lower energy state. On average, it releases the excess energy after a characteristic spontaneous emission lifetime (t_{sp}), which like the half-life of a radioactive isotope measures the time it takes half the excited atoms to drop to a lower state. No outside intervention is required.

Albert Einstein realized that there was another possibility. A photon with the same energy as the transition could stimulate the excited atom to release its energy as a second photon with exactly the same energy. That second photon would be a wave with pre-

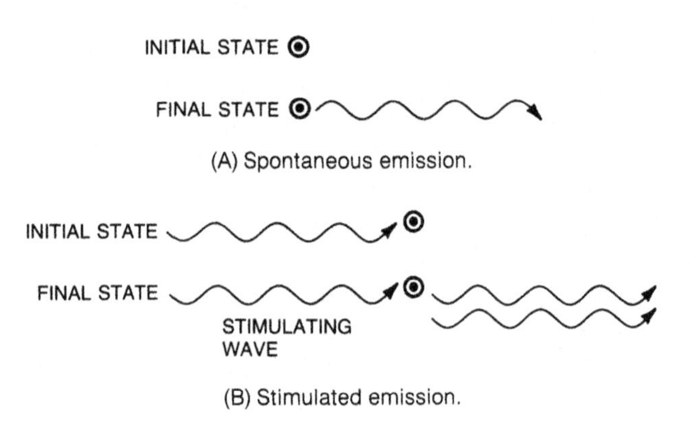

Figure 2-7. Spontaneous and stimulated emission.

cisely the same wavelength as the first, and would be precisely in phase with the stimulating photon, as shown in Figure 2-7B.

You can visualize the process better if you think of the stimulating photon as a wave that tickles the excited atoms so it oscillates at the photon frequency, which matches the transition energy. This oscillation increases the chance the atoms will release the extra energy as a second photon.

Now we have come almost to the point of having a laser. You learned earlier that "laser" was coined as an acronym for "Light Amplification by the Stimulation Emission of Radiation." We have seen that, like a laser beam, stimulated emission is all at the same wavelength and in phase, or coherent. But we have not made a laser beam yet. It took decades for physicists to clear the crucial hurdle needed to amplify stimulated emission.

2.2.6 Populations and Population Inversions

The problem with stimulated emission is that it does not work very well when matter is in *thermodynamic equilibrium,* a state in which atoms and molecules tend to be in the lowest possible energy level. It is a condition that physicists traditionally considered to be the normal state of matter. The atoms are not all in the ground state because they always have some thermal energy as long as the temperature is above absolute zero.

However, if you look at the number of atoms in each energy level, you find that the population is lower in the higher energy levels. Figure 2-8 shows an example, with the number of spots showing the relative population.

The ratio of the numbers of atoms or molecules in states 1 and 2 in thermodynamic equilibrium is given by

$$\frac{N_2}{N_1} = \exp\left[\frac{-(E_2 - E_1)}{kT}\right] \tag{2-16}$$

where N_1 and N_2 indicate the number of atoms in the two states, E_1 and E_2 are the energies of the two states, k is a number called the Boltzmann constant, and T is the temperature in degrees Kelvin. Under normal conditions, if state 2 has higher energy, this is ratio is quite small for transitions at optical wavelengths.

You can see how small it is by plugging in the numbers. The Boltzmann constant is 1.38×10^{-23} joule/Kelvin, or 8.6×10^{-5} elec-

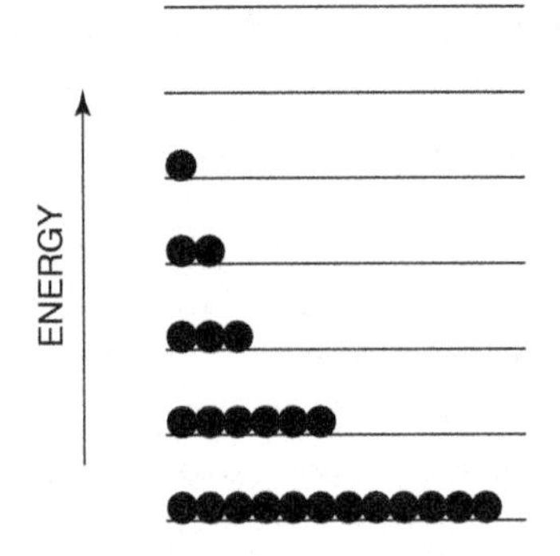

Figure 2-8. Relative populations of energy levels as a function of energy above the ground state at thermal equilibrium.

tronvolt/Kelvin. For a transition of 1 eV (with wavelength slightly longer than 1000 nm) at room temperature (300 K), you get a ratio of 1.5×10^{-17}. That means that virtually all atoms are in the lower state at room temperature for a visible transition.

Why does this make stimulated emission difficult? Because the best way to stimulate emission from an excited atom is with spontaneous emission from an identical excited atom. That would seem easy if you had a lot of excited atoms but, in fact, you would not at thermodynamic equilibrium. You would have trillions of atoms in the ground state ready to absorb the spontaneously emitted photon before it could stimulate emission from another excited atom. Physicists were able to show that stimulated emission existed, but they could not stimulate enough emissions to be of any practical use.

For stimulated emission to dominate, the majority of atoms must be in the excited state, so any spontaneously emitted photons are more likely to stimulate emission than be absorbed by atoms in the ground state. This is called a population inversion, because the population is inverted from the normal situation in which more atoms are in lower levels than in higher levels. Stimulated emission can continue as long as the population inversion exists, but stops once the ground-state population becomes larger and photons are more likely to be absorbed than to stimulate emission.

However, physicists in the 1920s and 1930s did not know how to produce a population inversion. They called such a condition a "negative temperature" because the only way to make the population larger in the higher-energy state shown in Equation 2-10 was to make the temperature T a negative number. Only after World War II did physicists realize they could produce a population inversion for more than a fleeting instant. You will learn more about that Chapter 3.

2.3 INTERACTIONS OF LIGHT AND MATTER

So far, we have examined light in isolation and interacting with individual atoms in fairly abstract terms. However, what matters in most practical situations is how light interacts with matter, and how those interactions can be used to manipulate light. That is the science of optics, and you are going to need to understand something about optics in order to understand lasers.

2.3.1 Optics and Materials

So far, we have seen how individual atoms can absorb or emit light. But when light interacts with matter in other ways, it interacts with bulk material rather than individual atoms. That is because the wavelength of visible light (400 to 700 nm) is more than a thousand times larger than an average atom (around 0.1 nm). How light interacts with a material thus depends on the bulk properties of the material.

We can sort objects into three groups according to how they interact with light:

1. Transparent objects (e.g., glass) transmit light
2. Opaque objects (e.g., dirt or rocks) absorb light
3. Reflective objects (e.g., mirrors) reflect light

However, nothing is perfectly transparent, opaque, or reflective. Thin slices of objects we consider to be opaque actually transmit a little light. Hold the pages of this book together, and they're opaque, but hold a single page up to a bright light and some light will pass through. That is because absorption builds up quickly. If each page transmits 10% of the light striking it, 10 pages stacked together allow just 10^{-10} or one ten-billionth of the original light to get through because each of them only transmits 10%.

Everything reflects some light, whether we consider it transparent, opaque, or reflective. We see the light that objects reflect, and our eyes interpret the amount of reflection as dark or light. Sometimes our eyes can fool us. A bright full moon reflects only about 6% of the incident sunlight, but it looks bright because we see it against the black night sky. Surfaces look white because they reflect light uniformly across the visible spectrum. The best metallic mirrors reflect about 99% of the incident light. The difference between white and shiny is the smoothness of the surface. The white surface is rough, so it scatters light in all directions; the shiny surface is very smooth so it reflects light directly to our eyes and we see reflections.

Transparent materials are in many ways the most important for optics, because light can pass through them. It turns out that nothing is perfectly transparent (other than a perfect vacuum, which is, of course, really nothing). As Section 2.3.6 explains, the transparency varies with wavelength, so materials block some colors but transmit others, as do colored sunglasses. Another complication is that all materials attenuate light a bit, and that the amount of attenuation increases with the thickness of the material, as described in Section 2.3.7. There is also some reflection at the surface of any transparent material, visible as reflections on window panes when you look outside at night.

However, we will start by assuming that materials are reasonably transparent, and deal with those details as we go along.

2.3.2 Refractive Index

The velocity of light in a vacuum, denoted by the letter c, is a universal constant, precisely defined as 299,792.458 kilometers per second, but usually rounded to 300,000 km/sec or 186,000 miles/sec. However, light slows down in matter, even in something as tenuous as air. This effect is measured by the *refractive index,* which equals the speed of light in vacuum divided by the speed of light in the material:

$$n_{\text{material}} = \frac{c_{\text{vacuum}}}{c_{\text{material}}} \tag{2-11}$$

Because the speed of light is faster in a vacuum than in a material, the refractive index normally is greater than one. (There are a few very interesting exceptions on the cutting edge of optical science, but they are not important at this stage of your learning about lasers.) Table 2-5 lists the refractive indexes of some common materials.

The speed of light in a transparent material equals the wavelength times the frequency, as it does in vacuum. Frequency remains constant, but the wavelength equals the wavelength in vacuum divided by the material's refractive index n_{material}:

$$\lambda_{\text{material}} = \frac{\lambda_{\text{vacuum}}}{n_{\text{material}}} \tag{2-12}$$

Lenses are normally used in air, which has a refractive index of 1.000278 at standard pressure, much smaller than that of glass or plastic optics. For most practical purposes, the refractive index of air is treated as equal to 1.000, simplifying calculations by avoiding the need to consider the index of air.

2.3.3 Refraction

This reduction in wavelength arising from the slowing down of light causes an effect known as refraction, which bends the path of light as it enters a material, as illustrated in Figure 2-9. The closely spaced parallel lines in the figure represent the peaks of successive waves, one wavelength apart in the direction in which the

Table 2-5. Refractive indexes of common materials for wavelengths near 500 nm, except where noted

Material	Index
Air (1 atmosphere)	1.000278
Water	1.33
Magnesium fluoride	1.39
Fused silica	1.46
Zinc crown glass	1.53
Crystal quartz	1.55
Optical glass	1.51–1.81*
Heavy flint glass	1.66
Sapphire	1.77
Zircon	2.1
Diamond	2.43
Silicon	3.49 @ 1.4 μm
Gallium arsenide	3.5 @ 1 μm

*Denotes depends on composition. Standard optical glasses have refractive indexes in this range.

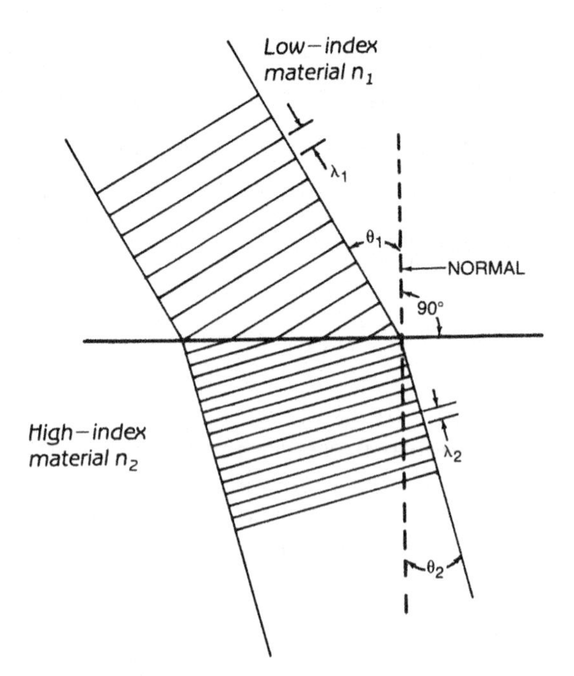

Figure 2-9. Refraction of light waves as they pass from a low-index medium with refractive index n_1 to higher-index medium with index n_2.

light travels. In the low-index material at the top, the peaks are far apart, but the wavelength shrinks when the light enters the higher-index material at the bottom. The peaks stay in phase, so the change in refractive index bends the light in the direction shown. If the light wave passes into the same low-index material on the other side of high-index material, the light waves become longer, and the light bends back. You can see how if you turn the page up-side down.

Refraction only occurs at the border between two transparent materials, not within the materials themselves. The amount of refraction is measured by comparing the direction of the light with the *normal angle,* perpendicular to the surface where refraction takes place, as shown in Figure 2-9. If light traveling in a medium with refractive index n_1 strikes the surface of a material with index n_2 at an angle of θ_1 to the normal, the direction of light in the second material is given by θ_2 by Snell's Law:

$$n_1 \sin \theta_1 = n_2 \sin \theta_2 \tag{2-13}$$

You can also rewrite this to directly calculate the angle of refraction θ_2:

$$\theta_2 = \arcsin\left(\frac{n_1 \sin \theta_1}{n_2}\right) \tag{2-14}$$

where the arcsine is the inverse of the sine.

Suppose, for example, that light in water (n_1 = 1.33) strikes the surface of crystal quartz (n_2 =1.55) at 30 degrees to the normal. The formula tells us that θ_2 = 25.4 degrees, meaning that the light is bent closer to the normal. If the light were going in the other direction, it would be bent further from the normal. If the difference in refractive index were larger, as when going from air into quartz, the change in angle would be larger.

Something different happens if light goes from a high-index material into a low-index material at a steep angle. Suppose, for example, that light in a quartz crystal hits the boundary with air (n = 1.0003) at a 50 degree angle to the normal. Plug the numbers into the equation above, and you find that θ_2 equals the arcsine (1.19). That is impossible because the sine of an angle cannot be greater than one, and your calculator will tell you it is an error if you try to calculate its sine. Try an experiment, and you will find that the light does not escape into the air; it is all reflected back into the quartz, a phenomenon called *total internal reflection* that is the basis of light guiding in optical fibers and that makes diamonds sparkle.

2.3.4 Transparent and Translucent Materials

You may have been confused at some point by the distinction between transparent and translucent materials. Transparent materials are clear, like a window. Translucent materials transmit light, but are cloudy, like milk, wax paper, or ground glass. The difference is that transparent materials let light pass straight through, whereas translucent materials scatter the light rays, blurring them so you cannot see clearly. This is similar to the difference between white and shiny metal surfaces; the white surfaces are rough, and the shiny surfaces are smooth and reflect light directly.

2.3.5 Reflection

Reflection also can redirect light, as we see every day in mirrors. The law of reflection is simplicity itself:

$$\text{Angle of Incidence} = \text{Angle of Reflection} \tag{2-15}$$

This means that if light strikes a mirror at a 50 degree angle to the normal, the mirror reflects it at the same angle.

Note that this law applies specifically to mirror-like or *specular* reflection, from a surface that is smooth on the scale of the wavelength of light. Such a smooth surface reflects light back at the angle of incidence, just as we expect a mirror to do. Because the wavelength of visible light is only 400 to 700 nm, a specular surface must be very smooth. However, metal and glass surfaces can be made that smooth.

A surface that is rough on a small scale reflects light *diffusely.* Each point reflects light at the angle of incidence, but the surface is so rough that the overall effect is to scatter light in all different directions. You can compare diffuse reflection to balls bouncing off a pile of rocks, and specular reflection to balls bouncing off a flat floor. The paper of this book reflects light diffusely.

Household mirrors reflect light from a metal film on their back side, so light is refracted going back and forth through the glass to the reflective surface. Mirrors used in high-performance optical systems are made to reflect from their front surface. Such *front-surface mirrors* require more care than household mirrors, because their reflective surfaces are directly exposed to the environment. However, they reflect more light, and are much more accurate optically. Front-surface mirrors may be made of solid metal, or made of glass coated with a reflective surface coating.

The surface of a transparent material also reflects some light due to an effect called Fresnel reflection. People learn to recognize this reflection as a subtle cue to avoid walking through glass doors, although a bit of dirt also helps. The amount of surface reflection depends on the difference in refractive index between two materials, n_1 and n_2, which can be reduced in optical systems by coating glass with a lower-index material. The fraction of light that a surface reflects, R, is given by

$$R = \left(\frac{n_1 - n_2}{n_1 + n_2}\right)^2 \tag{2-16}$$

2.3.6 Wavelength-Dependent Effects

The optical properties of materials depend on the wavelength of light. For example, the refractive index of silicon dioxide glass decreases slightly as wavelength increases, as listed in Table 2-6. The overall variation is small, so it might seem an esoteric matter. However, this small variation causes glass prisms to refract different colors at slightly different angles, displaying the full spectrum of colors. Rainbows arise from the variation of the refractive index of water in tiny droplets in the air.

The transmission of light through transparent materials also can vary with the wavelength. This effect is not visible in glasses used for windows or optics because those materials were selected to be reasonably transparent across the visible spectrum. You can see it in optical filters; for example, a red filter transmits only red light to the eye. It is also important if you work in other parts of the spectrum. Glasses transparent to our eyes block many infrared and ultraviolet wavelengths, for example. Likewise, many materials transparent in the infrared look opaque to our eyes.

Reflection also depends on wavelength. We pick paints and inks because they reflect specific colors. These are things you do have to think about when you are working with lasers. When supermarket scanners were first introduced, a Boston-area company printed bar codes on its white milk cartons in ink that reflected the red light of the laser scanner. The symbol was clearly visible to the human eye, but the scanner could not see the difference because the red ink reflected almost as much red light as the white paper.

2.3.7 Optical Attenuation and Losses

No real optical material is totally transparent. It is possible to reduce losses, but not to eliminate them completely.

Two factors attenuate light passing through a transparent material—absorption and scattering. The amount of absorption de-

Table 2-6. Refractive index of silicon dioxide glass as a function of wavelength

Wavelength (nm)	Index
404.656 nm	1.46961
508.582 nm	1.46187
589.262 nm	1.45847
706.519 nm	1.45515
852.111 nm	1.45248
1013.98 nm	1.45025
1395.06 nm	1.44584

pends on the nature of the material and the wavelength. If the wavelength is close to a transition in the material, it is much more likely to be absorbed than if it is far from a transition. Scattering occurs when light bounces off atoms or molecules and goes in another direction. The chance of scattering increases as the wavelength becomes smaller, and the atoms look bigger to the light. Light scattering is what makes the sky blue; the shorter blue wavelengths are scattered much more than the longer red lengths.

The losses caused by absorption and scattering add together as attenuation, the reduction in light power per unit distance as it travels through a material. If the material is uniform, the same fraction of the input light is lost each unit distance. It is easiest to calculate light transmission by using the fraction of light transmitted per unit distance, denoted α and equal to 1 minus the attenuation. If 90% of the light passes through a one-centimeter thickness (attenuation of 0.1/cm), the amount passing through 10 cm is the amount passing through one cm raised to the tenth power, $(0.9)^{10}$ or 35%. The general formula is an exponential equation:

$$I = I_0 \exp^{-\alpha D} \tag{2-17}$$

where I is the intensity of light that has passed through a distance D of the material, I_0 is the initial intensity, and α is a measure of the fraction transmitted through a unit distance.

2.3.8 Diffraction

Diffraction is an effect that arises from the wave nature of light. Look back at how light passes through a pair of slits in Figure 2-3, and you will see that the light waves radiate from each slit. The light does not travel in a straight line; it spreads out in waves from the edge. You can also see diffraction if one edge of the slit is removed, leaving the other edge to cast a shadow; the diffracted waves would spread toward the shaded area. You can see a similar effect if a barrier blocks waves on the surface of water.

When diffracted waves overlap, they interfere with each other, as Figure 2-3 shows in the two-slit experiment. The most striking optical effect we see in diffraction occurs when white light is diffracted from an array of many parallel lines. The angles at which the waves diffract depend on wavelength and spacing of the lines, and interference among the diffracted waves concentrates different wavelengths at different angles, spreading out a rainbow of light. The easiest way to see the effect is by reflecting light from a CD or DVD; the spiral patterns of data recorded on the disks act like a diffraction grating.

Diffraction effects are important in lasers because they put a lower limit on the spreading of a laser beam with distance, an effect called beam divergence, described in Chapter 4, Section 4.3.1.

2.4 BASIC OPTICS AND SIMPLE LENSES

The next step is to move from basic principles such as refraction and reflection to simple optical devices such as lenses, which focus light.

The way lenses work usually is described by imagining that incoming light is a number of parallel rays that are bent by the lenses. Optical designers use software that follows the paths that these rays take through complex optical systems. These rays do not really exist, but they are a convenient fiction because they show how optics manipulate light.

Refraction occurs when light passes between two materials with different refractive indexes. If you think of the path of light as a ray, it bends at the surface in a way determined by Snell's Law (Eq. 2-13), depending on the angle of incidence and the refractive indexes of the two materials. If the light is going through a window, with two flat surfaces parallel to each other, going from glass back into air simply bends the light rays back to their original angle. That is why a scene looks the same through a window as if the window were open. (Old window glass may distort the view if its two surfaces are not perfectly flat.)

Focusing light rays so they come to a point or form an image requires curved surfaces, either reflective surfaces on mirrors or refractive surfaces on transparent lenses.

Lenses have one or two curved surfaces. Figure 2-10 shows a sampling of important types of simple lenses and how they affect parallel light rays. Lenses that are thicker at the center than at the sides are called *positive lenses* and bend the light rays so they come together at a *focal point.* If the lens is thicker at the edges—a *negative lens*—it spreads out or diverges the light rays as shown. Properties of the lenses depend on their surface curvature, refractive index, and size.

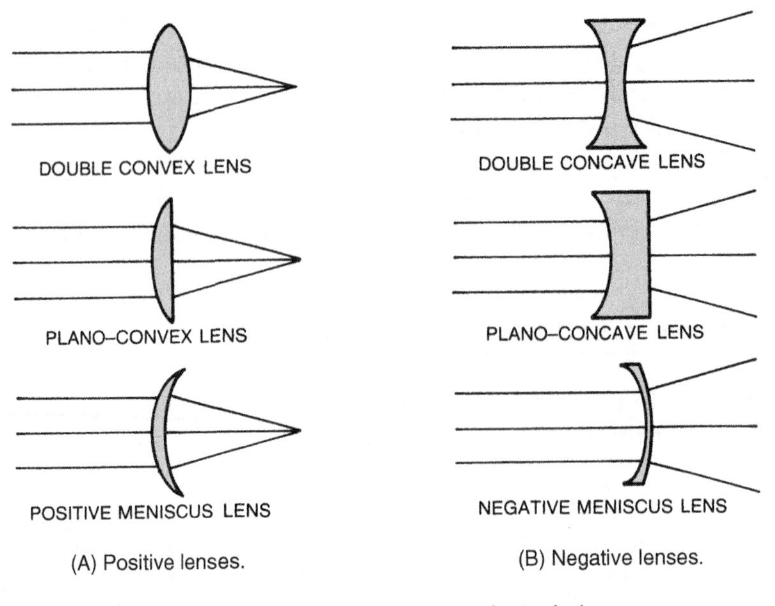

Figure 2-10. The six basic types of simple lenses.

2.4.1 Positive Lenses

The most important parameter of any lens is its *focal length.* For a positive lens, this is the distance from the center of lens to the point at which it focuses parallel light rays to a point, as shown in Figure 2-11. Most simple lenses have spherical surfaces, each with its own "*radius of curvature,*" defined as the distance from the

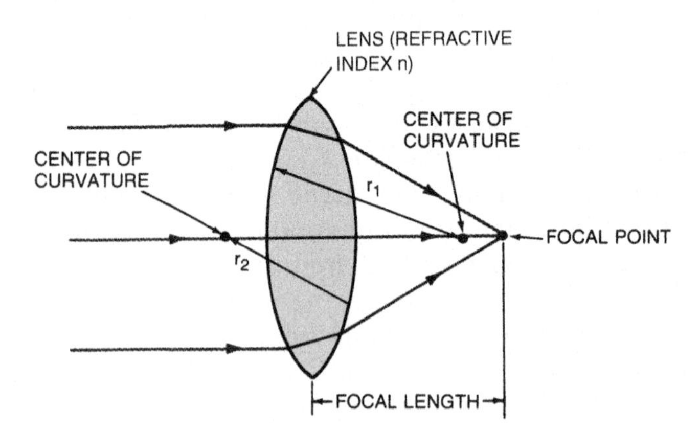

Figure 2-11. Important parameters of a positive lens.

spherical surface of the lens to the point at the center of the sphere. The two surfaces may have different curvatures, or one side may be curved and the other flat.

The general formula for the focal length f of a lens with radii of curvature of R_1 and R_2, and refractive index n is:

$$f = \frac{1}{(n-1)\left[\frac{1}{R_1} + \frac{1}{R_2}\right]} \quad (2\text{-}18)$$

That formula is easy to apply for double-convex lenses, in which both surfaces are like slices of the skin of a ball. If one surface is flat, as in a plano-convex lens, that radius of curvature is considered infinite, making the value of that $1/R$ term equal to zero. A surface that is concave, or curved inward, has a negative radius of curvature.

These examples assume that all light rays from an object are parallel, but that only happens if the object is extremely distant, like the sun. Thus, a lens can focus the rays to a small spot, producing enough energy to burn paper on a sunny day.

If an object is nearby, the light rays reaching the lens are diverging and the lens focuses them form an image of the object at a point beyond the focal length, as shown in Figure 2-12. The distance at which the real image forms depends on the focal length f of the lens and the distance D_o of the object from the lens. The distance of the image D_l is

$$\frac{1}{D_l} = \frac{1}{f} - \frac{1}{D_o} \quad (2\text{-}19)$$

or

$$D_l = \frac{1}{\left(\frac{1}{f} - \frac{1}{D_o}\right)} \quad (2\text{-}19)$$

You will note that the formula gives a negative number for the image distance if the object is closer to the lens than the focal length; that means you do not get an image.

The image size depends on the distance. The ratio of image height h_i to object height h_o is called the magnification ratio m. It also depends on the distance between the lens and the object:

$$m = h_i/h_o = D_i/D_o \quad (2\text{-}21)$$

As the formula indicates, the magnification is greater than one if the object is near to the lens and the image is far away, meaning that the image is larger than the object. If the image is nearer to the lens than the object, the magnification is smaller than one, and the image is smaller than the object. In Figure 2-12, the object and the image are at equal distances from the lens, and the magnification is one.

The image focused by the positive lens in Figure 2-12 is called a *real image* because you can focus it onto a surface and see it projected there. It is real because the light rays form the pattern of the image at that point. Optics specialists also speak of *virtual images,* which you can see with your eyes but cannot project onto a screen because the light rays do not actually come together in space to form them. The image you see in a magnifying glass is one example of a virtual image; your eyes can see it, but it cannot be projected onto a screen. Note that in this case the distance from the object to the lens is less than the focal length, a case for which the lens formula predicts you will not get a real image.

Both positive and negative lenses can produce virtual images. However, only positive lenses can bring light rays together to form real images.

2.4.2 Negative Lenses

The same optical laws apply for negative lenses as for positive lenses, but they work in somewhat different ways. The focal

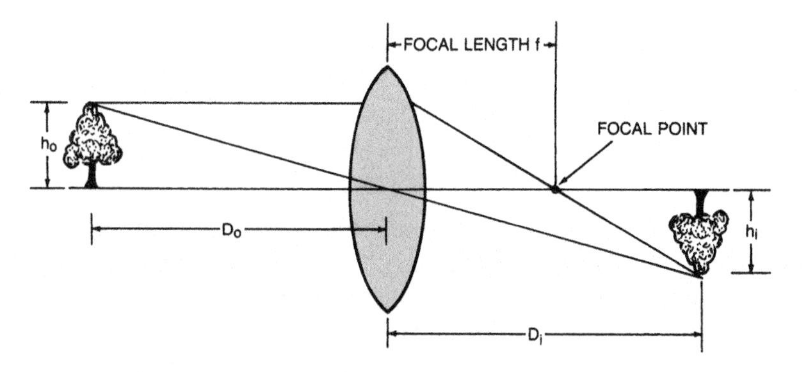

Figure 2-12. A positive lens forms a real image of an object.

lengths of negative lenses are negative. You can get this result mathematically by assigning negative radii of curvature to concave lens surfaces. Its physical meaning is that they do not focus parallel light rays to come together. Pass parallel light rays through a negative lens, and they seem to spread from a point behind the lens, as shown in Figure 2-13. The distance from the midplane of the lens to this point is the focal length; it has a negative value because it is on the opposite side of the lens from the focal point of a positive lens. (Optical designers use sign conventions to keep these things straight.) Because negative lenses do not bring parallel light rays together, they do not form real images, but they do form virtual images.

The equations giving magnification and image distance are not relevant for negative lenses because they apply only to real images.

2.4.3 Mirrors

Curved mirrors, like lenses, can focus parallel light rays or cause them to spread apart. Mirror surfaces, like lens surfaces, can be either concave or convex. A *concave mirror* seems hollow, like the inside of a bowl. Like a positive lens, a concave mirror with a spherical surface can form a real image or focus light rays to a point. A *convex mirror* has a rounded surface, like the outside of a ball; reflection spreads out light rays like a negative lens, and it cannot form a real image.

Like lenses, spherical mirrors have focal lengths, which indicate the distance from the mirror at which they focus parallel

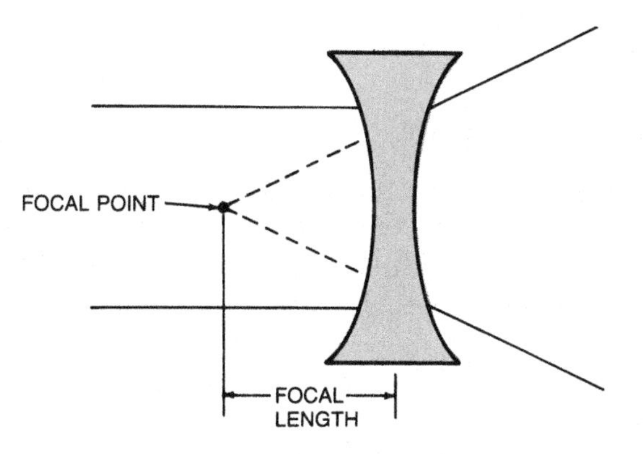

Figure 2-13. Important parameters of a negative lens.

rays. The focal length f is half the mirror's radius of curvature R:

$$f = R/2 \tag{2-22}$$

Following the same sign convention as for lenses, we assign a positive radius of curvature to a concave surface and a negative radius to a convex surface. This gives a concave mirror a positive focal length and a convex mirror a negative focal length.

As with positive lenses, the distance of a real image from a concave mirror depends on the object distance. If the object is a distance D_o away from a concave mirror with focal length f, the distance of the real image D_i is given by

$$D_1 = \frac{1}{\left(\frac{1}{f} - \frac{1}{D_o}\right)} \tag{2-23}$$

Likewise, the magnification m—the ratio of object height h_o to image height h_i—is given by

$$m = h_i/h_o = D_i/D_o \tag{2-24}$$

Look back at Equations 2-19, 2-20, and 2-21 for positive lenses, and you that see they look the same except for a rearrangement of terms. Some optics books use sign conventions that make the terms look different (by assigning a negative sign to distances measured in one direction), but the equations are otherwise the same because the same fundamental thing is happening—light rays are being focused.

2.4.4 Optical Complexities

Optics is a complex field that goes far beyond the realm of the simple lenses and mirrors we have described so far. If you are going to delve seriously into optics, you should get a book that deals specifically with optics. Appendix C includes some suggestions.

In this book, we treat optics as an adjunct to laser technology. That means we cover the basic elements of optics you need to know to understand lasers, but not the wealth of detail involved in optics per se. So far, we have given only a brief introduction to the field. In Chapter 5, we will talk more specifically about optics used with lasers.

2.5 WHAT HAVE WE LEARNED?

- Light acts both like waves and like massless particles called photons.
- Light is an electromagnetic wave, and can be characterized by its photon energy, wavelength, or frequency.
- Wavelength is the distance between wave peaks; frequency is the number of wave peaks passing a point per second.
- Wavelength times frequency equals velocity.
- The speed of light in vacuum is 300,000 kilometers per second.
- Photon energy equals Planck's constant (h) times frequency.
- The electromagnetic spectrum ranges from radio waves to gamma rays.
- Visible wavelengths are 400 to 700 nm.
- Optical wavelengths include parts of the infrared and ultraviolet.
- Adding the amplitude of light waves produces interference. It is constructive if the waves are in phase and destructive if the waves are 180 degrees out of phase and cancel each other.
- Quantum physics recognizes that energy comes in discrete chunks or quanta.
- Quantum numbers specify electronic energy levels.
- An electron loses energy if it drops to a lower energy level. It must absorb energy to make a transition to a higher level.
- Electronic transitions occur in atoms and molecules.
- Vibrational and rotational transitions occur in molecules and involve less energy than electronic transitions.
- Excited states can emit energy spontaneously or when stimulated by a photon with the same energy.
- Lasers generate stimulated emission.
- A population inversion occurs when more atoms or molecules are in a high-energy state than in one with lower energy, and is required for laser operation.
- Objects may transmit, absorb, or reflect light.
- The refractive index of a material equals the speed of light in vacuum divided by the speed of light in the material.

- Refraction bends light rays as they pass between transparent media. Snell's law shows how this depends on the refractive index and the angle of incidence.
- The angle of incidence equals the angle of reflection.
- The refractive index and light absorption of a material vary with the wavelength of light.
- Light transmission depends on transparency and thickness of a material; the intensity drops exponentially with distance traveled.
- Diffraction is the spreading of waves from an edge.
- A positive lens focuses parallel light rays to a spot. A negative lens causes parallel rays to diverge.
- The focal length of a positive lens is the distance from the lens to a point at which it focuses parallel light rays.
- A positive lens can project a real image.
- A virtual image cannot be projected, but can be seen by the eye.

WHAT'S NEXT?

In Chapter 3, we will use what we have learned about optics and physics to describe how lasers work in general. Later, we will learn about specific types of lasers, which differ greatly in detail.

QUIZ FOR CHAPTER 2

1. A carbon-dioxide laser has a nominal wavelength of 10.6 micrometers. What is its frequency?
   a. 300,000 hertz
   b. 2.8×10^{13} hertz
   c. 1.06 gigahertz
   d. 2.8×10^{10} hertz
   e. None of the above
2. What is the photon energy for an infrared wave with frequency of 10^{14} hertz?
   a. 10.6 micrometers
   b. 6.63×10^{-34} joule
   c. 6.63×10^{-20} joule
   d. 10.6×10^{20} joules
   e. About one joule

3. What is the wavelength of the infrared wave in Problem 2?
   a. 300 nanometers
   b. 3 micrometers
   c. 10.6 micrometers
   d. 30 micrometers
   e. 10.6 nanometers
4. How do two light waves of the same wavelength and amplitude interfere if they are 180 degrees out of phase?
   a. Destructively
   b. Constructively
   c. Partially, producing a wave with amplitude $\sqrt{2}$ times the two input waves
   d. Not at all
5. Calculate the wavelength of the transition in the hydrogen atom from the $n = 2$ energy level (the second orbit) to the $n = 3$ level. This is the first line in the Balmer series of spectral lines.
   a. 121.6 nm
   b. 91.2 nm
   c. 656 nm
   d. 632.8 nm
   e. 900 nm
6. An electron in the ground state first absorbs one photon at 500 nm, then a second at 1000 nm, to reach an excited state. What is the wavelength of the photon it would emit if it dropped all the way back to the ground state?
   a. 200 nm
   b. 225 nm
   c. 300 nm
   d. 333 nm
   e. 1500 nm
7. At thermodynamic equilibrium and room temperature (300 °K) what is the ratio of populations at the upper and lower levels of a transition with photon energy of 0.1 electronvolt? (That is, with $E_2 - E_1 = 0.1$ eV.)
   a. 0.0207
   b. 0.27
   c. 1
   d. 0.001
   e. 0.000009
8. What is the ratio in the above problem if the temperature is reduced to 100 °K?

a. 0.0207
b. 0.27
c. 1
d. 0.001
e. 0.000009

9. Light in a medium with refractive index of 1.2 strikes a medium with refractive index 2.0 at an angle of 30 degrees to the normal. What is the angle of refraction (measured from the normal)?
   a. 42 degrees
   b. 20.1 degrees
   c. 18.0 degrees
   d. 17.5 degrees
   e. 15.6 degrees
10. A material has an attenuation coefficient of 0.5 per centimeter and its scattering coefficient is negligibly small. What fraction of incident light can pass through a 2-centimeter thickness?
   a. 0.5
   b. 0.25
   c. 0.082
   d. 0.01
   e. None; the material is opaque
11. A positive lens with focal length of 10 centimeters forms a real image of an object 20 centimeters away from the lens. How far is the real image from the lens?
   a. 5 cm
   b. 10 cm
   c. 15 cm
   d. 20 cm
   e. 25 cm
12. What is the ratio of image size to object size for the case in Problem 11?
   a. 2
   b. 1.5
   c. 1
   d. 0.667
   e. 0.5

CHAPTER 3

HOW LASERS WORK

ABOUT THIS CHAPTER

In Chapter 2, we learned the basic physics and optics needed to understand how lasers work. In this chapter, we will learn what goes on inside lasers—how they are energized and how they produce beams. Here we will learn about the general principles of laser operation; Chapter 4 will describe the important characteristics of lasers.

3.1 BUILDING A LASER

The best way to understand how a laser works is to build one. In this book, we can only build a laser on paper, but we can follow the path of the original developers of lasers. They sat down, sketched out what needed to be done to make their idea work, then figured out how to proceed before they started building anything.

We have the obvious advantage that we know where we are going. Hindsight is always 20-20, so we do not have to pursue all the dead ends that wasted the original developers' time and energy, but reproducing the key steps will help us understand how the pieces go together. Each step is a separate section of this chapter.

The first step toward our conceptual laser is figuring how to produce the population inversion needed for stimulated emission, covered in Section 3.2.

Once we have the population inversion needed for stimulated emission to dominate, Section 3.3 describes how a resonant cavity helps amplify the light enough to make an oscillator.

Then Section 3.4 considers how oscillation creates resonant conditions and generates a laser beam.

Understanding Lasers: An Entry-Level Guide, Third Edition. By Jeff Hecht

Section 3.5 tells how the resonator and laser medium combine to select a laser wavelength.

Finally, Section 3.6 goes back and looks more closely at exciting the laser medium to produce a population inversion.

To keep the discussion simple, we will talk about atomic states rather than molecular states, but remember that the same principles apply to molecules.

3.2 PRODUCING A POPULATION INVERSION

The first requirement for making a laser is to be able to produce stimulated emission, which requires a population inversion. Specifically, we are going to consider a population inversion between generic energy levels. The *upper laser level* is the high-energy state from which an atom drops when it emits stimulated emission. The *lower laser level* is the lower-energy state into which the atom drops.

At thermodynamic equilibrium, most atoms are in lower energy levels, with the population decreasing as the energy increases, according to Eq. 2.10. That means that at equilibrium, more atoms are in the lower laser level than in the upper laser level. Thus, if an atom in the upper state happens to release a photon on that transition, it is more likely to encounter an atom in the lower state than an atom in the higher state, so it is more likely to be absorbed than to stimulate emission. Thus, you need a population inversion in order to get the cascade of stimulated emission needed for laser emission.

In principle, population inversions are *possible* on many different transitions. In practice, a population inversion is *necessary* only between a single pair of energy levels to demonstrate laser emission. (In fact, because it takes energy to create a population inversion, it is best to concentrate that energy on inverting one specific transition to make the laser more efficient, but that is beyond our scope here.)

Long before the laser was proposed, physicists had made extensive measurements of the wavelengths absorbed and emitted by various atoms, a field called *spectroscopy*. From this, they had deduced atomic transitions, energy levels, and the structure of electron shells. Laser developers used this spectroscopic data to identify potential laser transitions.

3.2.1 Excitation Mechanisms

It takes energy to excite atoms to produce a population inversion. Merely heating the material is not enough. Heating will increase the population in the higher level, but will not produce a population inversion, at least when things settle down to equilibrium. Equation 2-10 tells us that as long as matter is in equilibrium, more atoms will always be in a lower energy level than in a higher one. Heating the material increases the average energy but cannot produce a population inversion at equilibrium.

Producing a population inversion requires selectively exciting atoms so that more wind up in the upper laser level than in the lower laser level. To visualize what is necessary, think of energy levels as narrow flat steps on a staircase and the atoms as little marbles. Put a marble on a step, and soon it will roll to the edge and drop to a lower step, just as atoms drop to lower energy states by spontaneous emission. The only way to keep a number of marbles on the upper steps is to keep putting them there continually. In the same way, producing a population inversion requires adding energy to keep exciting atoms to the upper laser level.

Selective excitation is tricky. One approach is to illuminate the atoms with light at a wavelength that has enough energy to excite them to a high-energy state. Another is to transfer energy from electrons in an electric current passing through a gas or semiconductor. We will examine these ideas in Section 3.6, but for now let us consider excitation in general.

3.2.2 Masers and Two-Level Systems

The first major step on the road to the laser was the microwave-emitting maser. To produce the population inversion he needed, Charles Townes started with a beam of ammonia molecules that contained molecules in both the ground state and a higher-energy state. Then he passed them through a magnetic field, which bent the paths of the ground-state and excited-state molecules in different directions, separating them. He then directed the excited molecules into a chamber where spontaneous emission from some excited molecules stimulated emission of microwaves from other excited molecules to sustain oscillation on the 24-GHz ammonia transition.

Physically separating excited molecules or atoms from those in a lower state showed it was possible to produce a population inversion and enough stimulated emission to sustain oscillation. But it was a cumbersome, brute-force approach. More practical masers, and the later development of the laser, required ways to selectively excite atoms to particular states, and to see how that works, we need to look at the lifetimes of excited states.

3.2.3 Metastable States and Lifetimes

Normally, atoms in excited states release their extra energy by spontaneous emission within nanoseconds (billionths of a second) after they reach the excited state. That is so short that most excited atoms drop to a lower energy level before spontaneous emission from another atom can reach them to stimulate emission. What is needed is a longer-lived excited state.

Such states do exist. They are called *metastable* because they are unusually stable on an atomic timescale, although they may only last for a millisecond (a thousandth of a second) or a microsecond (a millionth of a second). They exist because the quirks of quantum mechanics affect the probability of atoms dropping to a lower energy level. For atoms to make transitions between energy levels, they must change their quantum states, and some changes are less probable than others. Fast transitions are ones that are easy for atoms; metastable transitions are less likely.

To go back to our mechanical analogy, imagine that short-lifetime states are steps with surfaces that curve down, so the marbles quickly roll off them. Metastable steps have surfaces that are flat or curve up a bit, so the marbles can stay in place longer.

A metastable state works very well as an upper laser level because atoms stay in that state long enough to build up a large population in the upper laser level. Some atoms will always drop to the lower energy level by spontaneous emission, but the longer the spontaneous-emission lifetime, the slower the leakage, and the easier it is to maintain the population inversion needed for stimulated emission to dominate.

3.2.4 Three- and Four-Level Lasers

Generally it is impractical to excite atoms directly into a metastable state, so the simplest approach uses three distinct energy levels, shown in Figure 3-1. In this case, the atoms start in the ground state, which also is the lower laser level. Energy from an

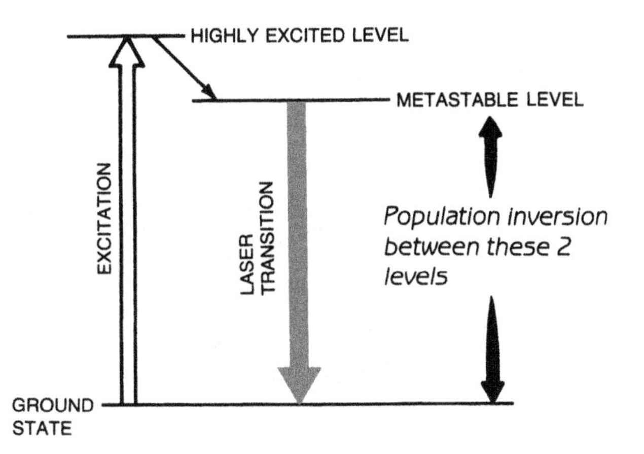

Figure 3-1. Energy levels in a three-level laser.

external source excites the atoms to a short-lived energy level slightly above the metastable level. Atoms quickly drop from that high level into the metastable level, which is the upper laser level, which typically has a lifetime a thousand times or more longer than the higher level. This process accumulates a large population in the metastable level, creating a population inversion between it and the lower laser level. The few photons spontaneously emitted by the metastable state are thus more likely stimulate emission from one of the abundant excited atoms than to be absorbed by one of the few atoms in the lower laser level. This allows stimulated emission to dominate.

Maiman's ruby laser is an important example of this three-level scheme. Although the system works, it is not ideal. One problem is that the ground state is also the lower laser level, so a majority of atoms must be excited out of the ground state to the upper laser level to produce a population inversion. For ruby, this requires an intense pulse of light, for which Maiman used a bright flashlamp. The population inversion is very difficult to sustain, so most three-level lasers operate in pulsed mode. (In a few cases, notably the erbium-fiber laser described in Section 8.8, such a laser can emit a continuous beam.)

Most practical lasers involve four energy levels, as shown in Figure 3-2. As in the three-level laser, the atom is excited from the ground state to a short-lived, highly excited level. The atom then drops quickly to a metastable upper laser level. Stimulated emission on the laser transition drops the atoms to the lower laser

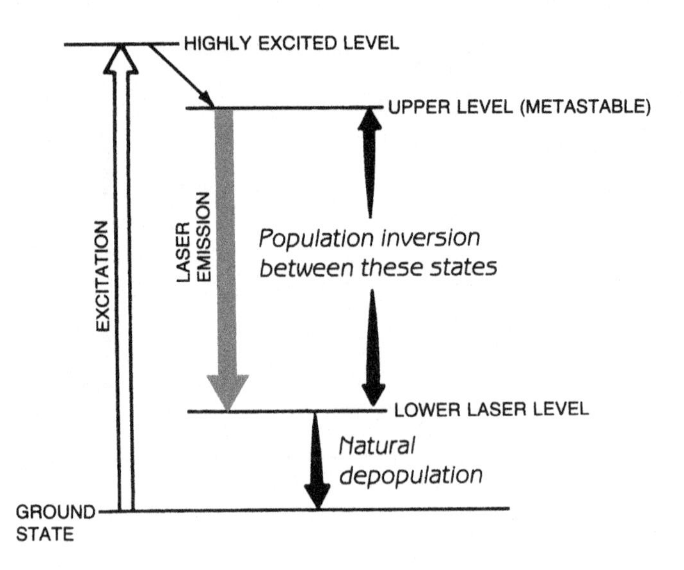

Figure 3-2. Energy levels in a four-level laser.

level, which in the four-level laser is a state above the ground level, as shown in Figure 3-2. The atoms then drop from the lower laser level to the ground state, releasing energy by spontaneous emission or other processes.

The key difference is that the lower level is not the ground state. Why is this so important? Because, normally, most atoms are in the ground state, so it has to be depopulated to produce a population inversion. That requires an intense light pulse to excite a majority of the atoms, like the bright pulses from Maiman's flashlamp. If the lower laser level is above the ground state so it starts with a low population, exciting a small fraction of the atoms to the upper laser level is enough to produce a population inversion. That is easier to do, and also makes it much easier to sustain continuous laser action.

A four-level laser does not automatically assure continuous laser operation because atoms can accumulate in the lower laser level, ultimately stopping laser emission when the population inversion runs out. To prevent that problem, the lower laser level should have a shorter lifetime than the upper laser level so it is steadily depopulated by transitions to the ground state or other lower levels. Then a laser can produce a steady beam (called *continuous-wave* emission). If atoms accumulate in the lower laser level, the laser is limited to pulsed operation, with an excitation pulse followed by stimulated emission and depopulation of the lower laser level before the excitation pulse is fired again.

If you look carefully at the operation of real lasers described in later chapters, you will see that the actual energy-level structures are more complex. Excitation is not always to a single high level; it may be to a group of levels, which decay to the same upper laser level. That improves efficiency by exciting atoms over a wider range of energies than would be possible if they could be raised to only a single state. The depopulation of the lower laser level also can be complex, and may involve a series of steps. The upper and lower laser levels can be much farther above the ground state than indicated in Figures 3-1 or 3-2, which makes it easier to control their populations, but limits overall operating efficiency.

Efficiency also can be improved by adding other types of atoms to the laser mixture. For example, the helium in a helium–neon laser captures energy from electrons passing through the gas, producing excited helium atoms that transfer their energy to neon atoms. That excitation process produces a population inversion in the neon gas and emission on transitions of neon. Other gases may be added to mixtures to depopulate lower laser levels.

In addition, one laser medium may be able to produce stimulated emission on two or more transitions. These may be closely spaced upper levels that drop to different lower levels and generate distinct wavelengths simultaneously. The choice of cavity optics may allow simultaneous operation or select a single wavelength. For example, the helium–neon laser emits at several lines in the visible and near infrared; the choice of cavity optics determines which line the laser emits. Other lasers, such as carbon dioxide, can simultaneously operate on many closely spaced wavelengths because the upper and lower levels are split up into many sublevels.

3.2.5 Natural Masers and Lasers

A major reason that decades passed between Einstein's prediction of stimulated emission and the first working laser was that physicists thought population inversion would be very difficult to achieve. They had no idea that stimulated emission might be visible in nature, but in the 1960s radio astronomers discovered that stimulation emission occurred naturally in outer space.

Natural masers occur in gas clouds near hot stars. Starlight excites simple molecules such as water or hydroxyl (OH) to high energy levels, and the molecules drop down the energy-level ladder to a metastable state. In a dense gas, the molecules would quickly lose the extra energy by collisions, but the gas is thin enough for them to retain the energy and produce stimulated emission at microwave and infrared frequencies.

Unlike their man-made counterparts, cosmic masers do not generate beams. Instead, they radiate stimulated emission in all directions, like other hot clouds of interstellar gas radiate light. In fact, without special instruments they look just like other gas clouds. Astronomers did not realize that the clouds were radiating stimulated emission until they carefully analyzed the infrared and radio spectra of the clouds, and spotted the anomalously bright emission lines.

3.3 RESONANT CAVITIES

Cosmic masers show that population inversions are feasible, and that stimulated emission can produce light on narrow lines. However, it takes more than a population inversion to make a laser. It takes a resonant cavity that helps build up stimulated emission.

The resonant cavity for a microwave maser was a metal box that resonated at the microwave frequency. Essentially, the cavity echoed, like sound waves bouncing off the walls of a shower stall if you hit the right note. Such resonances happen when a whole number of waves fits exactly into the box, or, more precisely, when an integral number of wavelengths equals the distance the wave has to travel to make a round trip in the box. Figure 3-3 shows the idea for a round trip between one pair of walls.

For a microwave maser, the wavelengths are measured in centimeters and the resonant cavity is no more than a few times the wavelength, as shown in Figure 3-3. Laser wavelengths are much shorter, 0.4 to 0.7 micrometer for visible light, so it is very hard to make little boxes for them to resonate in. Instead, we need to look at the workings of a laser in a different way, as the amplification of light contained in a different kind of cavity.

3.3.1 Amplification and Gain

Amplification is the process of increasing the strength of a signal. To amplify light, you generate more photons at the same wave-

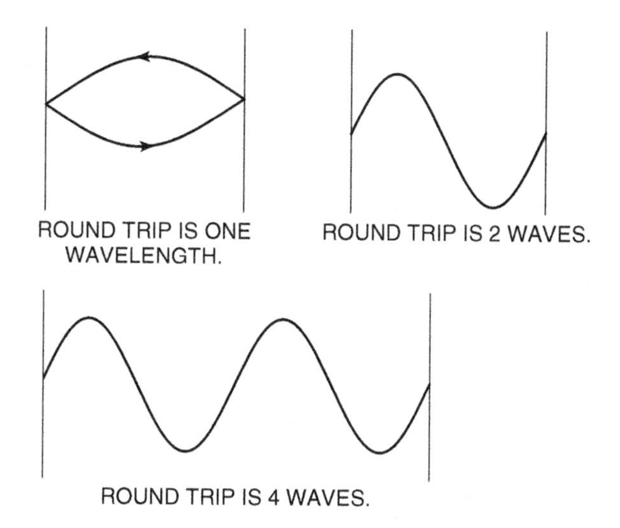

Figure 3-3. Waves resonate in a laser cavity when a round-trip is an integral number of waves.

length as the input signal. Stimulated emission can do this if you illuminate an excited medium with a population inversion at a wavelength that matches the input signal. (Remember that "laser" comes from light amplification by the stimulated emission of radiation.) An early application of the microwave maser was to amplify extremely weak microwave signals from astronomical objects. The maser could only amplify wavelengths that matched the energy of the inverted transition, but this worked fine if you matched the maser material to the wavelength being studied.

Things work differently if you are trying to generate your own signal rather than amplify one from outside. In this case, a population inversion comes with its own built-in source of photons—spontaneously emitted light is at just the right wavelength to stimulate emission from other excited atoms. You saw this in Figure 1-3. Spontaneous emission on the laser transition generates the first photon. When that photon encounters an atom in the upper laser level, it stimulates it to emit a second photon and drop to the lower laser level. As long as there is a population inversion, the photon is more likely to stimulate emission than to be absorbed by an atom in the lower laser level.

After the first stimulated emission, we have two photons of the same energy in phase with each other. Like the original spontaneous emission photon, each of them is more likely to encounter an atom in the upper laser level than in the lower level. Thus, they, too, are likely to produce stimulated emission. So are the stimulated-emission photons they produce. You can see the trend—the number of photons produced grows rapidly in a cascade of stimulated emission.

Laser physicists measure the amount of amplification as *gain,* the amount of stimulated emission a photon can generate as it travels a unit distance. For example, a gain of 0.05 per centimeter means that a photon generates an average of 0.05 stimulated emission photon each centimeter it travels. A gain of 2 per centimeter means that a photon generates two more photons each centimeter.

Power increases exponentially with gain because the photons produced by stimulated emission can themselves stimulate emission further along their path. So the power increases, rather like an investment increase with compounded interest, but the power compounds continually rather than incrementing at certain intervals.

The *amplification factor A,* which measures the increase in power through a length of laser medium L with gain G per unit length, is an exponential:

$$A = \exp(GL) = e^{GL} \tag{3-1}$$

For a gain of 0.05/cm, the calculated amplification factor is 1.0513 after 1 cm, 1.284 after 5 cm, and 1.6487 after 10 cm. That exponential increase in power cannot be sustained indefinitely because the gain term depends on conditions within the laser medium, including the populations of the upper and lower laser levels and the flux of photons stimulating emission, which can change.

The normal value given for gain is the *small-signal gain,* the gain when the light being amplified is still weak and plenty of atoms remain in the upper laser level. As the amplified power increases, the gain declines and the power produced eventually *saturates* because the excited atoms are producing stimulated emission just as fast as they can. In other words, the input light is stimulating emission from all available atoms in the upper laser level, so the power cannot be increased further.

The actual gain dynamics can be complex. When a laser switches on, the initial cascade of stimulated emission grows at the rate of small-signal gain. What happens next depends on the geometry of the laser medium and the design of the laser cavity, described in Sections 3.3.2 and 3.3.3.

3.3.2 Geometry of the Laser Medium

A photon produced by stimulated emission always goes in the same direction as the photon that stimulated the emission. Thus, when one photon triggers a cascade of stimulated emission, all the photons go in the same direction. The farther the light goes through the laser medium, the higher the power of stimulated emission (neglecting saturation effects).

Gas and solid-state lasers are cylindrically shaped to take advantage of this concentration of stimulated emission. The initial spontaneous emission can occur in any direction, and start a cascade of light in that direction, but the light emitted toward the sides of the cylinder quickly leaks out. The amplification of spontaneous emission is highest when the light passes through the longest length of laser material—along the length of the cylinder, as shown in Figure 3-4. If there are no mirrors on the rod, as in Figure 3-4A, this length effect concentrates stimulated emission into an angle θ, defined by an inverse sine function:

$$\theta = \arcsin\left(\frac{Dn}{2L}\right) \tag{3-2}$$

where D is the rod diameter, L is length, and n is refractive index.

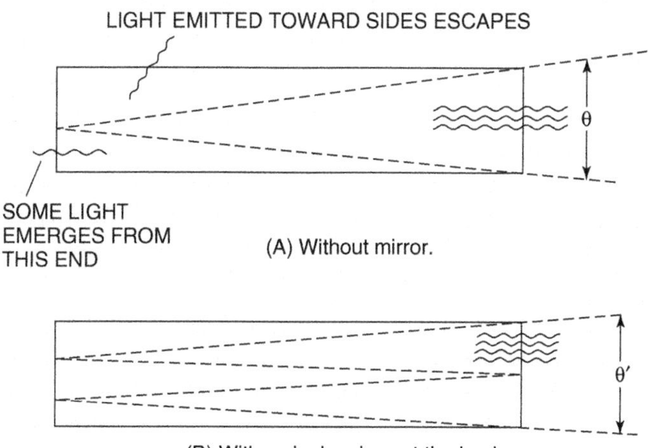

Figure 3-4. The process of stimulated emission makes light directional from a rod with no mirrors (top). Putting a rod on one end concentrates emission somewhat (bottom).

The effective length of the cylinder can be doubled by adding a mirror on one end, so stimulated emission emitted toward the mirror is reflected back in the other direction and out the other end. As shown in Figure 3-4B, this also concentrates the emitted light into a narrower angle θ':

$$\theta' = \arcsin\left(\frac{Dn}{4L}\right) \tag{3-3}$$

A few lasers with very high gain can operate in this way, including the excimer lasers in Section 7.7 and the edge-emitting semiconductor lasers in Section 9.6. However, most laser materials do not have high enough gain to build up useful amounts of stimulated emission when light makes a single pass back and forth through a laser. To generate useful powers, the light must bounce back and forth through the laser material many times. This is done by placing mirrors at each end of the laser rod or tube, with one mirror transmitting a small fraction of the light, which becomes the laser beam. This approach not only generates higher power; it also gives much better control over the emitted light, and gives lasers many of their special features such as coherence and single-wavelength output.

3.3.3 Mirrors, Resonant Cavities, and Oscillators

Putting a pair of parallel mirrors on opposite sides of a laser medium, or on the ends of a rod or tube, creates a *laser resonator, laser cavity,* or *resonant cavity.* The two mirrors serve the same function for light as the walls of a maser resonator serve for microwaves. The light waves bounce back and forth between the mirrors like sound waves in an echo chamber, but unlike the walls of an echo chamber or a maser cavity, the mirrors in a laser are separated by many optical wavelengths.

The two mirrors in a laser cavity are not identical. The *output mirror* reflects part of the laser beam back into the laser material, and transmits the rest to become the laser beam. The fraction of power transmitted depends on the type of laser, its internal gain, and the desired output. The rear cavity mirror normally reflects all light back into the laser material, although in some cases it may transmit a small fraction of the light to a power monitor or measurement system.

The laser itself is called an *oscillator*, because like an electronic oscillator it generates an electromagnetic wave that varies periodically, although light frequencies are hundreds of thousands of times higher than microwave frequencies. The idea that an oscillator generates a signal on its own is important. An amplifier normally boosts the power of a signal from an external source. An oscillator generates its own signal, determined by internal resonances. Optical amplifiers exist; like lasers, they generate power by the stimulated emission of radiation but, unlike lasers, they do not have a resonant cavity and cannot generate light on their own.

The seed that starts laser oscillation is the spontaneous emission of a photon along the length of the laser cavity, which stimulates emission that bounces back and forth between the mirrors. Figure 3-5 shows how one photon starts a cascade of stimulated emission that bounces between the mirrors, with some photons escaping through the output mirror to form the beam. The amount of stimulated emission grows on each pass through the laser medium until it reaches an equilibrium level. The oscillator must do a lot of amplifying because it must build up a handful of photons to produce observable power. In order to emit one milliwatt for one second, a helium–neon laser must generate 3.2×10^{15} photons.

3.3.4 Balancing Gain and Loss in a Laser

From the time a continuous-output laser turns on until it reaches a stable output power, the power bouncing back and forth between the mirrors grows a bit on each pass through the laser cavity. Then it reaches a point where the gain is balanced by losses so the output remains at a steady level, which is often called *continuous wave*. This means that the increase of power during a round-trip of the laser cavity must equal the sum of the power emerging in the beam and power lost inside the laser:

$$\text{Power Increase} = \text{Lost Power} + \text{Output Power} \tag{3-4}$$

Losses can be made low, but cannot be reduced to zero. No mirror can reflect every photon that reaches its surface. Material in the laser cavity also absorbs some light. These losses become important if the gain is low, as it is in many common types of continuous-wave lasers, and they must be considered in selecting transmission of the output mirror.

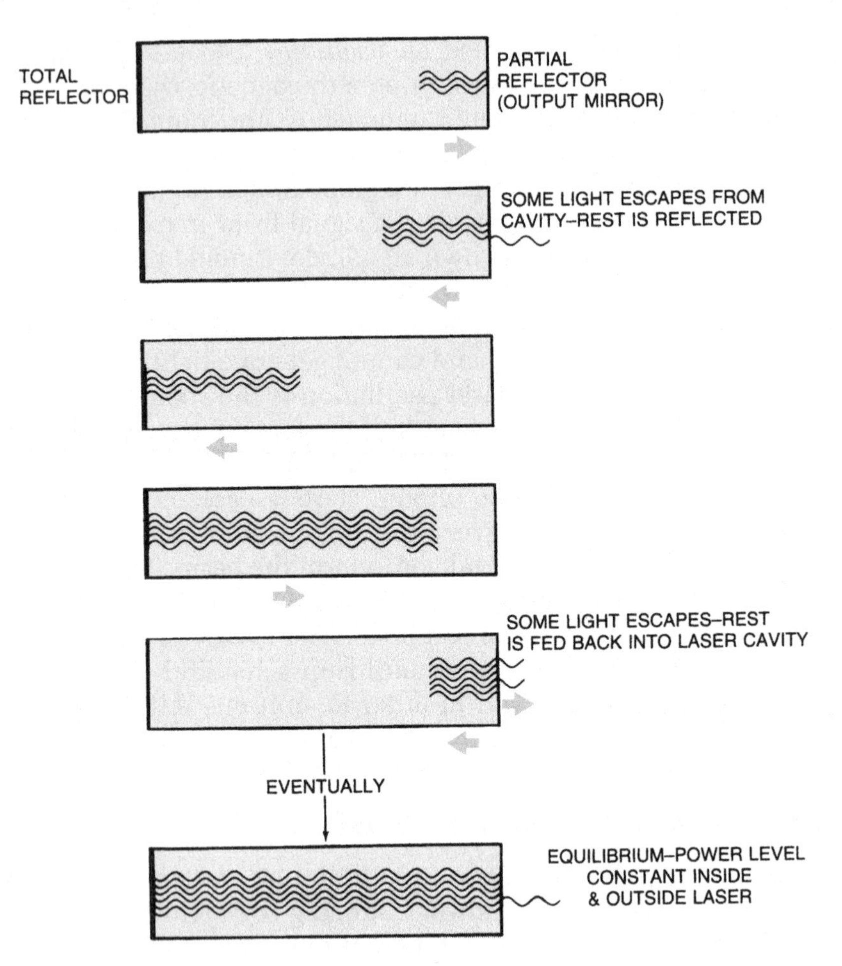

Figure 3-5. Growth of stimulated emission in a resonant laser cavity

For example, suppose that light is amplified 5% during a round-trip of a continuous-wave laser cavity. One percent of the light might be absorbed in the rear cavity mirror, and another 1% absorbed by the laser gas. The output mirror might transmit 2% of the light (the laser beam), absorb 1%, and reflect the remaining 97% back into the laser cavity.

The fraction of light transmitted in the output beam is also important in determining the power level both inside and outside the laser cavity. In this case, if the output power transmitted through the mirror is 2 mW and absorbed 1% of the incident light, the power reflected back into the laser cavity is 97 mW, 48.5 times higher than the output power. This difference would be smaller if the mirror transmission were higher. For example, if the output mirror transmitted 20% of incident light and reflected the rest, the reflected power in the laser cavity would be only four times higher than the transmitted power.

3.3.5 Energy Storage in a Laser

Another way to look at a laser cavity is in terms of the amount of energy stored in excited atoms. As in describing loss within the laser, the same considerations apply to both pulsed and continuous-wave lasers, but for energy storage they are simpler to describe for a pulsed laser.

Before stimulated emission begins, the laser cavity stores an amount of energy that depends on the number of excited atoms and their excitation level. How much energy can be stored depends on the how long the atoms can stay in the upper laser level before they start to release energy by spontaneous emission, which will trigger stimulated emission. The shorter the spontaneous emission lifetime, the less energy can be stored in the medium.

In practice, the excited-state lifetime is related to the likelihood of stimulated emission, which in turn is related to the laser gain. The higher the cross-section for stimulated emission, the higher the laser gain, and the shorter the lifetime of the upper laser level. This means that the higher the gain, the less energy can be stored in a laser medium. That may seem paradoxical at first, but it does make sense. High gain means that the laser transition occurs readily, so the laser tends to release energy quickly, making it hard to store energy in the upper laser level.

The amount of energy that can be stored can be increased by suppressing reflection that otherwise would build up stimulated emission in the laser cavity. For example, you could put a black slab between the laser material and the output mirror and exciting the laser medium would produce no output beam, until you remove the black slab, and the cascade of stimulated emission can bounce back and forth between the mirrors. Devices called Q switches, described in Section 5.5.2, exploit this principle.

3.4 LASER BEAMS AND RESONANCE

Now that we have "built" our laser on paper, let us go back and look more closely at some important details of laser operation that we have glossed over. We will cover some of these in more detail later, but a few definitions should help you now.

3.4.1 Beam Characteristics

You may think of a laser beam simply as a collection of parallel light rays that form a bright spot on a screen. However, if you drew a line across the laser spot and measured the light intensity along that line, you would find that intensity varies, peaking in the middle and dropping off at the sides. Because the power drops off gradually, it can be hard to define the edge of the beam, so, normally, the *beam diameter* is defined as the point at which power drops to a certain fraction of the central intensity (often $1/e^2$, where e is the root of natural logarithms).

A laser beam spreads with distance once it leaves the laser. Normally, this is not obvious over the scale of a room, but it is important over longer distances. If you know the spreading angle, called *beam divergence,* you can calculate the size of a laser spot at a certain distance from the laser. Normally, divergence, like beam diameter, is measured to points at which the beam intensity has dropped to a certain level. You will learn more about these laser characteristics in Chapter 4.

3.4.2 Cavity Resonances

Resonances arise from the wave nature of light. As Figure 3-3 showed, a round-trip of the cavity has to equal an integral number of wavelengths or, equivalently, the cavity length must equal an integral number of half-waves. In Figure 3-3, we just showed a few waves but, in reality, laser wavelengths are much shorter than the cavity, so many fit inside. Recall, as a starting point, that stimulated emission is coherent, so all light waves are in the same phase, meaning that they add in amplitude.

Figure 3-6 shows what happens if the waves fit exactly into a laser cavity, although to make the drawing clear the waves are shown much longer than they really are. (If we tried to show the real wavelength, you could not see it.) When the stimulated emission hits a cavity mirror, all the waves are in phase, and because they are in phase they add constructively. If light starts at a wave peak when it is reflected from the output mirror, it will travel an integral number of wavelengths before it reaches the output mirror

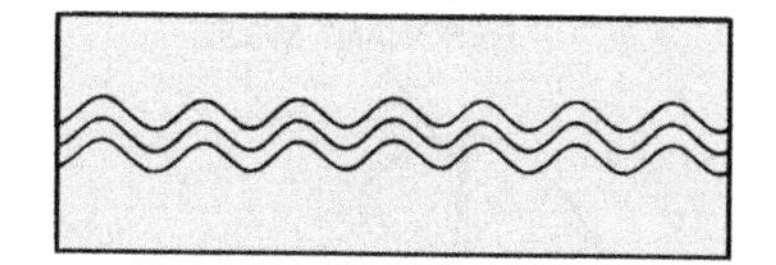

Figure 3-6. Light waves are resonant if twice the length the laser cavity equals an integral number of wavelengths.

again, where it again will be at a peak. So will the light waves stimulated by that wave. Thus, all of them will add in amplitude, by constructive interference, as long as twice the cavity length L is an integral multiple N of their wavelength in the laser cavity λ:

$$N\lambda = 2L \tag{3-5}$$

Suppose, however, that twice the cavity length is not an integral multiple of the wavelength. Then reflected waves will wind up out of phase with other light waves in the cavity, so destructive interference will reduce their strength. The waves do not have to be cancelled completely; all that is needed is to reduce the gain because, as Equation 3-1 shows, amplification depends exponentially on gain, so a comparatively small decrease in gain reduces the amplified output power by a much larger amount.

Suppose that the gain is 0.05/cm on resonance but only half that value, 0.025/cm, off resonance. In that case, the amplification factor over a meter will be 148 on resonance but only 12 off resonance. The longer light oscillates back and forth in the laser, the more the off-resonance wavelength will fade.

The need for precise resonance might seem to be very restrictive, but that restriction is offset by the fact that light wavelengths are much smaller than most laser cavities. For example, 30 centimeters, a typical round-trip distance in a small helium–neon laser, equals about 474,000 wavelengths of the laser's 632.8-nanometer red light. Resonance is possible not just at 474,000 wavelengths, but also at 474,001, 474,002, and so on. Each of these resonances is called a *longitudinal mode* of the laser, and each has a slightly different wavelength, with an integral number of waves fitting in the cavity, as you saw in Figure 3-6.

In addition, several effects combine to spread laser transitions over a range of wavelengths, called the *gain bandwidth,* which spans several individual resonant modes, as shown in Figure 3-7 for

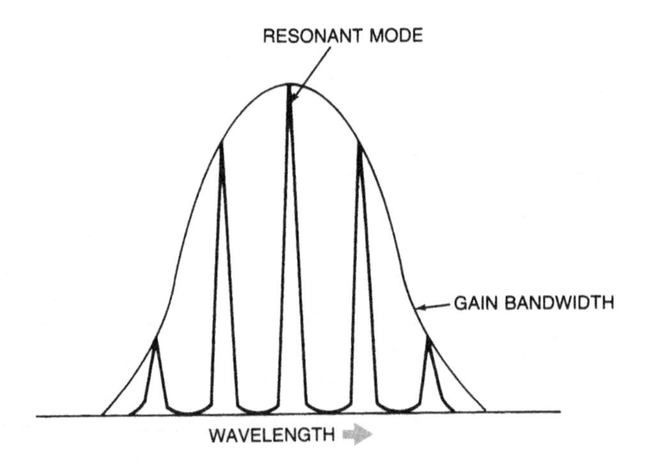

Figure 3-7. Several cavity resonances can fit within the gain bandwidth of a typical gas laser.

a helium–neon laser. Laser operation can be limited to one longitudinal mode, but multimode operation is fine for many purposes.

3.4.3 Types of Laser Resonators

So far, we have not said anything about the shape of the mirrors on the ends of a laser cavity. It is logical to expect both mirrors to be flat and parallel to each other, an arrangement called a Fabry–Perot resonator that is shown in Figure 3-8A. However, several other resonator designs are possible, as shown in Figure 3-8.

Although the Fabry–Perot resonator seems simple conceptually, it is hard to use with the many types of lasers with long cavities and low gain because slight misalignments of the mirrors cause photons to be reflected from one mirror to miss the other. If one mirror is only half a degree out of parallel with the other, a light ray aligned along the length of a 15-cm laser cavity would be reflected to a point 1.3 millimeters from the center of the other mirror, and would miss the edge of a cavity mirror 2 mm in diameter. Even with smaller misalignment, light bouncing back and forth between the two mirrors would soon be lost.

Such light losses can be avoided by using one or two curved mirrors in the resonator. Following the light rays in Figure 3-8B, C, D, and E shows how the focusing power of the curved mirrors directs light back into the cavity even if the light is not aligned

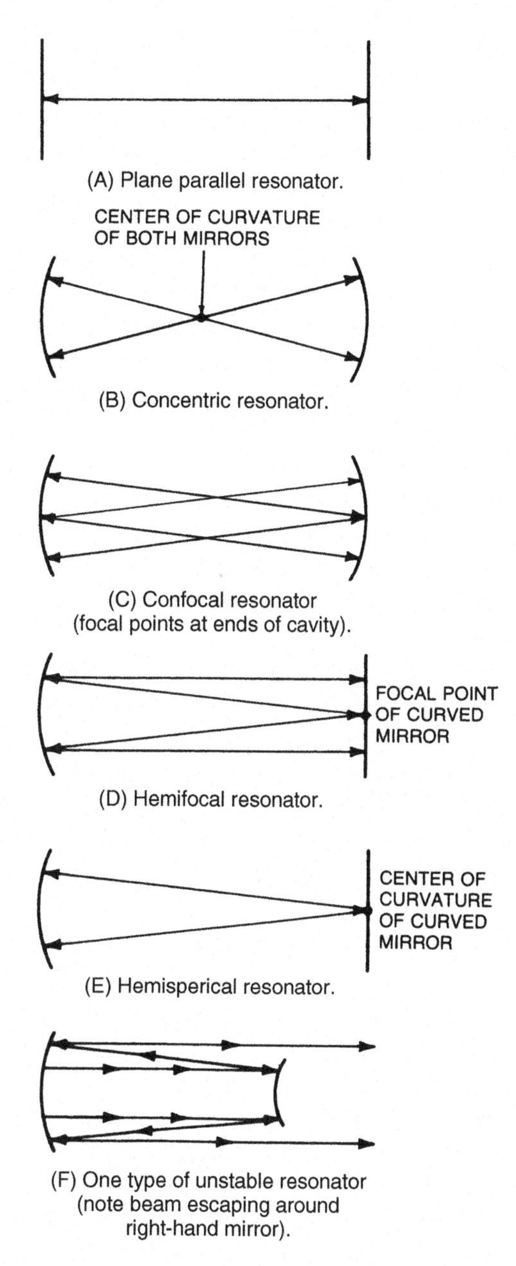

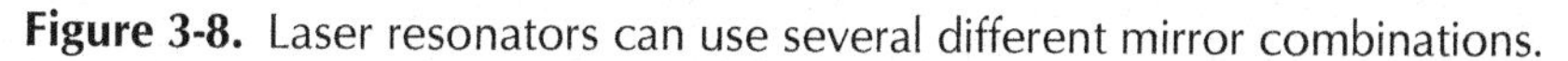

Figure 3-8. Laser resonators can use several different mirror combinations.

precisely with the resonator axis. Curved mirror configurations that reflect light back toward the other mirror are called *stable resonators* because light rays will keep bouncing between the mirrors indefinitely if there are no losses. Stable resonators are most attractive for low-gain lasers.

Other types of resonators work better in high-gain lasers, in which small losses of circulating power are not critical. Look carefully at Figure 3-8, and you can see that light rays reflected between cavity mirrors in some stable resonators do not pass through all of the excited laser medium. Without light passing through that volume to stimulate emission, the energy stored there cannot be collected.

One solution to this problem is the resonator shown in Figure 3-8F, which is called an *unstable resonator* because light rays bouncing back and forth between the mirrors eventually leak out the sides. Those losses are acceptable in high-gain lasers because they are offset by the advantage of collecting energy from a larger volume of laser medium.

The reason for this difference is that light does not need to make as many round-trips between cavity mirrors in high-gain lasers as does the light in low-gain lasers. Low-gain lasers require highly reflective mirrors and low cavity loss to sustain oscillation, so the light may be reflected 20 times before it exits the laser cavity in the beam. In contrast, high-gain lasers have less-reflective mirrors, so the average photon will be reflected at most a few times before being emitted.

A true plane-parallel mirror cavity is neither stable nor unstable, because it neither focuses nor diverges light rays. Although they do not work well for large lasers, their simplicity is very attractive for semiconductor lasers, which are very small and have high gain.

We will not go into detail on the theory of laser resonators. It bogs down in the sort of mathematical complexity you are reading this book to avoid, and many of the results are at best arcane. However, laser resonators do (quite literally) shape both the intensity distribution in the output beam and the rate at which the beam diverges.

3.4.4 Intensity Distribution and Transverse Modes

Longitudinal modes represent resonances along the length of the cavity, at wavelengths at which an integral number of waves fit into a round-trip. Lasers also have *transverse modes* that define the pattern of intensity distribution across the width of the laser beam. They arise from what are called the *boundary conditions* at the edges of the laser cavity; one boundary condition is that the electric and magnetic fields normally are defined as zero at the edges of the beam. Transverse modes may have one, two, or more intensity peaks in the central part of the beam.

The simplest, or "lowest order" transverse mode is a smooth beam profile with a peak in the middle, as shown at the top of Figure. 3-9. Its shape is that of a mathematical curve called a Gaussian curve, after the famed mathematician Carl Friedrich Gauss.

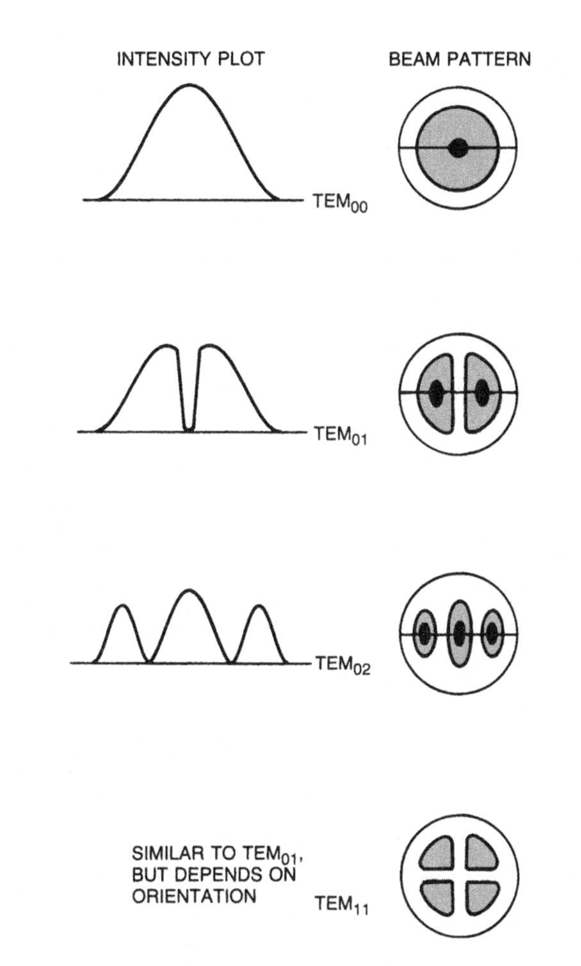

Figure 3-9. TEM_{00} mode beam and a sampling of other transverse beam modes.

This intensity distribution $I(r)$ as a function of distance r from the center of the laser beam is given by

$$I(r) = \left(\frac{2P}{\pi d^2}\right)\exp^{(-2r^2/d^2)} \quad (3\text{-}6)$$

where P is beam power and d is spot size measured to the $1/e^2$ intensity points. This first-order mode is called the TEM_{00} mode, where the T stands for transverse and the E and M stand for electric and magnetic modes, respectively. The numbers that follow are the number of zero-intensity points inside the pattern, with the first number giving the number of null points in the electric field and the second giving the number of nulls in the magnetic field. Recall that the electric and magnetic fields are perpendicular to each other.

As you might expect, there is a large family of TEM_{mn} modes, where m and n are integers denoting the number of nulls in electric and magnetic fields, respectively. Thus a TEM_{01} beam has a single minimum dividing the beam into two bright spots. A TEM_{11} beam has two perpendicular minima (one in each direction) dividing the beam into four quadrants, as shown in Figure 3-9, and so forth. The TEM_{00} mode is desirable because is suffers less spreading than higher-order modes. A low-gain laser medium with a stable resonator can readily produce this lowest-order mode; proper design adds losses to suppress the oscillation of higher-order modes. However, some stable-resonator lasers operate in one or more higher-order modes, especially when they are designed to maximize output power.

You may see similar modes in other situations, where electromagnetic waves are guided through long pipes or structures called *waveguides*. An optical fiber is an optical waveguide; metallic waveguides are sometimes used for microwaves.

Unstable resonators have fundamentally different mode patterns. You can see the basic reason why in Figure 3-10, which examines the unstable-resonator example from Figure 3-8F in more detail. In this simple unstable resonator, the output mirror is a solid metal mirror that reflects some light back into the laser medium while the beam escapes around it. Thus, near the laser the beam profile has a "doughnut" cross-section—a bright ring surrounding a dark circle where light was blocked by the mirror. Far from the laser, this intensity distribution averages out to a more uniform pattern.

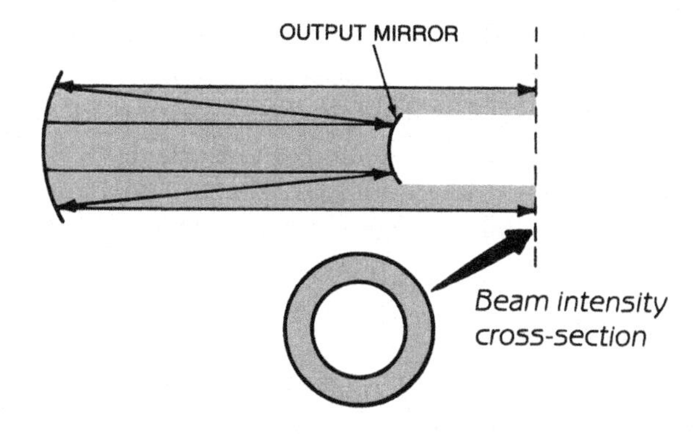

Figure 3-10. Near-field emission from an unstable resonator has a central minimum.

The details of unstable-resonator modes are even more complex than those of stable-resonator modes, but we can safely skip their many variations because they generally have little practical impact. In Chapter 4, we will learn how transverse modes affect the properties of laser beams.

3.5 WAVELENGTH SELECTION AND TUNING

The laser cavity and the laser medium combine to determine what wavelengths a laser emits.

Nominally, all laser transitions are at a sharply defined wavelength, but in practice they span a range of wavelengths, as shown in Figure 3-7 for neon in a helium–neon laser. In a gas, the random motion of gas molecules and their interactions broaden the line. In solids, the transition energy is spread by interactions among atoms in the solid. Gas molecules may simultaneously make both vibrational and rotational transitions.

As you saw earlier, the length of the laser cavity determines the precise resonant wavelengths. For helium–neon lasers and many other types, these resonances are narrower than the range of wavelengths over which the transition has gain, as you saw in Figure 3-7.

Another important factor is the variation of mirror reflectivity with wavelength. Although we normally think of mirrors as reflecting all wavelengths, their actual range is limited. Most metals reflect strongly at visible wavelengths, but absorb more light in other bands. Mirrors also can be made from combinations of materials in ways that make them strongly reflective in narrow wavelength ranges, as described in Section 5.3.3.

The overall result is that lasers oscillate at wavelengths at which the cavity and the laser transitions combine to give the highest gain. Laser designers use this to select emission wavelength. The details deserve more explanation.

3.5.1 Picking a Transition

So far, we have simplified our discussion of energy levels and transitions in one very important way—we have ignored the multitude of possible transitions. Atoms can absorb and emit light at many wavelengths, each corresponding to a different transition. Most of these transitions are not suitable for use in lasers but, in general, atoms have many possible laser transitions at different wavelengths. Some transitions are strong; many others are so weak that laser action is very difficult to achieve.

The potential strength of a laser transition depends on the properties shown in Figure 3-11:

- An efficient way to excite the atom to a high energy level
- Efficient transfer of those excited atoms to a metastable upper laser level where they accumulate to produce a population inversion
- The availability of a lower laser level, which is easily depopulated either naturally or by exciting the laser
- No other states that would strongly absorb stimulated emission on the laser transition

In many cases such as the helium–neon laser, it is possible to produce population inversions on two or more transitions at once. The initial excitation can populate two or three metastable levels in neon atoms, each of which can drop to more than one lower laser level. This could lead to simultaneous laser emission on multiple lines, but the nature of laser amplification makes the highest-gain transitions dominate. The differences in gain need not be great because the power depends exponentially on gain.

To understand how this works, consider a simple example: two transitions from the same metastable level with gains 0.04 and 0.05 per centimeter. Table 3-1 shows how the power grows exponentially on the two lines if we calculate total gain as a function of distance through the laser gas using Equation 3-1. The gain would

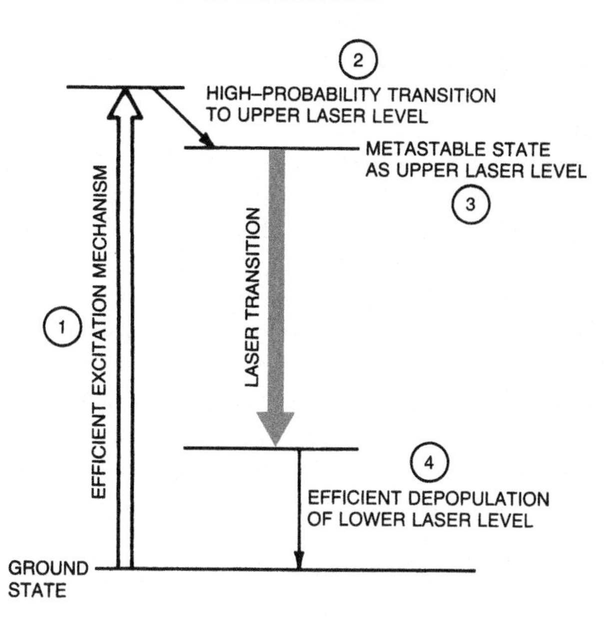

Figure 3-11. Requirements for efficient four-level laser action.

saturate before the power multiplied as much as in the calculations, but the difference in gain would still cause the strong line to overwhelm the weaker one.

3.5.2 Cavity Optics and Gain

A gain difference does not always suppress laser action on unwanted lines, particularly if the weak lines are the ones we want.

Table 3-1. Effect of gain on relative power

	Relative power		
Distance	Weak line	Strong line	Ratio
0.0 cm	1.00	1.00	1.00
1.0 cm	1.04	1.05	1.01
5.0 cm	1.22	1.28	1.05
10 cm	1.49	1.65	1.11
100 cm	54.60	148.00	2.71
1000 cm	2.35×10^{17}	5.18×10^{21}	22,043

The visible wavelengths that are preferred for most applications of helium–neon lasers have lower gain than less-desirable infrared lines, so gain must be suppressed on the undesired transitions.

Normally, this is done by using cavity optics with high losses at the undesired wavelengths. For example, red helium–neon lasers use mirrors which strongly reflect the 632.8-nm red line, but reflect little of the undesired infrared emission. The difference in mirror reflectivity more than offsets the difference in transition strength, suppressing the infrared line.

If a laser material has gain across a broad range of wavelengths, it is possible to tune the laser emission by putting a prism or diffraction grating into the laser cavity, as shown in Figure 3-12. Different wavelengths leave the prism or grating at different angles. Only one wavelength is bent at the proper angle to oscillate back and forth between the mirrors in the laser cavity. Shorter and longer wavelengths go off to the sides of the cavity and are lost. As a result, the net gain in the laser cavity is highest at the wavelength selected by the prism or grating. Turning the grating or prism changes the angles and, thus, tunes the laser wavelength.

3.6 LASER EXCITATION TECHNIQUES

Our general description of laser physics has given you an idea of how lasers work, but we have glossed over the actual excitation of

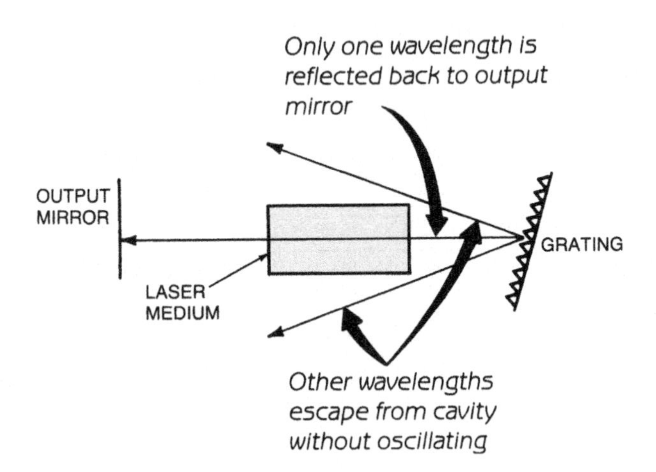

Figure 3-12. Turning a prism can tune laser wavelength.

atoms or molecules to produce a population inversion. There are three major approaches, plus others used only rarely. We will cover the basic ideas here, but the specifics will be covered in later chapters describing specific laser types.

3.6.1 Optical Pumping

Optical pumping is the use of photons to excite atoms or molecules to selected energy levels. It is conceptually simple and was the first approach demonstrated for lasers. The idea is to excite the lasing species to an energetic state, from which it drops down to the metastable upper laser level and emits light.

Figures 3-1 and 3-2 showed narrow transitions between isolated atomic energy levels, implying that the pump light must be at a narrow range of wavelengths to match that. Fortunately, this is not always the case. In solids in particular, many energy levels are broadened by interactions with neighboring atoms, so a range of wavelengths can excite the atoms to a band of energy levels, from which they drop to the metastable upper laser level. That allowed Theodore Maiman to use a photographer's flashlamp to excite the chromium atoms in his ruby laser. (Note, however, that some wavelengths may excite the material more efficiently than others.)

Optical pumping can be used with any laser medium that allows the pump light to reach the species that absorbs the pump light. In Maiman's ruby laser, the chromium atoms that absorbed the pump light were held in an aluminum-oxide matrix that transmitted the pump light from the flashlamp and the laser light emitted by the chromium atoms. Optical pumping is used mostly for crystalline solid-state lasers and liquid tunable dye lasers, either of which may be pulsed or continuous wave, but it also can be used with gases or even semiconductors.

3.6.2 Electrical Excitation of Gas Lasers

Most gas lasers are excited by an electrical discharge passing through the gas, a process somewhat like the excitation of fluorescent bulbs. In both cases, electrons in a high-voltage discharge transfer their energy to atoms in the gas inside a tube. In the fluorescent lamp, mercury atoms absorb the electron energy and spontaneously emit ultraviolet light, which excites the phosphors that emit visible light.

In a typical small gas laser, the electrons excite atoms in the laser tube. In some lasers, those atoms then drop into the metastable upper laser level, and are stimulated to emit laser light. In others, the excited atoms collide with other atoms, transferring energy and exciting the other atoms to the metastable upper laser level. One example is the helium–neon laser, in which helium atoms absorb energy from electrons and transfer the energy to neon atoms.

Electrically excited gas lasers may be continuous wave or pulsed. Some high-power gas lasers are excited by beams of electrons fired into the gas.

3.6.3 Electrical Excitation of Semiconductor Diode Lasers

Electrical currents also power semiconductor lasers, but their operation is so different from that of other electrically pumped lasers that they deserve separate mention here. Electrically, the semiconductor laser acts as a diode, in which current flows between materials with different compositions, exciting a population inversion in a narrow region called the *junction,* which is the boundary between regions where the semiconductor has different compositions.

Strictly speaking, what are excited in a semiconductor laser are current carriers—electrons and vacancies called (appropriately enough) *holes* that could accommodate electrons. The electrons and holes from the outer layers *recombine* in the junction layer, producing a metastable state which eventually releases its energy by spontaneous emission. Chapter 9 explains the process in much more detail.

3.6.4 Other Excitation Mechanisms

A few laser materials are excited in other ways. Rapid expansion of a hot gas into a cold region produces a short-lived population inversion in a *gas-dynamic* laser, described in Section 7.6.6. Chemical reactions can produce population inversions in reaction products, as described in Section 7.8. Stimulated emission also has been produced from beams of energetic electrons as described in Section 10.4.

3.6.5 Energy Conversion Efficiency

Like other light sources, lasers convert input energy into light with limited efficiency. Their efficiency varies widely, and depends on several factors including the nature of the laser transition, the excitation mechanism, and the fact that excitation is never 100% efficient.

The nature of the laser transition sets a ceiling on maximum efficiency, as you can see in the simple four-level laser diagram of Figure 3-2. The pump energy raises the laser species from the ground state to a higher energy level, but only part of that input energy is released on the laser transition. The diagram makes it look like more than half of the energy is released on the laser transition, and in some cases that is true. Optical pumping always requires photons with more energy than the laser transition, but sometimes the difference is not huge. For example, a semiconductor laser emitting at 808 nm can pump a neodymium laser emitting at 1064 nm, so about three-quarters of the energy from a pump photon emerges in a laser photon. Other types of pumping can be quite inefficient. For example, the laser transition in argon-ion gas lasers is far above the ground state, so the energy needed to excite the argon to that state is many times larger than the energy released on the laser transition. Those and other internal inefficiencies limit argon lasers to less than 0.1% efficiency.

Excitation efficiency is an additional issue, because all the pump energy does not get absorbed, and some of what is absorbed does not get converted into a laser beam. These effects also cause energy losses.

Overall, "wall-plug" efficiency compares electrical power in to optical power out, and varies widely. Semiconductor lasers can be quite efficient, with the best results having wall-plug efficiency above 50%. The wall-plug efficiency of optically pumped solid-state lasers is limited by the efficiency of the pump source, so it cannot match semiconductor lasers. Typically, gas lasers emitting at visible wavelengths, such as helium–neon and argon-ion, are the least efficient, converting less than 1% of input power into laser light.

3.6.6 Confining Energy

Laser excitation is most efficient when the excitation energy is confined to a small volume. The more concentrated the excitation energy is, the more atoms are likely to be excited, and the stronger the population inversion. It is like heating an object; the smaller the volume being heated, the more its temperature will increase.

Optical confinement of stimulated emission works similarly, by concentrating stimulated emission in the zone where a population inversion exists to amplify it further. The overall result is that confinement of excitation energy and stimulated emission both tend to improve laser operation and efficiency. This is critical for some types of lasers, such as ruby, in which more than half the chromium atoms that produce the laser light must be excited to produce a population inversion.

Confinement plays an important role in enhancing the efficiency of two important classes of lasers—fiber and semiconductor lasers. In fiber lasers, both stimulated emission and optical pump light are confined to the central core region, which guides light along the fiber. In semiconductor lasers, the confinement of current flow through the laser and of light produced within the semiconductor enhance efficiency, and are vital to continuous operation at room temperature. You will learn more about confinement in fiber and semiconductor lasers in Chapters 8 and 9, respectively.

3.7 WHAT HAVE WE LEARNED?

- A population inversion is needed for stimulated emission to dominate on a transition between an upper laser level and a lower laser level.
- Atoms must be selectively excited to the upper laser level to create a population inversion.
- Upper laser levels normally are metastable states with unusually long lifetimes.
- In three-level lasers, the laser transition is between the metastable upper laser level and the ground state, which is the lower laser level.
- In four-level lasers, the laser transition is between a metastable level and a level above the ground state, so a population inversion is easier to produce than in a three-level laser.
- Most continuous-wave lasers are four-level lasers.
- A spontaneously emitted photon starts laser oscillation by stimulating emission from an excited atom or molecule.
- Resonant cavities help build up stimulated emission in a laser oscillator.
- The simplest resonant cavity for a laser is a pair of parallel mirrors called a Fabry–Perot resonator on opposite ends of the laser.

- Laser amplification makes the power in a laser increase exponentially.
- The gain in laser power as the light oscillates in a resonant cavity must equal the power lost plus the power emitted from the laser.
- Beam diameter is defined by the point at which power drops to $1/e^2$ of the central intensity.
- The length of a round-trip within a laser oscillator must equal an integral number of wavelengths of the light generated.
- The output mirror transmits some light to form the laser beam.
- Lasers may oscillate simultaneously on two or more wavelengths within the gain bandwidth of the laser, which resonate in the cavity.
- Laser resonators with curved mirrors of a certain shape are called stable resonators because the mirrors keep reflecting the light rays between them.
- Transverse modes define the intensity pattern across the laser beam. A TEM_{00} mode is the simplest transverse mode.
- The laser cavity and the laser medium combine to determine what wavelengths a laser emits.
- The strongest laser line dominates emission unless the cavity optics select against it.
- Optical pumping uses light to excite atoms or molecules in solids, gases, or liquids.
- An electrical discharge can excite a gas laser.
- An electric current can excite a semiconductor laser.
- Like other light sources, lasers do not convert all input energy into light, and some lasers can be quite inefficient.

WHAT'S NEXT?

Now that we have looked at the basic mechanisms behind laser operation, it is time to look at important laser characteristics. Later on, we will look at individual types of lasers.

QUIZ FOR CHAPTER 3

1. What is the major advantage of a four-level laser over a three-level laser?

a. More levels to excite atoms to.
b. The lower laser level is not the ground state.
c. More metastable states.
d. No advantage

2. What is a metastable state?
   a. A long-lived energy state that can serve as the upper laser level.
   b. An energy state that is short-lived and unstable.
   c. An energy state of an radioactive atom.
   d. The ground state of a laser transition.
   e. A state in which an atom can easily be excited.
3. How does stimulated emission increase laser power?
   a. Linearly, each photon produced by stimulated emission produces one additional photon.
   b. Geometrically, each stimulated-emission photon produces exactly two additional photons.
   c. Exponentially, each stimulated emission photon can stimulate the emission of a cascade of additional photons.
   d. Stimulated emission triggers spontaneous emission, which produces more photons.
4. The gain per unit length of a laser medium is 0.01 per centimeter. What is the amplification factor after 20 centimeters?
   a. 0.02
   b. 0.20
   c. 1.001
   d. 1.020
   e. 1.221
5. The round-trip loss of helium–neon laser light in a cavity is 2%. The output mirror lets 1% of the light escape in the beam. When the laser is operating in a steady state, what is the round-trip amplification in the laser, measured as a percentage?
   a. 1%
   b. 2%
   c. 3%
   d. 5%
   e. 99%
6. The helium–cadmium laser has a wavelength of 442 nm. What is the round-trip length of a 30-cm long cavity measured in wavelengths?
   a. 13,260
   b. 135,700

c. 474,000
d. 679,000
e. 1,357,000

7. The 632.8-nm line of the helium–neon laser covers a wide enough range of wavelengths to have multiple resonances in a 15-cm laser cavity. What is the difference between the wavelengths of two adjacent longitudinal modes in that cavity? (Hint: Think of the wavelengths for which the cavity is N and N + 1 wavelengths long.)
   a. 0.001 nm
   b. 0.011 nm
   c. 0.055 nm
   d. 0.110 nm
   e. 0.111 nm
8. How many internal minimum-intensity points are there in a TEM_{01} mode beam?
   a. None
   b. 1
   c. 2
   d. 3
   e. 6
9. Which of the following excitation mechanisms cannot produce a population inversion?
   a. Optical pumping with an external laser.
   b. Passing an electrical current through a semiconductor.
   c. Heating a gas.
   d. Passing an electrical discharge through a gas.
   e. Optical pumping with light from a flashlamp.
10. Which of the following factors does not harm power-conversion efficiency in a laser?
    a. Atmospheric absorption.
    b. Excitation energy not absorbed.
    c. Problems in depopulating the lower laser level.
    d. Inefficiency in populating the upper laser level.

CHAPTER 4

LASER CHARACTERISTICS

ABOUT THIS CHAPTER

Now that we have seen what goes on inside lasers, it is time to look at the important characteristics of lasers and laser light. In this chapter, we will learn about coherence, wavelength, directionality, beam divergence, power, modulation, polarization, and related properties. We also will learn about the factors that determine laser efficiency.

4.1 COHERENCE

Coherence probably is the best-known property of laser light. Light waves are coherent if they are in phase with each other, that is, if their peaks and valleys line up, as shown in Figure 4-1. Coherence requires that the waves start in phase and that their wavelengths match to keep the waves from drifting out of phase with distance. Laser light starts out coherent because stimulated emission has the same phase and wavelength as the light wave or photon that stimulates it. The stimulated photon, in turn, can stimulate the emission of other photons, which are coherent with both it and the original wave.

In reality, laser light is not perfectly coherent. Not all stimulated emission is produced from a cascade triggered by a single spontaneously emitted photon, and even the stimulated emission generated by a single original photon is not perfectly in phase or exactly identical in wavelength. Differences in gain and spontaneous emission rates between types of lasers lead to differences in their coherence, which affects how the lasers are used. Let us look closer.

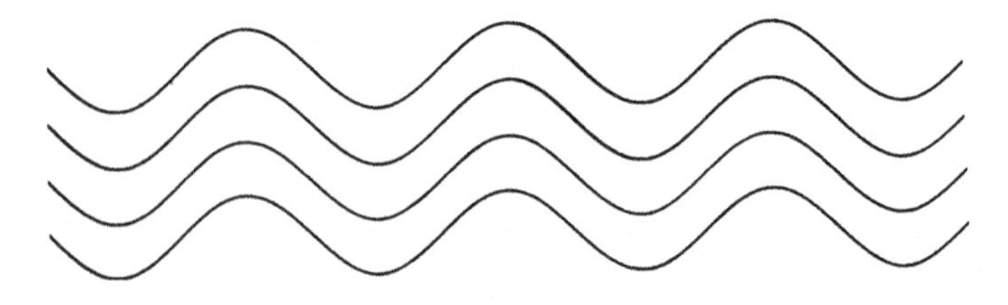

Figure 4-1. Coherent light waves.

4.1.1 Coherence of Laser Light

There is no such thing as "perfect" coherence. Although our simple descriptions may make it seem that a single photon could produce a cascade of stimulated emission that accounts for all the light in a laser beam, things are more complicated. Spontaneous emission can occur at any time, triggering another cascade of stimulated emission. Although the extra cascades of stimulated emission are on the same transition, they are not in phase with the initial light.

In addition, the photons produced by stimulated emission are not perfectly identical to the originals; their wavelengths can differ slightly, so the photons in the beam can drift out of phase over long distances. Tiny fluctuations within the laser, such as thermal gradients or vibrations, also can degrade coherence.

The degree of coherence differs widely among lasers. The more times light bounces back and forth through the laser cavity, the more coherent the light. That means high-gain lasers are less coherent, because their photons make only a few round trips of the laser cavity. Low-gain lasers tend to emit a narrower range of wavelengths, making them more coherent. Lasers oscillating in a single longitudinal mode have a narrower wavelength range than those oscillating in multiple modes, so they tend to be more coherent.

Coherence also depends on the duration of laser emission. The uncertainty principle causes the range of wavelengths in a pulse to increase as the pulse length decreases, so the shortest pulses tend to have the broadest wavelength ranges and be the least coherent. On the other hand, lasers that emit continuously can have the most narrow emission bands and the greatest coherence. Add this all together and you find that the most coherent beams come from continuous-wave, low-gain lasers operating in the lowest-order TEM_{00} mode.

4.1.2 Coherence Types and Measurement

There are two kinds of coherence: temporal and spatial. *Temporal coherence* measures how long light waves remain in phase. The term *temporal* is used because coherence is compared at different times, but temporal coherence also measures the distance that light travels during a period of time. All light has some temporal coherence over a characteristic *coherence length,* which is very close to zero for ordinary light bulbs but can be many meters for lasers. Coherence length is important for applications such as holography that require coherent light.

The coherence length can be calculated from the light's nominal wavelength λ and the range of wavelengths $\Delta\lambda$ it contains. Equivalently, it can be calculated from the speed of light in vacuum c and the light source's bandwidth in frequency units, $\Delta\nu$:

$$\text{Coherence length} = \left(\frac{\lambda^2}{2\Delta\lambda}\right) = \frac{c}{2\Delta\nu} \qquad (4\text{-}1)$$

A few simple calculations show what this means for some representative sources. Light bulbs emit light in the visible and near infrared, from about 400 to 1000 nm, with an average wavelength of 700 nm. Plugging in those numbers gives a coherence length of only 400 nm. An inexpensive semiconductor laser has a wavelength of 800 nm and wavelength range of 1 nm, giving a coherence length of 0.3 mm. An ordinary helium–neon laser has a much narrower linewidth of about 0.002 nm at its 632.8-nm wavelength, corresponding to a coherence length of 10 centimeters. Stabilizing a helium–neon laser so its emits in a single longitudinal mode limits the linewidth to about 0.000002 nm, and thus extends the coherence length to about 100 meters.

Spatial coherence, on the other hand, measures the area over which light is coherent. Strictly speaking, it is independent of temporal coherence. If a laser emits a single transverse mode, its emission is spatially coherent across the diameter of the beam, at least over reasonable propagation distances.

4.1.3 Interference and Coherence Effects

The interference effects described in Chapter 2 require some degree of coherence. If you superimpose many incoherent light waves of different wavelengths, interference effects average out.

This is why we do not see interference effects in everyday scenes illuminated by light bulbs or the sun. Adding light waves from a few sources with some degree of coherence does produce interference effects, as in the two-slit experiment you saw in Figure 2-3, where the amplitudes cancel at some points to produce dark zones, and add at other points to produce bright zones.

These interference effects are desirable in many measurement applications, because they let us measure distance by counting in units of wavelength. Because light waves are so small, this lets us measure distance very precisely, a practice called interferometry. Interference effects also manifest themselves in other ways.

Coherence is not always desirable, however. Small-scale turbulence in air causes slight shifts in the paths of coherent light passing through it, producing shifting patterns called *speckle,* which you can see when you look closely at a laser spot on the wall. Those patterns are an array of light and dark zones produced by interference among light waves that take slightly different paths through the air. They contain information on the quality of the air and the surface, but for most practical purposes that information is functionally noise, which makes laser illumination uneven across the surface.

4.2 LASER WAVELENGTHS

Laser light is called *monochromatic,* meaning single-colored, and it is monochromatic compared to light from other sources. However, as you learned in Chapter 3, stimulated emission generates a range of wavelengths on each transition, and that range depends on both the nature of the transition and the optics of the laser cavity.

To understand the nature of laser linewidth, we need to take a closer look at what happens inside lasers and how it affects laser output.

4.2.1 Gain Bandwidth and Amplification

The bandwidth of a laser transition can be measured in two distinct ways: as the gain of the laser as a function of wavelength and as the range of wavelengths in the beam. What makes them different is the amplification of stimulated emission inside the laser.

The wide, low curve in Figure 4-2 shows the *gain bandwidth,* the range of wavelengths over which stimulated emission in the laser medium produces gain. The curve shows that stimulated emission is most likely to occur for wavelengths at the middle of the transition, and that the probability of stimulated emission drops relatively gradually at higher or lower wavelengths.

The sharp, narrow curve in Figure 4-2 shows the *laser bandwidth,* the range of wavelengths in the beam emerging from the laser. The peak is not to scale; if it were it would not fit on the page.

The difference between the two arises from the difference in amplification across the gain bandwidth. Suppose, for example, that the gain is 0.1/cm at the center of the transition and 0.01/cm near the side. If we use Equation 3-2 to calculate the amplification factor in 100 cm of laser medium, we find that the power on the higher-gain line is increased 8100 times more than the power on the weaker line. That exponential dependence of output power on the gain makes the curve for laser emission much steeper and higher than for gain.

The laser intensity is highest at the wavelength at which the gain is the highest if the cavity optics do not affect the wavelength. However, laser action can occur at any wavelength where gain is large enough to overcome losses in the laser cavity, so the cavity optics can select other wavelengths, allowing tuning within the gain curve.

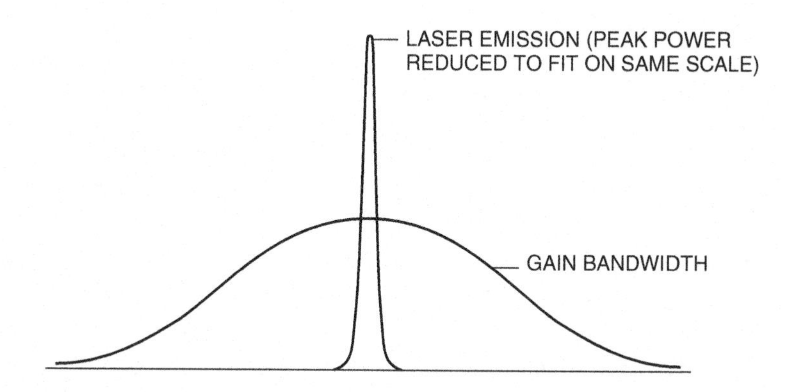

Figure 4-2. Gain bandwidth and laser emission on a laser transition, drawn to fit on the same scale. The laser peak is much higher.

4.2.2 Line Broadening

Several effects broaden the linewidth of a laser transition.

The quantum world is inherently uncertain. The likelihood of a transition is a quantum-mechanical probability function, which peaks at the nominal transition wavelength, and is lower, but not zero, at both higher and lower wavelengths.

The continual random motion of atoms and molecules in a gas also causes line broadening by a process called the Doppler effect. If an atom moves toward you, the wavelength of the emitted light becomes shorter because the atom comes closer between wave peaks. If the atom moves away, the wavelength is stretched, or red-shifted, toward longer wavelengths. The atoms in a laser gas are moving randomly, so the motion of some atoms Doppler shift the emitted wavelength toward the red, while the motion of others shift it toward the blue, spreading out the gain bandwidth.

Line broadening also comes from interactions between excited atoms and their neighbors. In gases, collisions between atoms and molecules cause broadening, which increases with gas pressure because collision frequency increases as more atoms and molecules are squeezed together at higher frequencies. Most gas lasers normally operate at low pressures, but some gas lasers operate at much higher pressures, at which collisional broadening can merge separate laser emission lines into a continuous spectrum. In solids, atoms can transfer vibrational energy to close neighbors, which also affects emission linewidth.

4.2.3 Wavelength Selection by a Cavity

The gain bandwidth and laser cavity combine to select the range of wavelengths at which a laser oscillates. Recall that in Section 3.4 you learned that a laser cavity of length L resonates at many wavelengths, described by Equation 3-6. If the laser cavity is much longer than the wavelength (which is the usual case), one or more wavelengths should fall within the gain bandwidth of the desired transition, as shown in Figure 4-3. Each wavelength is a distinct longitudinal mode.

Section 3.4 specified resonances in terms of the wavelength in the laser cavity, but Equation 3-6 was an oversimplification because it assumed that the refractive index of the laser medium was very close to one. That is not always the case. To make the equa-

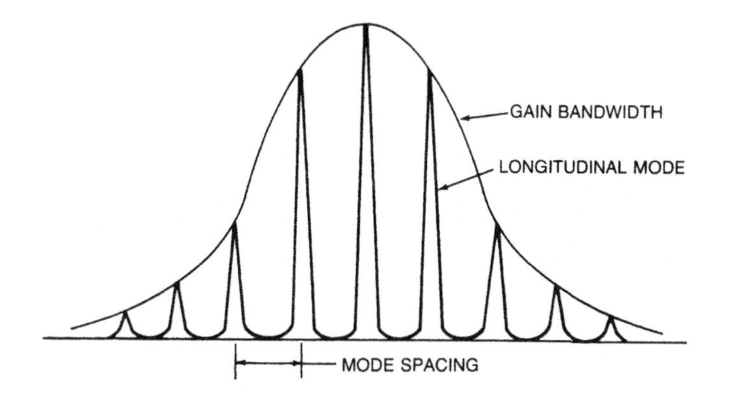

Figure 4-3. Multiple longitudinal modes fall within the gain bandwidth of a gas laser.

tion for cavity resonance exact, we should include the refractive index n, along with wavelength λ, cavity length L, and an integer N:

$$2nL = N\lambda \tag{4-2}$$

Refractive indexes of air and other gases are very close to one, so Equation 3-4 is usually good enough for gas lasers. However, as you saw in Table 2-5, solids and semiconductors have refractive indexes starting from about 1.4 to 3.5, so Equation 4.2 is needed for those materials.

The actual profile of wavelengths emitted by a laser is the product of the gain curve and the envelope of longitudinal oscillation modes. The profile in Figure 4-3 is based on typical emission from a gas laser, which can oscillate in a few longitudinal modes. In some other lasers, one central longitudinal mode dominates, with much lower power emitted at one or two peripheral modes.

Adding components to the laser cavity can limit the laser to oscillating in a single longitudinal mode, which has a linewidth of about one megahertz, which is equivalent (for helium–neon lasers) to 0.000001 nm. The most common technique is to place a device called a Fabry–Perot etalon into the resonant cavity. As shown in Figure 4-4, an etalon is a structure with a pair of parallel reflective surfaces, which is tilted at an angle to the axis of the laser cavity. At most wavelengths, the etalon reflects some light

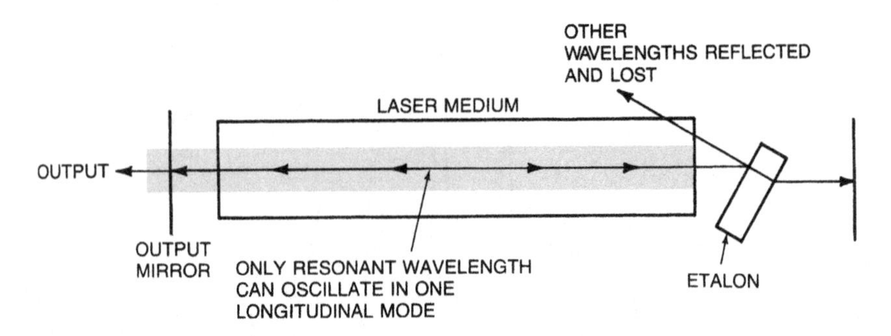

Figure 4-4. A Fabry–Perot etalon limits laser oscillation to a single longitudinal mode.

away from the cavity mirrors, increasing loss and suppressing laser oscillation, but in a narrow band interference effects eliminate the reflection, suppressing the loss and allowing laser oscillation. This limits the laser to operating in a single longitudinal mode, the wavelength of which can be varied by adjusting the spacing between the Fabry–Perot reflectors, or by turning the device in the laser cavity.

As you learned in Section 3.5, optics inserted in the laser cavity can tune the emission wavelength. Usually, the extra components deflect the unwanted wavelengths out of the laser cavity, but allow oscillation at the desired wavelength.

4.2.4 Single- and Multiwavelength Operation

The natural line narrowing caused by laser amplification, described earlier, does not prevent lasers from emitting simultaneously on two or more different transitions.

A few lasers can emit on a family of closely spaced transitions at the same time. In the carbon dioxide laser, the main transition occurs when the molecule shifts between two vibrational modes. (There are actually two such transitions close to one another, one centered near 10.5 μm, the other near 9.4 μm.) As it shifts vibrational states, the CO_2 molecule also changes rotational energy levels, which are spaced much more closely than vibrational levels. Figure 4-5 shows the family of lines emitted by these lasers, each one a transition between different rotational sublevels during a vibrational transition. At high pressures, they blur together to form a continuum.

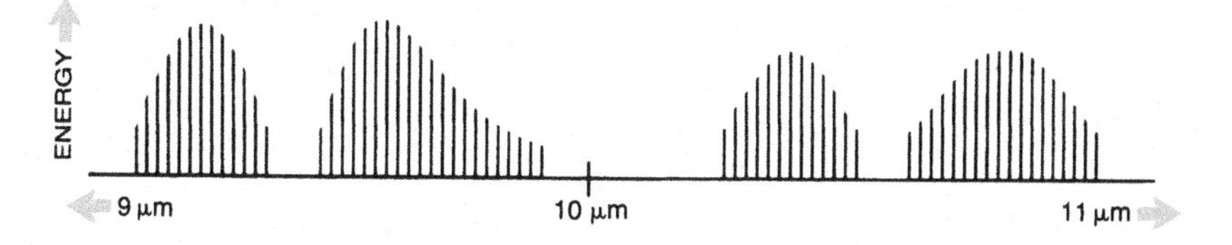

Figure 4-5. The many wavelengths emitted by a carbon dioxide laser near 10 μm.

Other lasers emit on two or more discrete transitions, typically involving different upper and/or lower laser levels. The argon- and krypton-ion lasers described in Chapter 7 are important examples that can simultaneously emit on different lines.

4.3 BEHAVIOR OF LASER BEAMS

We think of laser beams as tightly focused and straight, but their behavior is more complex than that. In general, beams spread or diverge with distance, but their precise behavior depends on distance from the laser.

4.3.1 Beam Divergence

The spreading angle of a laser beam is called its *divergence,* and usually is measured in milliradians, thousandths of a radian, a unit of circular measurement. A full circle equals 2π radians, and one milliradian (mrad) equals 0.0572958°, or about 3.5 arcminutes. Radians are convenient because the tangent of a small angle roughly equals its size in radians, making calculations easy. Most continuouswave gas lasers have divergence of around 1 mrad, but divergence usually is larger for pulsed lasers, and much larger for many semiconductor lasers which lack external beam-focusing optics. (Rapidly diverging beams are usually measured in degrees.)

Beam divergence depends on the nature of the resonator, the width of the emitting area, and diffraction, which you learned in Chapter 2 arises from the wave nature of light. The theoretical lower limit on beam divergence, called the *diffraction limit,* arises from diffraction at the edge of the emitting aperture or output mirror. It is lowest for TEM_{00} beams. The diffraction limit roughly equals wavelength divided by the diameter D of the output aperture.

$$\text{Diffraction limit(radians)} = \lambda/D \tag{4-3}$$

Focusing optics can change beam divergence. Lenses are used to focus the rapidly diverging beam from a red semiconductor laser into the narrow beam required for laser pointers.

4.3.2 Near-Field Conditions

In the near field close to the laser, a narrow beam normally shows little spreading and remains essentially a bundle of parallel light rays. The distance over which the light rays remain parallel is sometimes called the *Rayleigh range,* and depends on the beam diameter exiting the laser, D, and the wavelength λ:

$$\text{Rayleigh range} = D^2/\lambda \tag{4-4}$$

For a visible beam with 500-nm wavelength, the near-field range is 2 m for a 1-mm beam, and 50 meters for a 5-mm beam.

4.3.3 Far-Field Beam Divergence

In the far field beyond the Rayleigh range, beam divergence becomes the critical parameter in calculating beam diameter, as shown in Figure 4-6. Divergence angle normally is measured from the center of the beam to the edge. Beam intensity drops off gradually at the sides, so the edge of the beam usually is defined as where intensity drops to $1/e^2$ of the maximum value.

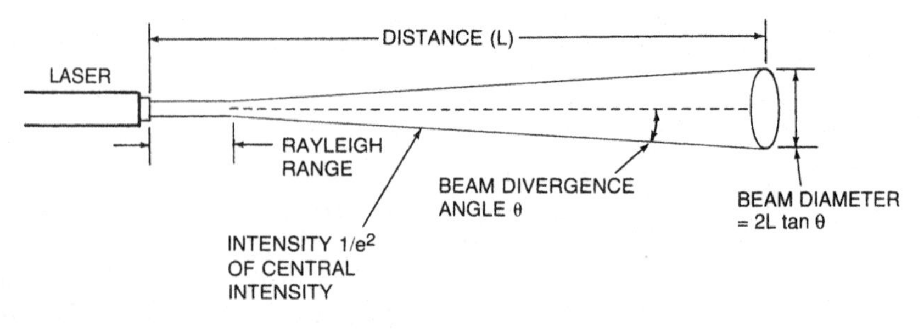

Figure 4-6. Divergence of a laser beam, exaggerated to make it visible.

Calculating beam size is a matter of trigonometry. Multiplying distance by the tangent of the divergence angle θ gives the beam radius, which must be doubled to get the diameter:

$$\text{Diameter} = 2 \times \text{distance} \times \tan\theta \tag{4-5}$$

If divergence is under about 0.1 radian (6 degrees), you do not have to bother calculating the tangent; the angle in radians is a good enough approximation. That is the case for most lasers except semiconductor types.

4.3.4 Beam Waist and Divergence

Things get more complicated if the resonator mirrors are curved, which produces a narrow point or "waist" of the beam inside the laser cavity. In this case, the far-field divergence θ depends on the beam waist W rather than the emitting aperture. In the far field, we can approximate the divergence angle with the formula

$$\theta = \frac{\lambda}{\pi W} \tag{4-6}$$

The beam waist diameter itself depends on laser wavelength, cavity length, and design. For a simple confocal resonator (with two mirrors both having their focal points at the midpoint of the cavity) of length L, the beam waist diameter W is

$$W = \sqrt{\frac{L\lambda}{2\pi}} \tag{4-7}$$

or about 0.17 millimeter for a 30-centimeter (one-foot) long helium–neon laser emitting at 632.8 nanometers. Note that this is smaller than the output spot size, typically about 1 mm for such lasers.

4.3.5 Output Port Diameter and Beam Divergence

The diffraction limit mentioned above is a fundamental rule of optics, which relates to how much an optical system can focus light. It says that the minimum possible divergence angle θ roughly equals the wavelength divided by the diameter of the optics D.

The exact relationship includes a constant K, which depends on the distribution of light across the output port:

$$\theta = \frac{K\lambda}{D} \tag{4-8}$$

The constant K equals 1.22 if light is equally bright across the entire opening. It is closer to one if the intensity is highest in the center and drops off smoothly toward the edges, which is the case for most laser beams. The formula is general for any emitting aperture, so it works for laser output mirrors, focusing lenses, telescopes, or even radio antennas.

This formula tells us that the larger the output optics, the smaller the beam divergence. Thus, you need large optics to focus light onto a small spot. Conversely, the shorter the wavelength, the smaller the spot size. In practice, this means that you cannot have both tiny spot diameter and narrow divergence. If you start with a 1-mm beam from a helium–neon laser, for example, its minimum divergence is about 0.6 mrad.

Divergence is larger than these formulas indicate if the beam contains multiple transverse modes, which spread out more rapidly than a single-mode TEM_{00} beam. Far-field divergence increases roughly with the square root of the number of transverse modes ($N^{1/2}$) the laser emits, and the area of the illuminated spot is thus roughly proportional to the number of transverse modes.

4.3.6 Focusing Laser Beams

The diffraction limit also affects how tightly lenses can focus laser beams or other light, or how finely optical systems can resolve detail in distant objects. The minimum diameter of a focal spot S formed by a lens of diameter D and focal length f with light of wavelength λ is

$$S = \frac{f\lambda}{D} \tag{4-9}$$

If you are familiar with photography, you may recognize the ratio f/D as the focal ratio or "f number" of a lens, so the formula also implies that the smallest focal spot for ordinary lenses is about a wavelength across.

4.3.7 Laser Modes

Discussions of laser modes can be confusing because lasers have two types of modes, longitudinal and transverse, but laser operation is often specified only as "single-mode" or "multimode." In most cases, descriptions of beam modes refer to transverse modes. Lasers oscillating in a single longitudinal mode often are called single-wavelength or single-frequency lasers, and have narrower linewidths than lasers oscillating in multiple longitudinal modes.

A laser operating in a single transverse mode can oscillate in two or more longitudinal modes. Many gas lasers emit a single TEM_{00} mode beam that contains two or more longitudinal modes, although special optics can restrict them to oscillating on a single longitudinal mode.

4.4 LASER POWER

Output power is a vital characteristic for most applications. Power turns out to be a more complex matter than you might expect because it can be measured and delivered in different ways. We will start with a general introduction, then shift to two critical areas: laser efficiency and the variation of output with time.

4.4.1 Intracavity and Beam Power

Normally, laser power refers to the power in the laser beam. However, there actually are two distinct levels of laser power: the *intracavity* power that circulates within the laser resonator, and the power in the beam itself.

The intracavity power is the stimulated emission generated inside the laser that bounces back and forth between the laser mirrors. On each pass of the cavity, some fraction of the light is extracted to form the laser beam. How much light exits the cavity depends on the output mirror transition. For example, if the output mirror transmits 1% of the light striking it, the beam power is roughly 1% of the circulating power inside the cavity. For such a laser to generate a continuous 1-mW beam, it has to have a circulating power of about 100 mW. The ratio is not as simple to calculate in a pulsed laser, but the point remains that the power inside the cavity may be much higher than in the laser beam.

Because intracavity power is much higher than the power in the beam, certain devices that require high power may be inserted into the resonant cavity to take advantage of the extra power. For this to be practical, one or both of the mirrors must be separate from the laser medium, as shown in Figure 4-7. Chapter 5 describes some devices that may be used inside a laser cavity.

4.4.2 Laser Output Power and Scaling

The output power in the laser beam depends on a number of factors including the gain, the nature of the laser cavity, and the inherent properties of the laser medium.

Many types of lasers are inherently feeble because their internal physics makes it impossible to reach high powers. One example is the red helium–neon laser, which can generate powers from a fraction of a milliwatt to about 50 mW, but becomes impractical at large sizes.

Other lasers are far more adaptable, and can be scaled over a much larger range. Commercial carbon dioxide lasers can generate continuous beams from a fraction of a watt to more than 10 kilowatts, but no single laser covers that whole range. Different versions must be designed to cover different power ranges. Small size, low cost, and easy use are important at low power; at higher power, the dissipation of waste heat becomes crucial, so much effort must be devoted to control gas flow and heat transfer.

No single version of any type of laser can span the whole range of possible output power. Helium–neon lasers have to be en-

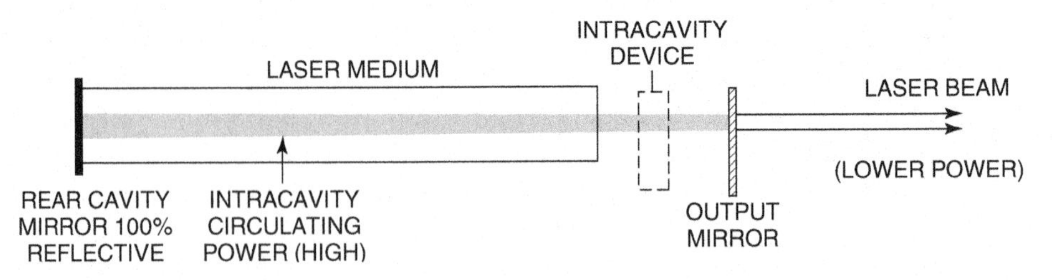

Figure 4-7. Circulating power inside a laser cavity is higher than beam power, so devices that need high power may be put between the mirrors and the laser medium.

gineered differently to emit 50 mW than to emit only 1 mW. Carbon dioxide lasers change even more dramatically. Low-power versions have sealed tubes; high-power versions flow gas through the tube and may use refrigeration or flowing water to remove waste heat. You will learn more about power ranges when we cover individual laser types.

4.4.3 Power Measurement and Time

Power in a laser beam normally is measured in watts, which is the amount of energy flowing through the beam. One watt equals energy flow of one joule per second. This measurement is straightforward if the beam is continuous.

Because power is an instantaneous measurement, things get more complicated if the laser beam varies with time. The output of pulsed lasers may be an average over a number of pulses repeated at a regular interval, or a peak power in an individual pulse. Pulses also may be characterized by the energy they contain.

Alternatively, power delivered to a surface can be measured in watts per unit area, typically per square centimeter. If the spot is close to the laser, or optics focus a large beam into a small area, this number can be quite high. Strictly speaking, optical power per unit area is defined as *irradiance* rather than *intensity,* which is defined as power per unit solid angle. However, this fine distinction is often ignored, and power per unit area is often called intensity.

4.4.4 Modulation

Many laser applications require modulation of the laser output power. For example, if a laser beam is to carry a signal in a fiber-optic communication system, the beam must be switched on and off to carry digital ones and zeros.

Lasers can be modulated internally (also called directly), by switching the power to the laser off and on, or externally, by passing light through a device that transmits varying fractions of the light. The ease and effectiveness of direct modulation depends on the type of laser; it works much better for semiconductor lasers than for gas or solid-state lasers. External modulation can be used with any laser.

4.5 LASER EFFICIENCY

Lasers do not produce energy; they convert energy from other forms into light. The efficiency of that energy conversion is important on several levels. The more obvious is producing the most possible power from the laser. The more subtle importance is that waste energy winds up as heat that the laser must dissipate, a particular problem for both high-power operation and inefficient lasers.

Laser efficiency can be measured in several ways. For practical purposes, the most useful measurement generally is the overall or *wall-plug efficiency,* the fraction of the energy delivered to the laser (through the wall plug) that emerges in the laser beam. That overall efficiency, in turn, depends on a number of other efficiency terms. We will start by looking at overall efficiency, then turn to its components, and finally mention a few cases in which other measures of efficiency are more important.

4.5.1 Elements of Overall Efficiency

In considering the components of wall-plug efficiency, we will start by looking at a gas laser, but the principles are the same for all types of lasers. Electrical power enters the laser through a power supply, which converts much (but not all) of the input energy into drive current that is passed through the laser gas. Much (but not all) of that electrical energy is deposited in the gas (energy-deposition efficiency). Much of the deposited energy excites the laser medium and produces a population inversion. Much of the energy in the population inversion is converted into a laser beam.

The central point is that wall-plug efficiency is the product of the efficiencies of several processes. For the gas laser, these are

Power supply efficiency
× Energy-deposition efficiency
× Excitation efficiency
× Fraction of energy available on laser transition
× Fraction of atoms that emit laser light
× Cavity output-coupling efficiency

= Overall efficiency

Plug in some numbers and you can see why lasers have low wall-plug efficiencies. If the power supply, energy deposition, and excitation each are 80% efficient, the laser transition represents 50% of the excitation energy, half the atoms emit laser light, and 80% of the laser energy is coupled out of the cavity in the beam (all of which are optimistic assumptions for most gas lasers), the wall-plug efficiency is $0.8 \times 0.8 \times 0.8 \times 0.5 \times 0.5 \times 0.8 = 0.1024$, or 10%. Similar considerations apply to solid-state and semiconductor lasers.

Let us look at the major elements in the efficiency equation.

4.5.2 Power-Supply Efficiency

The power supply converts 120-volt alternating current (or higher voltages used for heavy-duty or industrial lasers) to the form needed to excite the laser. This is straightforward for continuous-wave gas lasers excited by low-current, high-voltage discharges, and for semiconductor lasers, which run on low-voltage direct current. However, no power supply is 100% efficient, so some energy is lost. More energy is lost when the electrical input has to be converted to high-voltage pulses to drive pulsed gas lasers.

The same concept applies to semiconductor and solid-state lasers. The power supply for a semiconductor laser must convert input electricity into the low-voltage drive current. The "power supply" in an optically pumped solid-state laser actually involves two stages: an electrical power supply that drives the pump source, and the light source which converts some of that energy into light. In this case, the efficiency of generating pump energy is the product of the electrical power supply efficiency and the efficiency of converting electrical energy into light. Both are low for pumping with pulsed flashlamps, which are driven by high-voltage electrical pulses. Semiconductor lasers generate light much more efficiently, with more than half the electrical energy converted into light in some cases.

4.5.3 Excitation Efficiency

All the excitation energy that goes into the laser medium does not excite atoms to the upper laser level. The medium does not absorb all the energy, and some of the energy does not excite atoms to the proper states for laser action.

You can understand the principles by thinking about a laser rod that is pumped by light from a flashlamp. Inevitably, some light will pass through the rod and escape, even if the rod and flashlamp are contained in a reflective cylinder. Likewise, the rod will absorb some pump light that is not at the right wavelengths to excite atoms to the upper laser level, so that energy will heat the rod rather than produce laser light.

4.5.4 Laser Transition Efficiency

Some losses are inherent in the laser transitions. Exciting a four-level laser raises an atom to an excited level above the upper laser level. After stimulated emission drops the atom to the lower laser level, it then drops further to the ground state. You can see this in Figure 4-8, which shows a hypothetical atom with energy levels indicated in energy units—electron volts. Initially, absorption of 4 eV raises the atom to an excited state, from which it drops to the upper laser level at 3 eV. The laser transition takes the atom to the lower laser level at 2 eV, from which the atom drops to the ground state at 0 eV. It takes 4 eV to produce a 1 eV laser photon, meaning that the transition is at most 25% effi-

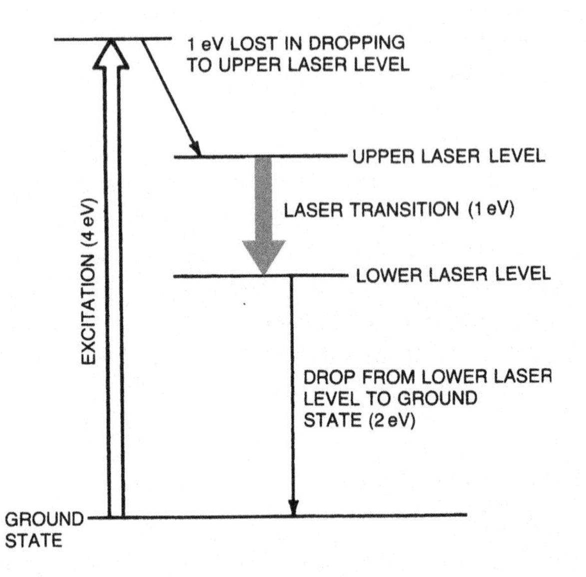

Figure 4-8. Only part of the energy that excites a four-level laser winds up in the laser beam. In this example, 4 eV of excitation produces only 1 eV of laser energy.

cient. Some real laser transitions are better than our example, but many are much worse.

Another problem is that laser emission cannot start until energy has been pumped into the laser medium to produce a population inversion. The *threshold* for laser action is not reached until the population inversion is large enough for stimulated emission to offset other losses. Above that point, light emission increases dramatically. As shown in Figure 4-9, each increment of input power raises output power by an amount proportional to the increase in input; the rise in output divided by the rise in input is the *slope efficiency.* This is an important concept because it shows the dramatic change in laser efficiency above the threshold. For example, if you need ten watts of input power to pass the threshold, that much input may produce only 0.001 W of output, or 0.01% efficiency. However, if slope efficiency is 10%, adding another watt of input power adds 0.1 W of output power, raising total power to 0.101 watt, for 0.9% efficiency.

4.5.5 Energy Extraction Efficiency

Energy also is lost in extracting light from the laser cavity. Figure 4-10 illustrates the problem by showing how the mirrors of a con-

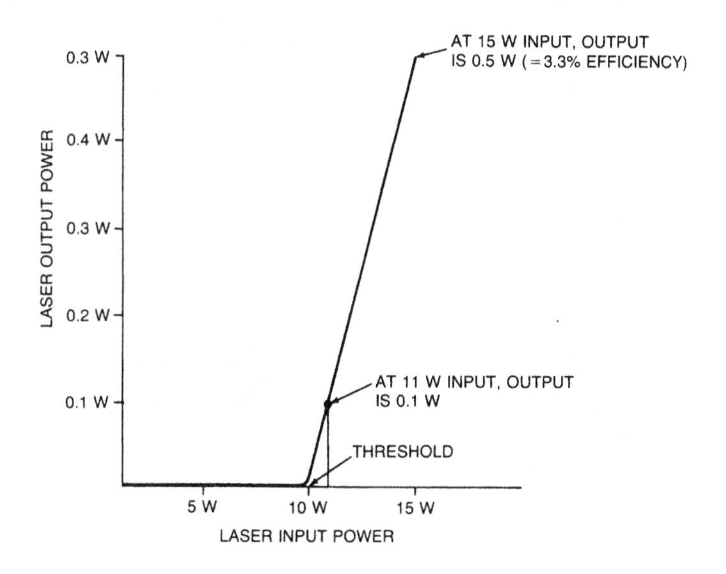

Figure 4-9. Laser emission does not emerge until the input energy has passed a threshold at 10 W, then it rises with 10% slope efficiency.

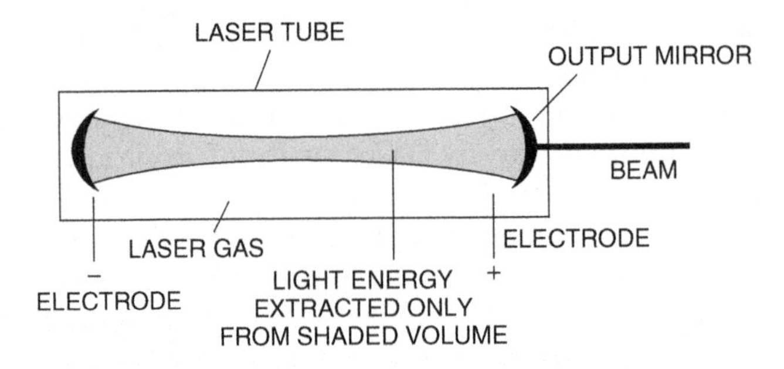

Figure 4-10. Electrical discharge passes through a laser tube, but the resonator only extracts energy from the gas in part of the tube.

centric resonator collect stimulated emission from only part of the laser medium. Although the entire laser medium is excited, light from part of the volume does not contribute to the laser beam. In general, unstable resonators can collect light from more of the laser cavity than stable resonators.

4.5.6 Waste Heat and Thermal Issues

The energy lost inside the laser does not vanish; most of it winds up as heat. Small and efficient lasers, like semiconductor laser pointers, can easily dissipate the heat by convection cooling into a room. But as shown in Table 4-1, many lasers have wall-plug efficiency less than 1%, so they can produce substantial amounts of heat at low powers. For example, a 10-watt argon laser that is 0.1% efficient produces 10 kilowatts of heat.

As the waste heat increases, more effort must be made to keep the laser from overheating. Modest-power lasers include fans for

Table 4-1. Efficiency ranges of commercial lasers

Type	Wall-plug efficiency
Argon-ion	0.001–0.1%
Carbon dioxide	5–20%
Copper vapor	0.2–0.8%
Excimer (rare-gas halide)	1.5–2%
Helium–cadmium	0.002–0.014%
Helium–neon	0.01–0.1%
Neodymium	0.1–over 10%
Semiconductor diode	1–over 50%

forced-air cooling. Higher-power leasers require flowing-water cooling or closed-cycle refrigerators. A few lasers require cooling below room temperature, but they are rarely used outside the laboratory.

4.5.7 Pump Conversion Efficiency

Efficiency also can be measured as the efficiency of converting energy input to the laser into light. This is common when the laser is packaged separately from the power supply or pump source.

For semiconductor lasers, pump conversion efficiency compares the input electrical drive power to the output optical power. This number can be over 50% for high-power semiconductor diode lasers.

For optically pumped lasers, conversion efficiency compares the input of optical pump power to the laser output. This number can be in the 80% to 90% range when some lasers are pumped with external lasers matched to strong absorption in the laser medium, particularly in fiber lasers (see Section 8.8).

4.6 DURATION OF EMISSION

Like wavelength and power level, the duration of laser emission is critical for laser applications. Lasers fall into two basic categories: pulsed or continuous wave. Pulse duration can vary across a wide range, and pulses may be repeated at a steady rate or generated one at a time. How a particular laser operates depends both on the nature of that type of laser, and on specific design details.

Many lasers are limited to pulsed operation because they can sustain a population inversion for only a limited time. In the three-level ruby laser, the lower laser level is the ground state, which quickly fills up as stimulated emission drops atoms to the ground state. Some four-level lasers cannot sustain a population inversion because atoms drop out of the lower laser level slowly, causing it to fill up and halt laser operation. Both types can be pulsed repeatedly.

Excitation techniques also can limit lasers to pulsed operation. Excimer gas lasers are excited by intense electrical discharges that pass through a mixture of gases, but are stable for less than a microsecond. The end of the discharge stops the laser pulse.

The need to dissipate waste heat also can limit lasers to pulsed operation. High temperatures can reduce gain by making it harder to sustain a population inversion. In solid-state lasers, uneven temperature distribution can induce strain into the glass or crystal laser medium, causing perturbations that increase internal losses, reduce output power, and degrade beam quality.

Steady-state continuous-wave operation requires a combination of favorable circumstances: energy levels and transitions that allow continuous operation, a stable way to supply energy continuously to the laser medium, and an efficient way to remove waste heat. Some important types of lasers can emit continuous waves at low power or with active cooling, but are limited to pulsed emission at high power or without cooling.

Excitation also plays an important role in the timing of laser emission. Most lasers start emitting almost immediately after excitation begins, and stop almost immediately after excitation ends, but rise and fall times are not instantaneous. Some continuous-wave lasers require relatively long times for emission to stabilize, particularly gas lasers. Metal-vapor lasers may require significant warm-up times before operation.

You will learn about the characteristics of individual laser types in later chapters, but here we will introduce measurements and fundamental limits.

4.6.1 Power and Energy Versus Time

Pulsed and continuous-wave lasers differ in how their output power varies with time. Before we go on, we should look carefully at these variations and how they are measured.

Figure 4-11 shows a fairly typical laser pulse, which rises slowly at first, then rapidly to its *peak power,* then drops off symmetrically. Pulses have characteristic *rise and fall times,* which are measured at the points where laser power is between 10% and 90% of the maximum value (not from zero to peak power). Some lasers may emit a small prepulse (before the main pulse), and/or have a long tail dropping off slowly after the main pulse, which are shown as dashed lines in Figure 4-11. Duration of the whole pulse typically is defined as "*full width at half maximum,*" the interval from the time the rising pulse passes 50% of the peak power to the time the power drops below that 50% point.

The actual pulse shape may vary from quite rounded, as in Figure 4-11, to nearly rectangular, but even digital pulses do not

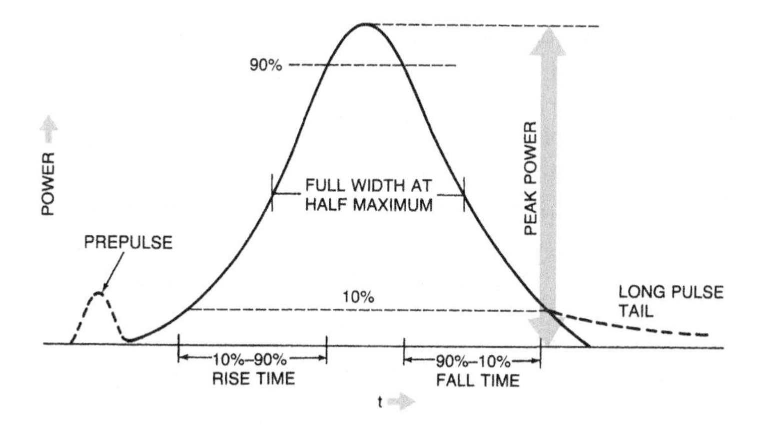

Figure 4-11. Power variation as a function of time (dashed lines show prepulse and long tail).

turn on and off instantaneously. Although the peak power may be reached only briefly, for many purposes it is assumed that the laser emits at peak power for the full width at half maximum duration of the pulse duration. This greatly simplifies calculations of pulse power and energy.

Another vital parameter is the *repetition rate,* or number of pulses per second. This is a frequency, so it is measured in units of hertz, like the frequency of an electromagnetic wave. In both cases, the frequency is the number of peaks or pulses per second.

It is often useful to calculate energy in a laser pulse. To do so rigorously requires calculating the energy delivered as a function of time, equivalent to measuring the area under the pulse in Figure 4-11. (Using calculus, you integrate the power as a function of time.) However, you can make a simple approximation by multiplying the full-width, half-maximum power P by the pulse duration Δt to get pulse energy E:

$$E = P \times \Delta t \tag{4-10}$$

Using this formula, we find that a one-megawatt pulse lasting one microsecond delivers 1 J. If you know pulse energy and duration, you can flip the formula to calculate the pulse power:

$$P = E \times \Delta t \tag{4-11}$$

If you have a series of equally spaced pulses, you can modify this formula to calculate the *average power* emitted during a period in which the beam is both off and on. In that case, you sum energy of the pulses emitted over a period of time and divide by the time interval. If all the pulses have the same energy E and the laser is emitting N pulses per second, the average power is

$$\text{Average power} = NE \tag{4-12}$$

If the laser emits 100 pulses per second, each with energy of 1 J, the average power (over a one-second interval) is 100 W.

Another useful number is the *duty cycle,* the fraction of the time the laser emits light. If a laser switches from "on" to "off" once every microsecond, the duty cycle is 50% because it is on half the time. If the laser emits N pulses per second, each lasting a time Δt, the duty cycle is a dimensionless number, the product of repetition rate times pulse length:

$$\text{Duty cycle} = N \times \Delta t \tag{4-13}$$

If the pulse length is 1 microsecond and the repetition rate is 100 pulses per second, the duty cycle is 0.0001 or 0.01%.

4.6.2 Limits on Pulse Duration and Bandwidth

The Heisenberg Uncertainty Principle sets fundamental limits on the product of pulse duration and the range of wavelengths a laser can emit during a pulse. This limit means that you need a very wide bandwidth to produce a very short pulse, or a continuous beam to generate a very narrow bandwidth. In other words, very short pulses cannot be very narrow because you cannot precisely measure both the location and energy of a photon. This is called the *transform limit,* because it is derived from the Fourier transform between time and frequency domains.

Numerically, the product of the pulse length Δt (in seconds) times the bandwidth Δn (in hertz) must equal at least 0.441 for pulses with the Gaussian shape of Figure 4-11. This means that the bandwidth $\Delta \upsilon$ of a laser pulse can be no smaller than

$$\Delta \upsilon = \frac{0.441}{\Delta t} \tag{4-14}$$

However, the bandwidth does not become infinitely narrow for a continuous beam. Turn the formula around, and you see that a laser must have a broad bandwidth to generate short pulses.

$$\Delta t = \frac{0.441}{\Delta \upsilon} \qquad (4\text{-}15)$$

To understand just how broad a bandwidth you need, let us calculate the requirement for generating pulses lasting only 5 femtoseconds (5×10^{-15} second), which is only a few wavelengths long. First we find $\Delta\upsilon$ in frequency terms, which comes to 88.2×10^{12} Hz or 88.2 terahertz (THz). To see what that means in relative terms, the visible spectrum runs from 400 to 700 nm, equivalent to frequencies of 430 to 750 THz, a bandwidth of 320 THz. Thus, the wavelengths in such a pulse would have to span more than a quarter of the visible spectrum.

4.7 POLARIZATION

Polarization is an important property of electromagnetic waves that can appear in laser beams. Recall that light waves consist of oscillating electric and magnetic fields that are perpendicular to each other, as you saw in Figure 2.1. Light waves are polarized when their electric (or magnetic) fields are aligned with each other.

The simplest type of polarization to understand, and the most common in introductory laser physics, is *linear polarization.* The polarization direction normally is defined for the electric field, so linearly polarized light waves all have their electric fields aligned in the same plane. One peculiar property of light waves is that they always can be separated into two components with linear polarizations that are perpendicular to each other. You can think of this as projections onto the X and Y axis of a graph, with one component vertically polarized (along the Y axis) and the other horizontally polarized (with electric field along the X axis).

Normally, most lasers emit unpolarized light. However, a laser can generate linearly polarized light when an optical component is added that separates the two perpendicular polarizations. The usual choice is a Brewster window, a surface which is aligned at a critical angle θ_B, defined in air as the arctangent of the refractive index n of the window material:

$$\theta_B = \arctan n \qquad (4\text{-}16)$$

For optical glass with refractive index 1.5, Brewster's angle is about 57 degrees.

As shown in Figure 4-12, the Brewster window transmits all light and reflects none if the polarization is in the plane of the page in Figure 4-12. However, it reflects about 15% of the light if the plane of polarization is perpendicular to the page.

Placing a Brewster window between the laser medium and one mirror in the reflective cavity increases losses for the plane of polarization perpendicular to the page. That means that light polarized in the plane of the page is amplified more, and if the gain is high, only light polarized in that plane will appear in the output beam. This produces a linearly polarized output beam with the same power that an unpolarized beam would have had. (Suppressing the other polarization concentrates laser power in the plane of the page, but it does not "throw away" power.)

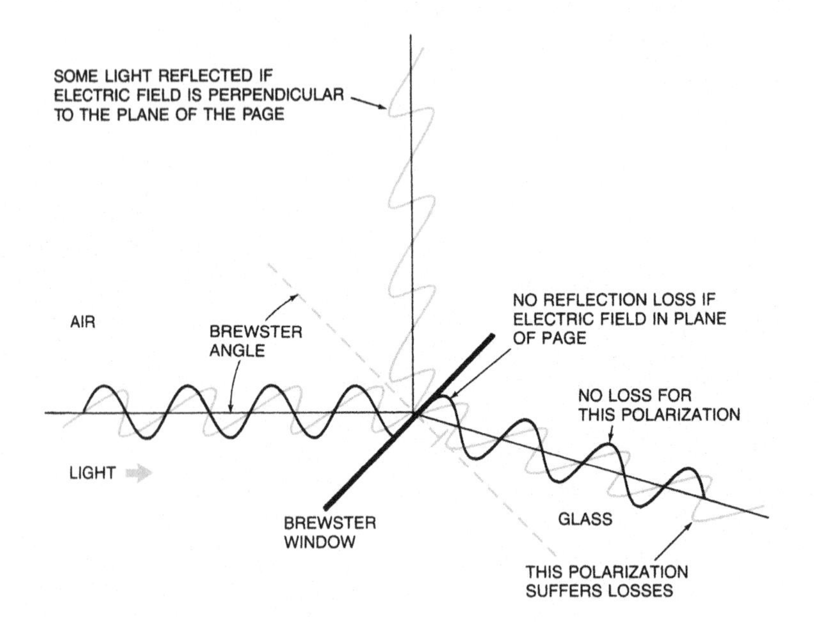

Figure 4-12. Operation of a Brewster window in a laser.

4.8 WHAT HAVE WE LEARNED?

- Coherent light is in phase and has the same wavelength.
- Temporal coherence indicates how long light waves remain in phase; spatial coherence is the area over which light is coherent.
- Coherence length equals the speed of light divided by twice the bandwidth.
- Interference effects require some degree of coherence.
- Coherent effects generate grainy patterns called speckle.
- Laser light is nominally monochromatic, but spans a finite wavelength range.
- Amplification narrows the width of laser lines so they are narrower than the gain bandwidth.
- The thermal motion of gas molecules broadens the linewidth of lasers.
- Multiple longitudinal modes of a laser cavity typically fall within the gain bandwidth of a laser transition.
- An etalon can restrict a laser to oscillation in a single longitudinal mode.
- Lasers can emit on single wavelengths or multiple transitions, depending on the laser cavity.
- Beam divergence decreases as the diameter of the output optics increases, and increases as wavelength increases.
- The diffraction limit is the lower limit on beam divergence.
- Focusing optics can change beam divergence.
- Beam behavior differs in the near field and far from the laser.
- Lasers oscillating in a single longitudinal mode are often called single-wavelength or single-frequency lasers.
- Lasers emitting a single transverse mode may oscillate in multiple longitudinal modes.
- Intracavity power circulating in the laser resonator normally is higher than output power.
- Power measures the amount of energy flowing past a point.
- The peak power of a laser pulse comes when its amplitude is highest.
- Wall-plug efficiency measures how much input electrical power a laser needs to generate light. It is the product of the efficiencies of several processes.
- Efficiency varies widely between types of lasers. Semiconductor diode lasers are the most efficient.

- Excitation efficiency measures how much of the input energy excites the laser medium.
- Slope efficiency is the increase in laser output per unit increase of input power when the laser is operated above threshold. It is higher than total efficiency because the laser must reach threshold before it starts emitting light.
- Waste energy not converted to light becomes heat, which must be dissipated.
- Pump conversion efficiency measures how much input pump energy is converted into laser emission. It can be very high for diode-pumped solid-state lasers.
- Some lasers are limited to pulsed operation by internal physics or by excitation techniques.
- Laser pulses have a characteristic rise time, pulse duration, and fall time.
- Pulse duration normally is measured as full width at half maximum of power.
- The repetition rate is the number of pulses per second.
- Energy is the power received during an interval.
- Average power is the total energy divided by the time.
- The uncertainty principle limits how precisely we can measure wavelength and pulse duration. Very short pulses have very wide bandwidths; very narrow bandwidths require continuous emission over a long time.
- A Brewster window can polarize laser output.

WHAT'S NEXT?

In Chapter 5, we will learn about major laser accessories and how they are used.

QUIZ FOR CHAPTER 4

1. A semiconductor laser emits 820-nm light with a bandwidth of 2 nm. What is its coherence length?
   a. 822 nm
   b. 168 μm
   c. 8.4 mm
   d. 840 mm
   e. 12.8 m

2. A single-frequency dye laser has bandwidth of 100 kilohertz (10^5 Hz) and wavelength of 600 nm. What is its coherence length?
   a. 168 μm
   b. 8.4 mm
   c. 840 mm
   d. 12.8 m
   e. 1.5 km
3. Which of the following contributes to the broadening of laser emission bandwidth?
   a. Doppler shift of moving atoms and molecules
   b. Amplification within the laser medium
   c. Coherence of the laser light
   d. Optical pumping of the laser transition
   e. None of the above.
4. How many longitudinal modes can fall within a laser's gain bandwidth?
   a. 1 only
   b. 2
   c. 3
   d. 10
   e. No fixed limit; depends on bandwidth and mode spacing
5. What is the distance over which light from a typical helium–neon laser, with 1-mm beam diameter and 632.8-nm wavelength, shows no divergence?
   a. No such distance; beam starts diverging immediately
   b. 0.23 ms
   c. 1.58 m
   d. 6 m
   e. 1.5 km
6. A helium–neon laser beam has divergence of one milliradian. What would its diameter be at the 38,000-km altitude of geosynchronous orbit?
   a. 1 m
   b. 38 m
   c. 76 m
   d. 1 km
   e. 76 km
7. Suppose we direct a helium–neon laser through a one-meter telescope to limit its divergence. What then is the beam diameter at the 38,000-km altitude of geosynchronous orbit? (As-

sume that the telescope is diffraction limited to a divergence of λ/D.)

a. 24 m
b. 48 m
c. 76 m
d. 94 m
e. 3.8 km

8. A semiconductor diode laser has output wavelength of 650 nm and beam divergence of 20 degrees without external optics. If you pointed it at a wall 2 m away, how big would the laser spot be?
a. 1 mm
b. 14.6 mm
c. 14.6 cm
d. 1.46 m
e. 2 m
9. Suppose you can buy a lens that can focus the diode laser beam to make it diffraction limited to a divergence of λ/D. How big would the lens have to be to focus the light onto a 1-mm spot on a wall 2 meters away.
a. 650 μm
b. 1 mm
c. 1.3 mm
d. 2.6 mm
e. 650 mm
10. What is the maximum wall-plug efficiency for a laser with a power supply that is 70% efficient with an energy-deposition efficiency of 25%, in which the laser transition represents 60% of the excitation energy and half the excited atoms emit laser light?
a. 1%
b. 5.25%
c. 10%
d. 25%
e. None of the above
11. A diode laser emitting at 980 nm pumps a ytterbium-doped fiber laser emitting at 1080 nm. Assuming all the pump light is absorbed, all the emitted light is collected, and everything else goes perfectly, what is the maximum possible conversion efficiency?
a. 10%
b. 50%

c. 89.6%
d. 90.7%
e. 98.0%

12. A pulsed laser generates 500-kilowatt pulses with full width at half maximum of 10 nanoseconds at a repetition rate of 200 hertz. Making simplifying assumptions about pulse power, what is the average power?
    a. 1 watt
    b. 2 watts
    c. 5 watts
    d. 10 watts
    e. 20 watts

CHAPTER 5

OPTICS AND LASER ACCESSORIES

ABOUT THIS CHAPTER

Many types of optical devices help lasers perform useful functions. This chapter is a brief introduction to the field, running the gamut from simple lenses to emerging concepts such as photonic crystals. It explains basic concepts but not all the details. We will start with lenses and simple optics, then describe special-purpose optics, nonlinear devices, modulation, pulse generation, measurement, and some emerging technologies

5.1 CLASSICAL OPTICAL DEVICES

In Chapter 2, you learned the basic concepts of how lenses and mirrors refract and reflect light. These are the foundation of the field called classical optics because it is based on concepts developed long ago. In this chapter, you will learn more about how optical devices can manipulate laser beams, both as individual optical elements and as components inside more complex optical equipment and instruments. The coverage is brief because the subject of classical optics would require a book in itself.

We will start from where we left off in Chapter 2. Our descriptions largely consider laser beams as bundles of parallel rays of light, as if they were emitted by an object an infinite distance away.

Understanding Lasers: An Entry-Level Guide, Third Edition. By Jeff Hecht

5.1.1 Simple Lenses

Simple lenses are made of a single piece of glass or other transparent material, with one or both sides curved to focus light. A single positive lens focuses parallel light rays to a small spot at its focal point. The size of that spot depends on the wavelength of the light and the size and curvature of the lens. The minimum spot size for a lens with focal length f and diameter D at a laser wavelength λ is the same as the minimum spot size for a laser beam (Equation 4-8):

$$\text{Spot Size} = \frac{f\lambda}{D} \tag{5-1}$$

The minimum practical spot size possible is comparable to the wavelength.

Lenses focus light rays the same way, no matter which direction they travel, so the same optics that focus parallel light rays to a point also can bend light from a point source to form a parallel beam. This is how a flashlight lens or searchlight mirror forms light from a bulb into a narrow beam.

A negative lens does not focus light to a point, but it can deflect parallel light rays so they spread out, as if they were coming from a point source.

5.1.2 Compound Lenses

Compound lenses are assembled from two or more simple single-element lenses. Camera and projector lenses are good examples. Compound lenses are more complex and expensive than single-element lenses, but they do a better job of focusing light because careful design can compensate for the inherent imperfections of simple lenses over a wide field of view, long distances, or a large depth of focus. Some compound lenses have fixed elements, but many have elements that move back and forth to bring light into focus, particularly in cameras.

5.1.3 Collimators, Beam Expanders, and Telescopes

Though it may seem amazing today, some two millennia passed between the discovery that lenses or mirrors could focus light and the discovery of the telescope. The ancient Greeks knew about burning glasses, but the telescope was invented about 1600 A.D. It seems unlikely that either discovery was the outcome of a systematic investigation. Legend tells us that children playing with lenses in the shop of Dutch spectacle maker Hans Lippershey discovered the Galilean telescope, which is made of a large positive lens with a smaller negative lens that serves as the eyepiece. Replacing the negative lens with a positive lens makes a more powerful telescope, but turns the image upside down.

Telescopes do more than make distant objects look closer. As shown in Figure 5-1, a simple refracting telescope made from a pair of positive lenses also can expand the bundle of parallel light rays in a laser beam, enlarging the beam diameter. A telescope that serves this purpose is called a *beam expander* or *collimator*. If you reverse the direction of the light, the collimator optics can shrink the laser beam.

5.1.4 Optical Aberrations

Simple optics do not bring light to a perfect focus because they suffer from *aberrations* that blur light even if the lenses are polished perfectly to the ideal shape specified by the optical designer. These aberrations arise from limitations inherent in the nature and design of simple lenses.

One important type is *chromatic aberration,* which occurs because the refractive index of transparent materials varies with the wavelength. In Chapter 2, you saw that the focal length of a positive lens with radii of curvature of its surfaces R_1 and R_2 has a focal length f that also depends on refractive index n:

$$f = \frac{1}{(n-1)\left[\dfrac{1}{R_1} + \dfrac{1}{R_2}\right]} \tag{5-2}$$

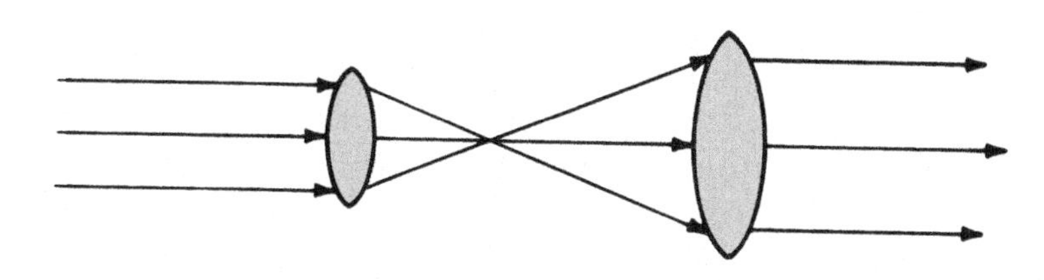

Figure 5-1. A refracting telescope serves as a beam expander or collimator for a laser beam.

Refractive index does not change a large amount, but it changes enough to affect the focusing of visible light. In one common type of optical glass, n is 1.533 at 434 nm in the violet and 1.517 at 656 nm in the red. This means that the focal length in the red is 1.03 times that in the violet, or 3% longer. The effect is most obvious near the edge of a lens or optical system, where it may give white objects a red fringe on one side and a blue fringe on the other. It is evident in cheap binoculars, or in thick plastic eyeglass lenses.

Chromatic aberration is not a big problem with lasers that emit only one wavelength. However, compensation may be needed if the laser emits multiple wavelengths, or if the same optical system is used by the human eye. The simplest way to compensate for chromatic aberration is to glue together components made of different glasses, which have refractive indexes that change in different ways with wavelength. Careful selection of glass compositions can largely cancel the variations in refractive index with wavelength, although the correction is not perfect. Such lenses are called *achromatic lenses.*

Standard lenses have surfaces that are spherically curved, that is, the surfaces are shaped like parts of a perfect sphere. That shape is used for most optics because it is easy to manufacture precisely, but it does not bring light passing through all parts of the lens to the same focus, an effect called *spherical aberration.* The focal length is slightly shorter in the outer parts than in the central zone, an effect that is stronger in lenses with strongly curved surfaces. Using only the central part of the lens or using only lenses with long focal length eases the impact of spherical aberration. Optical surfaces also can be made with aspheric (nonspherical) shapes that focus all light rays passing through the lens at the same distance, but they are harder to manufacture.

5.1.5 Cylindrical Optics

Most lenses are radially symmetric, with spherically curved surfaces, so they refract the same way if you turn them around their centers without tilting them. Light reaching such a lens at the same angle and the same distance from its center is always bent in the same way.

Spherical lenses are fine for most purposes, but not if your laser beam is diverging much more rapidly up and down than side to side. That is the case with many semiconductor lasers, as you will learn in Chapter 9. In that case, you need to focus the beam in the direction in which it is diverging most rapidly, but not in the perpendicular direction. (In practice, the resulting divergence is still too large to form a pencil-like beam, so an additional spherical lens is added in devices such as laser pointers.)

To do that, you need a lens that bends light in only one direction, as shown in Figure 5-2. This is called a *cylindrical lens* because the surfaces are slices of a cylinder, not a sphere, and, as shown in the figure, it focuses light vertically but not side to side. That is, if you look at it in a vertical plane, it acts like a normal lens. If you look at in a horizontal plane, it is just a flat piece of glass, and light is not bent at all in that plane.

5.1.6 Mirrors and Retroreflectors

As you learned in Chapter 2, curved mirrors can focus light like lenses. Flat mirrors are widely used to change the direction of laser beams, but unlike a household mirror, they reflect light from their front surface rather than the back surface. *Front-surface mirrors* are used for precision optics because they avoid undesirable effects such as the double reflection from the front and back surfaces of a household mirror.

Although many laser applications use standard flat or curved mirrors, some use a special type of mirror called a *retroreflector* or *corner cube.* This consists of three flat mirrors at right angles to one another, like the corners of a cube, as shown in Figure 5-3.

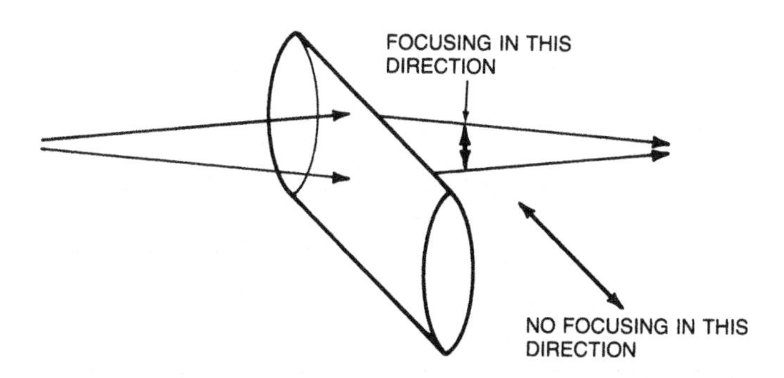

Figure 5-2. A cylindrical lens focuses light in one direction, but not in the perpendicular direction.

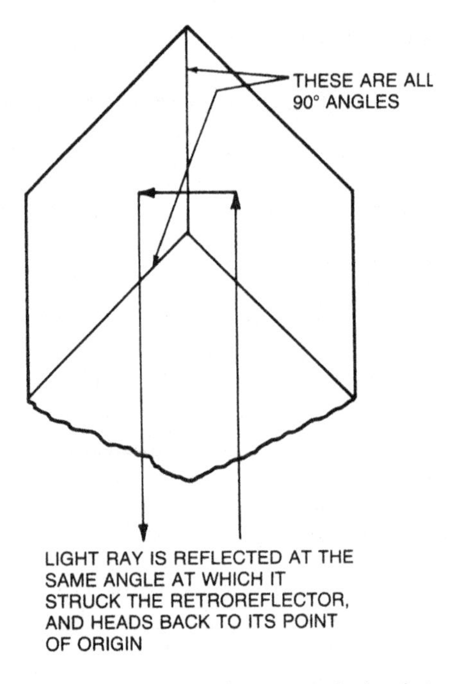

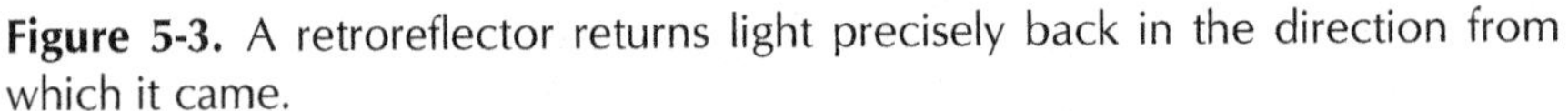

Figure 5-3. A retroreflector returns light precisely back in the direction from which it came.

Some retroreflectors are "hollow," made up of three front-surface mirrors. Others are prisms with reflective rear surfaces.

What is special about retroreflectors? A flat mirror reflects an incident beam at an angle equal to the angle of reflection, so it goes off in another direction unless the laser beam is perfectly perpendicular to the beam. A retroreflector bounces light directly back to the source.

5.1.7 Dispersive Optics

Often, it is useful to spread out or disperse different wavelengths of light. Two types of optical components, *prisms* and *diffraction gratings,* rely on completely different principles to do the same task.

A prism is a slab of glass with flat polished faces. Prisms are made in many shapes, and many of them are used essentially as mirrors to redirect light. Solid retroreflectors are really prisms designed to function as retroreflectors. Other prisms are used inside binoculars to fold the optical path and show the image right side up.

Dispersive prisms are triangular in cross section, usually equilateral, and are designed to take advantage of the variation of refractive index with wavelength to spread out the spectrum. Passing light through the prism bends the light at an angle that depends on the refractive index. The refractive index is higher at short blue wavelengths than at longer red wavelengths, so the blue wavelengths are bent the most, as shown in Figure 5-4. If light left the glass through a face parallel to the surface through which it entered, refraction at the exit face would cancel refraction at the entrance, and all wavelengths would emerge at the same angle. However, because the prism's output face is not parallel to the input face, different colors emerge at different angles, forming a spectrum.

A diffraction grating is a flat piece of glass or plastic with a series of parallel grooves in its surface. The grooves diffract (or scatter) light at all wavelengths from the surface across a wide range of angles. However, the diffracted light waves interfere with each other. For any one wavelength, the interference is constructive at only one angle and destructive at the other angles, so all the light

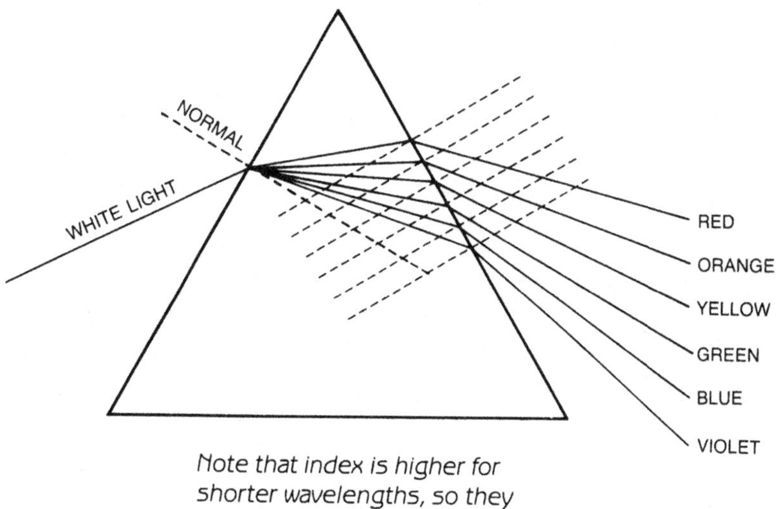

Figure 5-4. A prism forms a spectrum by bending different wavelengths at different angles.

at that wavelength is directed at the one angle. The diffracted angle depends on wavelength, and the effect spreads out a rainbow spectrum. You do not need a special diffraction grating to see the effect; the tightly wound spiral of data bits on a CD-ROM or DVD acts in the same way, diffracting light and spreading out a rainbow, which you can see if you hold the disk in the light and look for its reflection. The width of the spectrum depends on the spacing of the lines for both a standard diffraction grating and an optical disk.

Some important optical instruments contain diffraction gratings or prisms to spread out a spectrum that separates light by its wavelength. The *spectroscope* is an instrument that spreads out the spectrum of light from an external source, such as a laser or fluorescent lamp.

A more elaborate instrument called the *monochromator* spreads out the spectrum from an internal light source in order to generate a narrow range of wavelengths. In a simple monochromator, a single slit transmits a narrow slice of the spectrum to the outside world. The monochromator was invented before the laser, and its output is often called "monochromatic," although most lasers emit at a narrower range of wavelengths.

5.1.8 Fiber Optics and Total Internal Reflection

Light waves normally travel in straight lines, but optical fibers can guide them along curved paths. The main use of optical fibers is to carry communication signals, but they also can be used in other types of optical systems.

Figure 5-5 shows how light is guided in a simple type of optical fiber. Light at one end enters a core of glass with high refractive index n_1 surrounded by a cladding layer with a lower refractive index n_2. If you think of light as a ray that is directed along the length of the fiber, it eventually hits the interface between core and cladding at a glancing angle. If the angle is small enough, an effect called *total internal reflection* bounces the light back into the core.

Total internal reflection is a variation on the normal process of refraction. In Figure 2-9, you saw light going from a low-index medium into a higher-index material. Total internal reflection occurs when you are trying to go in the other direction, from high-index to low-index materials. In this case, the formula for angle of

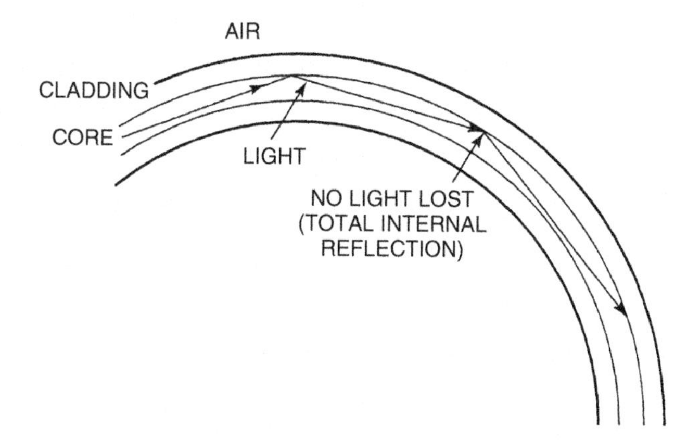

Figure 5-5. Total internal reflection guides light through the high-index core of an optical fiber clad with a lower-index cladding.

refraction is the same as you learned in Chapter 2, but you are trying to find the angle of refraction θ_1 (measured from the normal) into the low-index material:

$$\theta_1 = \arcsin\left(\frac{n_2 \sin \theta_2}{n_1}\right) \tag{5-3}$$

Because n_2 is greater than n_1, things work differently, and you are multiplying the angle from the normal in the core by a number larger than one. As θ_2 approaches 90°, its sine approaches one and, depending on the difference between n_2 and n_1, you eventually reach a point where you are trying to find the arcsine of a number greater than one. Your calculator will tell you that is an error. Nature simply does not allow the light to enter the lower-index material, and all the light is reflected back into the core, without the usual losses at a mirror surface.

This is a very simplified view of light guiding in optical fibers. Some fibers have a gradual boundary between core and cladding; others have a small core that transmits light in only a single transverse mode. Nonetheless, it conveys the essential aspect of how optical fibers guide light.

The principles of light guiding in optical fibers work for fibers of different sizes. Some guide light in a core less than 10 μm across, surrounded by a glass cladding that makes fiber diameter 125 μm. Others may have light-guiding cores a millimeter or more across. Stimulated emission, optical amplification, and laser action can occur inside fiber cores, as you will learn later.

Typically, single fibers are used for communications, but fibers also may be bundled together for imaging or illumination. Special-purpose optical fibers can be used as sensors.

5.1.9 Polarization and Polarizing Optics

As you learned in Chapter 4, light can be divided into two orthogonal polarizations, with their electric fields perpendicular to each other. In the simplest case, called *linear polarization,* the electric field is oriented in a particular direction. As shown in Figure 5-6, this polarization can be considered as a vector going from the origin (0, 0) to a point in the X–Y plane. You can break this vector down into horizontal and vertical components, as shown in Figure 5-6, or you can turn the coordinates so one direction matches the direction of polarization.

There are a number of types of polarization. Light is *linearly polarized* if its electric field remains continually in the same plane. If the plane of polarization changes with time but the amplitude is constant, the light is *circularly polarized,* because the polarization vector draws a circle in the X–Y plane. If the polariza-

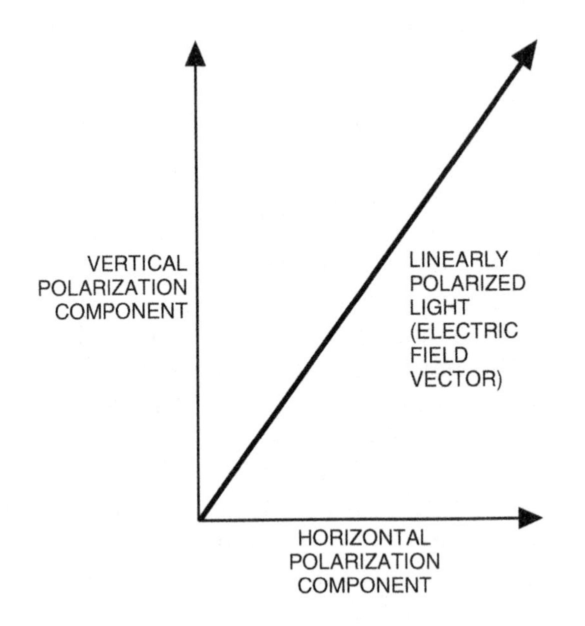

Figure 5-6. Direction of polarization as a vector.

tion vector changes both in direction and length, it draws an ellipse in the X–Y plane, and the light is called *elliptically polarized*. Light is called unpolarized if it is the sum of light of all different polarizations. (Random polarization, which might seem to be the same thing, is different, and means the direction of polarization changes randomly in time.) Laser light may be polarized or unpolarized.

A number of optical devices can manipulate light polarization.

Polarizers are optical devices that select or separate light of particular polarizations. Simple dichroic film polarizers (like those used for polarizing sunglasses) absorb light polarized along one axis, and transmit light linearly polarized along the other axis. For example, a polarizer might transmit light polarized along the Y axis in Figure 5-6 and absorb light polarized along the X axis. Higher-performance polarizers normally are made from *birefringent* materials such as calcite, which split light of different polarizations so it goes in different directions.

The refractive indexes of birefringent materials depend on the polarization of light, so they also can retard or rotate polarization. Polarization retardation is based directly on the difference between refractive indexes for the two linear polarization components of light. If horizontally polarized light experiences a larger refractive index, it takes longer to travel through a birefringent material than vertically polarized light. Thus, if the input light is unpolarized, the horizontally polarized component is retarded relative to the vertically polarized light.

How much the polarization is retarded depends on the thickness of the birefringent material. Typically, polarization retarders are designed so the delays are one-quarter or one-half of a wavelength, and are called *quarter-wave* or *half-wave plates*. One application is to rotate the phase of a linearly polarized input beam. If the input polarization is at an angle θ to the principal plane of the retarder, the polarization of the output beam is rotated by 2θ. Another application is to convert an input linearly polarized beam to circular polarization by passing it through a quarter-wave plate with its axis 45° from the input beam polarization.

5.1.10 Beam Splitters

One of the most important optical components used with lasers is the *beam splitter*. It does what its name says, splitting a beam into two parts. Normally, the beam splitter is at an angle to the incident beam, so it transmits some light and reflects the remaining light at an angle, as shown in Figure 5-7. (You need not align the beam splitter so the reflected beam is at a right angle to the incident and transmitted beam; that is just a convenient way to draw the beams.)

Beam splitters come in many types. Some reflect light of one linear polarization and transmit light of the orthogonal polarization. Others are insensitive to polarization. The reflectivity and transmission of some are very sensitive to wavelength; others vary little with wavelength. Two important things to note are that the beams do not have to be reflected and transmitted at right angles, and that the light is not always divided equally between transmitted and reflected beams.

5.2 TRANSPARENT OPTICAL MATERIALS

So far, we have said little about the materials from which optics are made. In Chapter 2, you learned that light transmission of any material varies with the wavelength. Materials we take for granted as transparent, such as air, water, and glass, are opaque in parts of the infrared and ultraviolet. Other materials that are opaque to visible light are transparent at some of those wavelengths. Reflectivity also varies with wavelength. As a consequence, different materials are used for optics in different parts of the spectrum.

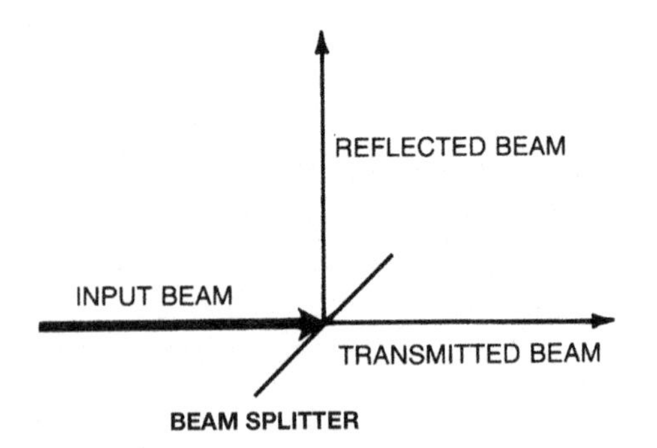

Figure 5-7. A beam splitter transmits part of the input beam and reflects the rest.

What we call glass—an impure form of silicon dioxide (silica or SiO_2)—is the most common transparent optical material for wavelengths in and near the visible. It is transparent, durable, and is supported by a well-developed technology. We can make clear glasses with a range of refractive indexes sufficient to meet most optical needs by adding different raw materials to the mixture that becomes glass. Some optics also are made from *quartz,* a natural crystalline form of silica. Optical fibers are made from an extremely pure form of silica known as *fused silica.*

Optics for visible light also can be made from transparent plastics, which are lighter than glass. Their biggest practical advantage is the low cost of making molded plastic lenses; light weight is secondary for many applications, but important in eyeglasses. However, plastics often have internal inhomogeneities that scatter enough light to cause problems in many laser applications. They also scratch more easily than glass lenses, and many plastics degrade when exposed to ultraviolet light. Although plastics are steadily improving, glass lenses normally are used for laser applications.

Many other materials are used for optics in other parts of the spectrum, particularly in the infrared. Figure 5-8 lists some important materials and the wavelengths at which they are transparent enough for use in transparent optics. Optical glasses are transparent enough for use with lasers at wavelengths to about 2 μm. Other materials are used at the 10-μm wavelength of carbon dioxide lasers.

Look carefully at Figure 5-8 and you will see some surprises. One is sodium chloride; purified crystalline table salt is highly transparent in the infrared and has good optical characteristics. However, it is also very soft and must be protected from atmospheric moisture.

5.3 OPTICAL SURFACES, COATINGS, AND FILTERS

Surfaces play a vital role in optical devices. Refraction changes the direction of light at surfaces. Reflection also occurs at surfaces, and not just on mirrors. Clear glass reflects a small fraction of the incident light, as you can see at night when you sit inside and look out a window, and see the bright objects in your room reflected by the window surface.

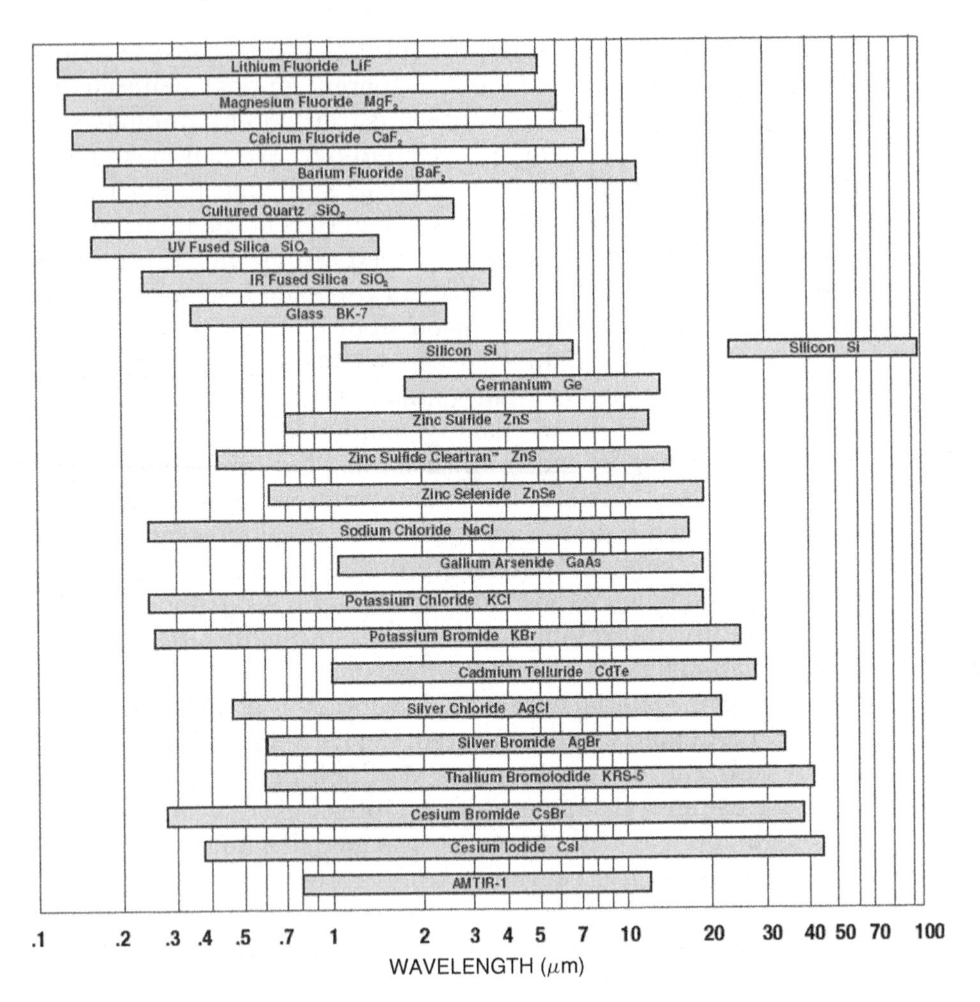

Figure 5-8. Important optical materials and their transmission ranges. (Courtesy of Janos Technology Inc.)

The high-performance optics used in laser systems normally are coated with thin layers of materials that affect their optical performance. Some coatings are single layers of metal or other materials, but most are made of many layers of transparent materials.

5.3.1 Metal Mirror Coatings

A thin layer of reflective metal can convert any surface into a mirror, but most metal-coated mirrors are made of glass. (Some special-purpose mirrors are solid metal.) Metal coatings are applied to the front surfaces of high-performance mirrors, so the light is reflected directly from the surface without having to pass through a slab of glass, as in bathroom mirrors. Metal mirrors may have an added protective layer deposited to protect the metal layer.

Although metal mirrors look highly reflective, the metal surface typically absorbs at least a few percent of the light. Aluminum is widely used for mirrors, but it typically reflects only about 90% of light in the visible spectrum. The absorption losses are not important for most optical applications, but can cause undesirable heating of a mirror that is reflecting a high-power laser beam.

5.3.2 Antireflection Coatings

Some reflection is inevitable any time light passes from air into glass or, in general, from a material with a low refractive index into another transparent material with a higher refractive index. The reflectance depends on the refractive indexes of the materials, the polarization of light, and the angle. If the light is incident perpendicular to the surface, the reflectance is

$$\text{Reflectance} = \frac{\left(\dfrac{n_{\text{high}}}{n_{\text{low}}} - 1\right)^2}{\left(\dfrac{n_{\text{high}}}{n_{\text{low}}} + 1\right)^2} \tag{5-4}$$

where n_{high} and n_{low} are the refractive indexes of the high- and low-index materials, respectively. For light in air, this yields a reflection loss of about 4%.

At different angles, the reflectance also depends on the polarization direction and the angle, but the overall dependence on refractive index difference remains. That reflective loss can be reduced by coating the glass with a thin layer intermediate in refractive index between the air and the glass. This creates two interfaces: between the air and the coating, and between the coating and the glass. However, the differences are smaller if the coating has an index of 1.25, halfway between glass and air. In that case, loss is 1.23% at the air–coating interface and 0.83% at the coating–glass interface. Total loss is about 2%, calculated by multiplying the fractions of light transmitted through the surfaces, not by adding reflective losses.

5.3.3 Multilayer Interference Coatings

More complex effects are possible by depositing many thin layers of material on optical surfaces to form *multilayer interference coatings,* also called *dielectric coatings* because they are made of materials that do not conduct electricity, called "dielectrics." These coatings consist of many alternating layers of two different materials, one with high refractive index and one with a lower index. Layer thicknesses are chosen so light reflected from one layer constructively (or destructively) interferes with that reflected from other layers, depending on the wavelength. This allows the designer to control how the coating transmits and reflects light across a range of wavelengths.

Proper selection of the layer materials and thicknesses lets coating designers meet demanding specifications. For example, they can make coatings that serve as filters and reflect or transmit only a narrow range of wavelengths. Such coatings can transmit (or block) a narrow range of wavelengths at a laser line. Thus, they can block white light from overhead lights while transmitting the red line of the helium–neon laser in the laser scanners used for automated checkout in supermarkets.

Multilayer coatings can give very high peak reflection or transmission at a particular wavelength, but this makes them very sensitive to wavelength. They also are very sensitive to the angle of incidence, because changing the angle changes the thickness of material the light has to pass through, and thus changes the wavelength at which interference effects occur. Tilt an interference coating that strongly reflects one color, and its apparent color will change with the viewing angle.

Coatings deposited on flat glass are used as filters to select particular ranges of wavelengths. Generally, most of the light is transmitted or reflected rather than absorbed. Table 5-1 lists some examples.

5.3.4 Color Filters

Wavelengths also can be selected by passing them through colored transparent glasses or plastics called *color filters.* The most common type of color filters are used for photography. Unlike thin-film filters, color filters absorb the light they do not transmit. That is fine for photography, but can cause heating if they are used to attenuate high-power laser beams.

Table 5-1. Some types of multilayer coatings

Mirror coatings	Strongly reflect selected wavelengths; used in laser cavities
Narrow-band filters	Transmit a narrow band of wavelengths, e.g., a laser line
Wide-band filters	Transmit a wide band of wavelengths, e.g., red light
Band-pass filters	Transmit a range of wavelengths: narrow, medium, or wide
Band-rejection filters	Blocks a band of wavelengths, e.g., a laser line

5.3.5 Neutral-Density Filters

Neutral-density filters reduce or attenuate light intensity uniformly at all wavelengths covered. Metallic neutral-density filters are coatings that attenuate light by reflecting the unwanted portion. Other neutral-density filters absorb the unwanted portion. Because their absorption is uniform across the visible spectrum, they look grey to the eye.

Attenuation of a neutral-density filter is normally measured as *optical density,* defined as

$$\text{Optical density} = -\log_{10}\left(\frac{\text{Output}}{\text{Input}}\right) \tag{5-5}$$

Thus, an optical filter that transmits 1% (0.01) of incident light has an optical density of 2. Reversing the formula gives the fraction of transmitted light:

$$\text{Transmission} = 10^{(-\text{optical density})} \tag{5-6}$$

5.3.6 Spatial Filters

A spatial filter transmits one part of a laser beam and blocks the rest. A common type of spatial filter is a pinhole in a sheet of black-painted metal. It transmits the center of the laser beam, but not the outer portions. Often, spatial filters are used to pick out the central, most uniform, part of a laser beam. Slits also are spatial filters.

5.4 NONLINEAR OPTICS

So far, we have described simple optics that have only a *linear* effect on the light passing through them. Linear in this context means that the output is proportional to the first power of the input:

$$\text{Output} = A \times (\text{Input}) \qquad (5\text{-}7)$$

Mathematically, both (Input) and (Output) terms are functions that give the instantaneous amplitude of an electromagnetic wave over time, such as $K(\cos \omega t)$, where ω equals 2π times the frequency ν.

Other types of devices are based on *nonlinear optics,* in which the output also depends on higher powers of the input:

$$\text{Output} = A(\text{Input}) + B(\text{Input})^2 + C(\text{Input})^3 + \ldots \qquad (5\text{-}8)$$

Nonlinear optics can do things impossible with linear optics, such as change the frequency and wavelength of light waves. The theory of nonlinear optics is the complex stuff of differential equations three pages long and Ph.D. theses, but some analogies will help you understand the basic idea.

You can think of nonlinear effects as a kind of distortion. When you play music in the normal range of a sound system, the output is linearly proportional to the input, like linear optics. Turn the sound up to the point where you hear distortion and the output is no longer linearly proportional to the input. Just as early rock musicians took advantage of the distortion in vacuum-tube amplifiers to generate new types of sound, nonlinear optics manipulate light in previously impossible ways.

As distortion in a sound system requires high audio power, nonlinear optical effects require high light intensities. Lasers achieve such intensities by concentrating light in the beam so the power density (in watts per square centimeter) is higher than that from ordinary light sources. The invention of the laser made possible the first observation of optical harmonic generation, which generates light at frequencies that are integral multiples of the input.

Nonlinear optical effects increase in strength more rapidly than linear effects because they depend on the square or some higher power of the input. Double the input power, and the output of the second harmonic increases by a factor of four, the square of the input increase. Raise the input a factor of five, and the output jumps by a factor of 25.

Some nonlinear optics can be even more complex. Let us take a look at the most important types of nonlinear optical devices and how they work.

5.4.1 Harmonic Generation

The simplest nonlinear optical effect is harmonic generation, the production of light waves at integral multiples of the frequency of the original wave. The second harmonic is at twice the frequency of the fundamental wave or, equivalently, at half its wavelength. The third harmonic is at three times the frequency or a third of the wavelength. In the laser world, the frequencies are very high, so the light waves normally are identified by wavelength, even though the process is still called harmonic generation. Thus, laser specialists say they frequency-double the output of a 1064-nanometer neodymium laser to get 532-nanometer light, but they never mention the frequencies involved (2.83×10^{14} and 5.66×10^{14} hertz, respectively).

The harmonics come from the higher-order terms in the formula for nonlinear output (Equation 5-8). Earlier, you saw that the input light wave can be described by $K(\cos \omega t)$. Another convenient way of describing the wave is as $Ke^{-i\omega t}$ where i is the imaginary square root of –1. (The formula is a fuller description than the cosine formula and gives useful results.) Using that description in the $(\text{Input})^2$ term gives you $K^2e^{-i2\omega t}$, which describes a wave with twice the frequency of the input wave. Including the K term, which also is squared, shows that second-harmonic amplitude increases with the square of the fundamental wave intensity, faster than the input intensity itself.

Although harmonic generation may sound esoteric, it has become an important practical tool. Most laser development has been concentrated on a few particularly efficient and practical types, leaving holes in the spectrum where no good lasers are available. Harmonic generation makes more wavelengths available. Frequency-doubling of the 1064-nm infrared neodymium laser to give 532-nm green light is the most important example, but there are many others.

The strength of nonlinear effects depends on the internal structure of the material. Nonlinear effects are very weak in most materials such as ordinary glass. Few materials have the high nonlinear coefficients needed for practical second-harmonic generation. High power density also improves harmonic generation, so pulses with high peak power produce higher harmonic powers than continuous beams.

Second-harmonic generation is most important in practice, and in some cases the third harmonic may be generated directly. Because the fourth-order nonlinear coefficient is much lower than the second-order coefficient, in practice the fourth harmonic is generated by doubling the frequency of the second harmonic.

5.4.2 Sum-and-Difference Frequency Generation

Nonlinear optics can also generate light waves at frequencies that are the sum or difference of two input light waves passing through a nonlinear material. (Harmonic generation produces the sum frequency of two identical waves.) The strength of the sum and difference frequencies are proportional to the product of the input intensities of the two waves.

It's also possible to essentially reverse the process of sum-frequency generation in a device called an *optical parametric oscillator* (described in more detail in Section 6.5). In this case, a strong pump beam at a frequency $\nu_1 + \nu_2$ is directed into a nonlinear medium placed between two mirrors that reflect one or both of ν_1 and ν_2. This creates a resonant cavity, similar to a laser cavity, which resonates at one or both frequencies ν_1 and ν_2. Which wavelengths are generated depends on the nature and orientation of the nonlinear material and the reflectivity of the mirrors.

5.4.3 Raman Scattering

Another important nonlinear interaction is a change in photon energy called Raman scattering, named after its discoverer, Indian physicist C. V. Raman. Normally, the energy of a photon does not change when an atom or molecule scatters it in another direction. Instead, the atom or molecule absorbs the photon, then almost instantly reemits it.

Raman scattering can occur when photons interact with atoms or molecules in a gas, solid, or liquid. The process transfers some energy to or from a vibrational or rotational energy state. Usually, some photon energy is lost to excite the other state, so the reemitted photon has less energy and a longer wavelength. More rarely, the photon picks up energy when the atom or molecule drops to a lower vibrational or rotational state, and emerges with shorter wavelength. The effect is strongest at higher power levels.

The Raman shift (measured in frequency or energy change of the photon) is an inherent characteristic of the material's vibrational or rotational state, so it is the same for any illumination wavelength. Raman scattering is a valuable tool for shifting laser wavelength to otherwise unavailable lines in the visible, near-infrared, or near-ultraviolet. As you will learn in Section 8.9, an effect called stimulated Raman scattering can serve as the basis for optical amplification and laser action.

5.4.4 Nonlinear Materials

Nonlinearity is an inherent property of materials that can be measured as a nonlinearity coefficient. Most materials are only weakly nonlinear, but some are highly nonlinear because of their crystalline structures. Several types are used with lasers:

- Beta barium borate (β-BaB_2O_4 or BBO)
- Lithium iodate ($LiIO_3$)
- Lithium niobate ($LiNbO_3$)
- Lithium triborate (LiB_3O_5 or LBO)
- Potassium dihydrogen phosphate (KH_2PO_4 or KDP)
- Potassium niobate ($KNbO_3$)
- Potassium titanyl phosphate ($KTiOPO_4$ or KTP)

5.5 BEAM INTENSITY AND PULSE CONTROL

A variety of techniques can be used to control laser beam intensity. Most fall into two distinct categories: modulators that vary the intensity in an arbitrary way, and pulse-generation techniques that cause the laser to emit a series of pulses at regularly spaced intervals.

The distinction is an important one. Modulators change the laser beam so it can carry a signal. The modulation may turn the laser beam on or off to convey a simple "yes" or "no" message, or it may switch a laser off and on 10 billion times a second to transmit high-speed computer data. Modulators are parts of communication systems, and can operate on either pulsed or continuous laser beams.

A series of regularly spaced repetitive pulses carries essentially no more useful information than an unchanging continuous beam. But many laser applications work better if the laser energy is concentrated into short, intense pulses than if it is delivered continuously. For example, short, intense laser pulses can punch holes through a sheet of plastic or metal much better than a continuous beam could.

Both modulation and pulse generation can be done in different ways. For convenience, we will look first at modulators, then at the three main types of pulse generation: Q-switching, cavity dumping, and mode locking.

5.5.1 Modulators and Modulation

There are several ways to modulate beam intensity, including mechanically blocking the beam, changing the input power that drives the laser, or passing the beam through a device with variable transmission.

Mechanical modulation typically is done with a shutter or beam chopper. *Shutters,* like those on cameras, open to let the beam through and close to block light. Shutters are used as safety devices on many lasers, to block the beam when it is not needed, or when the laser may not be operating correctly. Shutters are controlled manually or automatically and operate in a fraction of a second.

Beam choppers are mechanical devices that interrupt the beam repeatedly, and typically operate at higher speed. One example is a rotating disk with holes or slots that transmit the beam for selected intervals, as shown in Figure 5-9. When you align a laser beam with the holes in the disk, spinning the disk switches the beam off and on. Another example is a mask placed on the arm of a vibrating tuning fork, which swings in and out of the beam to block or transmit it. Note that both types of beam choppers generate a series of uniformly spaced pulses.

Direct modulation changes laser output power by changing the input power delivered to the laser. This is analogous to changing the output power of a light bulb by switching it on or off, or controlling it with a dimmer switch. It sounds simple, and works well for some lasers, but not for others. Direct modulation is easy for semiconductor diode lasers, which are small and respond very quickly to changes in their drive current. Most gas lasers are poor-

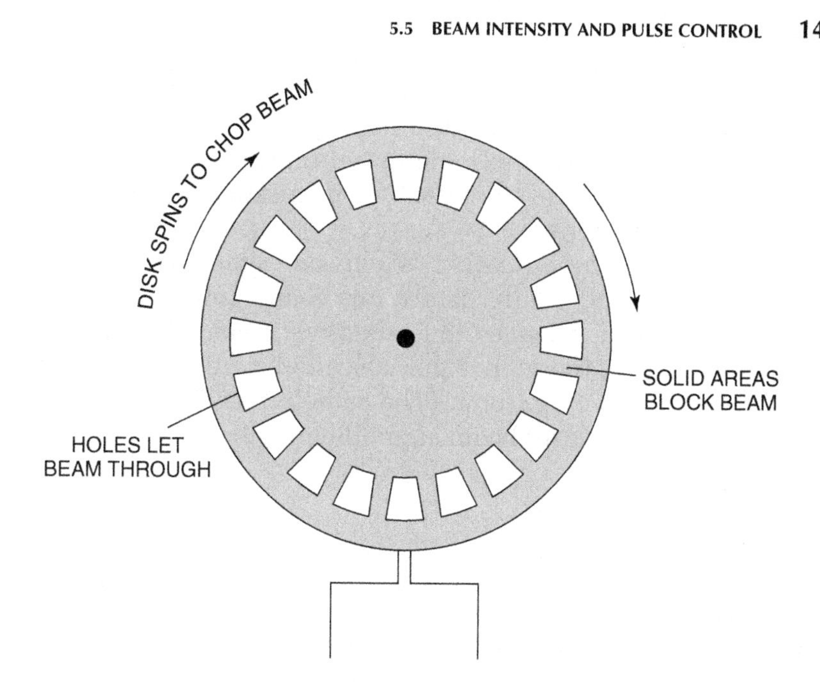

Figure 5-9. A beam chopper is a spinning disk with holes or slots that transmit a laser beam during intervals in which the solid part of the disk does not block the light. The faster the disk spins, the faster the chopper switches the light off and on.

ly suited for direct modulation because they take much longer to stabilize when their drive current is changed. Unlike beam chopping, direct modulation can vary output power continuously between zero and a maximum, rather than switching it between zero and maximum, with no intermediate levels.

External modulation passes the laser beam through a separate modulator, the transparency of which depends on a control signal applied to it. A number of types of external modulators have been developed for various applications:

- *Electrooptic phase modulators.* Application of an electric field across materials such as lithium niobate ($LiNbO_3$) causes a change in refractive index proportional to the voltage. This shifts the phase of the light, and can be used to modulate light amplitude if the phase-shifted light is combined to interfere with a portion of the light that has not been phase-shifted.
- *Electrooptic polarization rotators.* In other configurations, application of an electric field changes refractive index differently

for different polarizations of light, effectively rotating the polarization of the light. Combining a polarization rotation stage with a suitably oriented polarizer produces an amplitude modulator, because the polarizer's transmission changes with the polarization rotation.

- *Acoustooptic modulators.* When an acoustic wave passes through a solid, it alternately raises and lowers the pressure, which, in turn, changes the refractive index. The stronger the acoustic wave, the more light is scattered from the main beam, modulating its amplitude. (The same effect can be used to deflect part of the laser beam at another angle, as you will learn in Section 5.6.1.)

5.5.2 Q Switches

Q switches generate a series of short, intense pulses by modulating the *quality factor* or Q of a resonant laser cavity.

The Q of a cavity measures internal loss within the laser; the higher the Q, the lower the loss. Loss is not necessarily bad, because a high-loss cavity can store more energy than one with lower loss. Suppose, for example, you block one laser mirror, then excite the laser medium. Excitation puts many atoms into the upper laser level, but with one mirror blocked the cavity loss is too high for the laser to oscillate, so you can build up a larger population inversion than if the mirror were not blocked. If you suddenly remove whatever was blocking the mirror, the population inversion stored in the laser cavity will be much larger than needed for laser action. Stimulated emission then will drain the energy stored in that large population inversion in a short pulse with peak power much higher than the laser could otherwise produce, as shown in Figure 5.10.

This process is called *Q switching,* and it is based on a *Q switch* that quickly shifts from absorbing light to transmitting it, causing a sudden drop in cavity loss when it becomes transparent so light can oscillate between the two cavity mirrors. The duration of a Q-switched pulse depends on output-mirror reflectivity R and on the time t that laser light takes to make a round-trip of the laser cavity, and is roughly

$$\text{Pulse duration} = \frac{1}{(1 - R)} \tag{5-9}$$

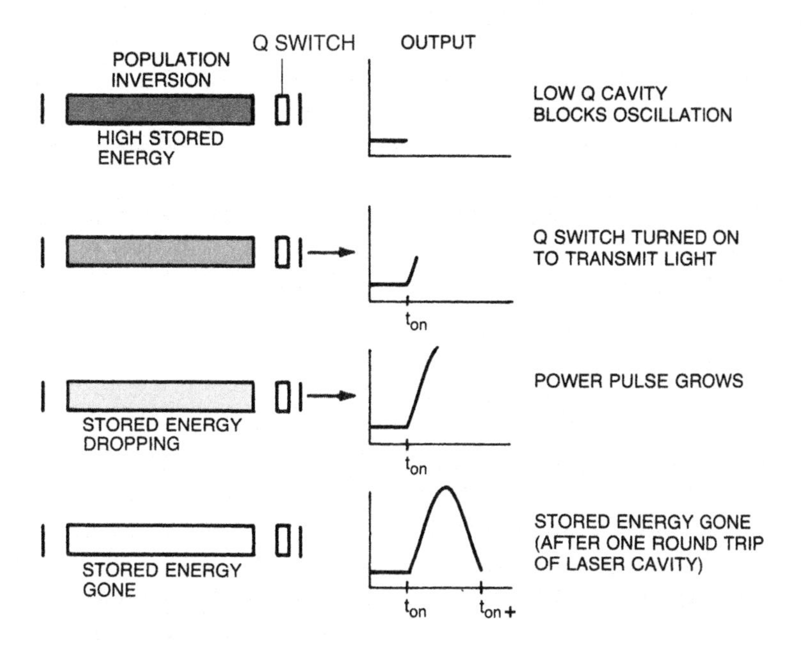

Figure 5-10. Q switching generates a laser pulse.

Typical pulse lengths are around 10 nanoseconds. If you insert the value of cavity round-trip time—twice the cavity length L divided by the speed of light in the laser medium (c/n) you have a more useful estimate of Q-switched pulse duration:

$$\text{Pulse duration} = \frac{2Ln}{c(1-R)} \tag{5-10}$$

A Q switch can fire once to produce a single shot, or repeatedly to generate a series of regularly spaced pulses. Both electrooptic and acoustooptic modulators can serve as Q switches; so can a rotating glass prism with a number of flat faces. When none of the rotating faces is aligned perpendicular to the laser axis, the Q is too low for laser oscillation, but when one face rotates into position, it raises the Q suddenly so the laser generates a pulse.

It is also possible to build a passive Q switch using a cell containing a dye that normally absorbs light. Absorbing a photon puts the dye into a higher-energy state that cannot absorb light at the same wavelength, where it remains for a while. If stimulated emission from the laser medium is strong, it raises all the dye molecules to the excited state, "bleaching" the dye so it no longer absorbs the laser wavelength. This makes the dye transparent, raising cavity Q by exposing a mirror behind it, and thus generating a Q-switched pulse.

Q switching is widely used, but it requires that atoms can stay in the upper laser level long enough to generate the Q-switched pulse. Some lasers do not meet this criterion.

5.5.3 Cavity Dumping

Like Q switching, cavity dumping works by releasing energy stored in the laser cavity, but the details differ. Normally, fully reflecting mirrors are on both ends of a cavity-dumped laser, with no output mirror. In this configuration, power builds up steadily within the laser cavity until another mirror switches into place and dumps the circulating laser energy out of the cavity in a single pulse. This pulse lasts only as long as light takes to make a round-trip of the laser cavity:

$$\text{Pulse duration} = \frac{2Ln}{c} \tag{5-11}$$

where L is cavity length, n is refractive index of the laser medium, and c is the speed of light in a vacuum. For a gas laser 30 centimeters long, the pulse lasts two nanoseconds, significantly shorter than a Q-switched pulse.

Cavity dumping generates lower energy pulses than Q switching, and it is not used as often. However, lasers that cannot be Q switched can be cavity dumped.

5.5.4 Mode Locking

Mode locking generates the shortest pulses that can be produced in a laser cavity. From a theoretical viewpoint, the idea is to lock many longitudinal modes together so that the laser simultaneously oscillates on all of them. This can be done by inserting suitable components into the laser cavity. However, it may be simpler to visualize the process by imagining that the mode-locking device operates on a clump of photons bouncing back and forth between the laser mirrors. Every time the photon clump reaches the mode locker, it "opens up" to transmit the photons. Each time the clump hits the output mirror, some photons pass through and the laser emits a very short pulse. If you were sitting outside the laser, you would see one laser pulse each time the clump of photons made a round-trip of the cavity.

Mode-locked pulses can be very short, so their peak power can be very high during that brief pulse. The minimum duration of a mode-locked pulse depends on the range of wavelengths (or, equivalently, frequencies) that the laser emits. This frequency range $\Delta\nu$ also measures the range of modes in the mode-locked pulse. The lower limit on pulse duration Δt is roughly $1/\Delta\nu$. To give a numerical example, a mode-locked laser with bandwidth of 1000 GHz could generate pulses as short as 1 picosecond (10^{-12} second). This is a consequence of the uncertainty principle, which tells us that we cannot pin down both the frequency of a light wave and the duration of a pulse with indefinite precision. This fundamental limit is called the *transform limit* because the relationship is considered a transformation between time and frequency domains.

Laser oscillation inherently limits the range of wavelengths in a laser pulse, because the amplification process concentrates emission on the strongest lines in the laser's limited gain bandwidth. Further extension of the bandwidth requires stretching the bandwidth of laser pulses beyond their normal limits. The details are beyond the scope of this discussion, but the results make it possible to generate pulses of light lasting less than five femtoseconds, equivalent to just a couple of waves of visible light.

5.6 BEAM DIRECTION AND PROPAGATION

Chapter 4 described the divergence and propagation of a laser beam without outside interference. Many laser applications require modifying the beam after it leaves the laser. This chapter has already described important concepts of beam-focusing optics, but simple optical focusing is often not enough. For example, applications such as reading bar codes for retail-store checkout require a way to scan the beam across objects. This section will introduce a number of concepts that do not fit readily with conventional focusing optics.

5.6.1 Beam Scanners

Several techniques are used to scan laser beams over an area.

A simple concept is spinning a polygonal mirror around a central axis. Each mirror face turns as the laser beam strikes it, scanning the reflected beam at an angle dictated by the rule that the angle of incidence equals the angle of reflection. When a new facet turns toward the laser, it repeats that scan. Holographic scanners are similar in concept, but instead of reflecting the laser beam off a flat mirror they reflect it from holographic optical elements, allowing complex scan patterns.

Resonant and galvanometer scanners also rely on moving mirrors, but they twist flat mirrors back and forth across a limited angle, so the laser beam always strikes the same flat mirror surface. The twisting mirror sweeps the laser beam across a line.

Mechanical motion is not required in solid-state acoustooptic beam deflectors. As in acoustooptic beam modulators, a sound wave passing through a solid creates zones of varying refractive index, which deflect some of the light passing through the solid. Acoustooptic scanners use the deflected beam, which emerges at an angle proportional to the ratio of the acoustic wavelength to the light wavelength. The strength of the deflected light depends on the acoustic power. The absence of moving parts is an advantage for some applications.

5.6.2 Waveguides and Integrated Optics

Light can also be directed by guiding it through an optical *waveguide,* which confines the light optically so it follows the path set by the waveguide. An optical fiber is one type of waveguide, in which total internal reflection confines the light to a high-index core surrounded by a cladding with a lower refractive index. Optical waveguides also come in other forms, including hollow tubes and optical nanowires smaller than a micrometer across, which guide light in different waves.

Planar waveguides are fabricated from thin layers of material deposited on a flat substrate. They are rectangular in cross section. Most designs confine light by total internal reflection, because they are surrounded partly by air and partly by other material with lower refractive index than the waveguide.

Other planar optical components can be integrated with planar waveguides to form *integrated optics,* which, like integrated electronic circuits, combine many discrete components on a single substrate. Integrated optics are used in applications such as communications that require processing low-power light signals.

5.6.3 Adaptive Optics

The interaction of laser beams with air is more complex than mere transmission and refraction. Turbulence causes air to move about. Look across a warm car hood or a warm parting lot on a sunny day, and you can see ripples in the background, caused by moving air. Look at the stars at night, and you can see them twinkle as air currents move in the atmosphere. If you aim a powerful laser beam through the atmosphere, air molecules will absorb a tiny fraction of the beam, warming the air so that it tends to defocus the beam.

To cope with these perturbations, optical engineers have invented devices they call *adaptive optics.* The idea is to measure the perturbations, and use that information to adjust the surface of a mirror so it focuses light in a way that compensates for the perturbations. The changes are small, usually a fraction of a wavelength, but they can greatly improve transmission. Most of the world's largest telescopes use adaptive optics to see the stars more clearly than they otherwise could through the atmosphere.

Adaptive optics also can compensate for the perturbations arising from heating of the air by high-energy laser beams, so the lasers can be focused more sharply onto their targets. In fact, military research paid for most of the development of adaptive optics, but so far astronomers have reaped most of the benefits in better viewing from the ground.

5.7 MOUNTING AND POSITIONING EQUIPMENT

Walk into any well-equipped laser laboratory, and the first thing you are likely to notice is a massive table, often painted flat black, and often with screw holes regularly spaced across its metal surface. This behemoth is called an *optical table* and is shown in Figure 5-11. On it you will find an array of special mounting equipment that holds lasers, lenses, and other optical components.

Figure 5-11. An optical table on vibration-isolation legs.

Optical tables and mounts are not optics per se, but they are used with optics and lasers. Optical tables and massive linear rails called optical benches serve as foundations. They generally are mounted on shock-absorbing legs to isolate optical equipment from vibrations that can affect precision instruments and holography. Vibrations need not be large to affect optical measurements, only a fraction of the wavelength of light, and they can be caused by a person walking through the room or a truck passing on the street outside.

Typically, individual optical elements and instruments are held in place by mounts with holes in their bases through which screws can be installed in threaded holes in the surface of an optical table. The mounts allow precision adjustment as well as holding components in place.

Optical mounts and positioning equipment sound like mundane technology compared to lasers, but they can be important to the success of precision measurements and experiments.

5.8 OPTICAL MEASUREMENT

Optical measurements use some of the same units, terms, and equipment as other measurements, but also require some special equipment and define some terms in different ways. We will introduce the basic concepts most important to understand when working with lasers, but we cannot cover the whole field of optical measurement.

5.8.1 Light Detection

The first step in measuring light is to detect it and convert it into an electrical signal that can be measured using electrical techniques. Generally, this is done by directing the light into a semiconductor *photodetector,* where it raises an electron into the valence band and causes current flow. An older alternative is a *vacuum photodiode* in which light frees electrons from a *photoemissive* metal surface, and the electrons flow through the vacuum to a positive electrode. A *photomultiplier tube* is a vacuum photodiode in which the photocurrent passes through a cascade of amplifying stages, making the tube extremely sensitive to light.

The wavelength response of a detector depends on its composition because a photon must have a threshold energy to free an electron from a valence bond or from a metal surface. Table 5-2 lists the response ranges of some important photodetectors. Some semiconductor detectors, called *avalanche photodiodes,* have internal amplification stages analogous to photomultipliers, although they operate differently.

Table 5-2. Wavelength ranges of detectors

Type	Material	Wavelengths (nm)
Photoemissive	Potassium–cesium–antimony	200–600
Semiconductor	Silicon	400–1000
Semiconductor	Germanium	600–1600
Semiconductor	Gallium arsenide	800–1000
Semiconductor	Indium gallium arsenide	1000–1700
Semiconductor	Indium arsenide	1500–3000
Semiconductor	Lead sulfide	1500–3300
Semiconductor	Lead selenide	1500–6000

Table 5-3. Radiometric units

Quantity and symbol	Meaning	Units
Energy (Q)	Amount of light energy	joules (J)
Power (P or ϕ)	Flow of light energy past a point at a particular time ($\Delta Q/\Delta t$)	watts (W)
Intensity (I)	Power per unit solid angle	watts/steradian
Irradiance (E)	Power incident per unit area	watts/cm^2
Radiance (L)	Power per unit angle per unit projected area	W/steradian-m^2

5.8.2 Radiometry, Photometry, and Light Measurement

Light measurement is often divided into two categories: radiometry and photometry. *Radiometry* is the measurement of the power and energy contained in electromagnetic radiation, regardless of wavelength. *Photometry* is the measurement of the amount of light visible to the human eye, with the contribution measured according to the eye's sensitivity across the visible range, 400 to 700 nm. Thus, photometry ignores infrared wavelengths, which are invisible to the human eye, and counts photons of green light (to which the eye is very sensitive) more than red or violet photons (to which the eye does not respond as strongly).

Virtually all laser-related measurements are radiometric. (See Appendix A for a brief introduction to laser eye safety.) Table 5-3 summarizes these units and lists common symbols for them, with brief descriptions of their meanings. As the table shows, these units are related to each other. *Power* is the rate of energy flow past a point in a given time; it is measured in watts. *Irradiance* is power per unit area; it is measured in W/cm^2. *Intensity* is power per unit solid angle, W/steradian. Similar units are defined for photometric power visible to the human eye.

Many instruments also measure power in decibels, a logarithmic ratio of two power levels, P_{in} and P_{out}:

$$dB = 10 \times \log_{10}\left(\frac{P_{out}}{P_{in}}\right) \tag{5-12}$$

The value is positive when the output power is higher than the input, and negative when the output power is low, that is, when some power has been lost. Power also can be measured in decibels relative to a predefined level, typically one milliwatt (1 mW) or one microwatt (1 μW). A positive number indicates that the measured power level is above the comparison, a negative number indicates it is lower.

If you are familiar with electronic measurements, you may have seen the logarithm of the voltage or current ratio multiplied by 20 rather than 10. This is because decibels are a power ratio, and power is proportional to the square of voltage or current, so the factor of 2 must be added.

5.9 WHAT HAVE WE LEARNED?

- A simple lens is made of a single piece of transparent material, with one or both sides curved to focus light.
- A lens can focus parallel light rays to a point or light from a point source into a parallel beam.
- A collimator is a simple telescope that expands the diameter of a laser beam.
- Variations of the refractive index with wavelength cause chromatic aberration, focusing light to different points depending on its wavelength.
- Achromatic lenses compensate for chromatic aberration.
- Cylindrical optics focus light to a line rather than to a point.
- A retroreflector bounces light directly back to its source.
- Prisms and diffraction gratings are dispersive optics that spread light out to show a spectrum.
- Total internal reflection confines light in optical fibers.
- Light can be linearly, circularly, or elliptically polarized.
- Silica glass is widely used for transparent optics at visible wavelengths. Other materials are used in the infrared and ultraviolet.
- Metal coatings on glass are used as mirrors.
- A low-index coating can reduce surface reflection by glass.
- Multilayer coatings use interference effects to control light transmission and reflection as a function of wavelength.
- A neutral-density filter absorbs the same fraction of light at all wavelengths in its operating range.
- Nonlinear optics alter light in a way proportional to the square of light intensity. The output of linear optics is proportional to the light intensity.
- Harmonic generation uses nonlinear effects to produce a second- or higher-order harmonic of the input laser frequency.

- Combining two input waves can generate new waves at the sum or difference of the frequencies of the waves.
- Raman shifting changes wavelength by adding or subtracting vibrational energy.
- Q switches generate a series of short, intense pulses by modulating the quality factor or Q of a resonant laser cavity.
- Mode locking generates the shortest pulses that can be produced in a laser cavity.
- An optical waveguide confines light inside itself and guides the light along a path.
- Adaptive optics compensate for atmospheric distortion.
- Radiometry measures electromagnetic power at all wavelengths.
- Photometry measures only light visible to the human eye.
- Power is the rate of flow of electromagnetic energy.

WHAT'S NEXT?

In Chapter 6, we will describe the major types of lasers in a variety of ways.

QUIZ FOR CHAPTER 5

1. A double-convex lens with two equal radii of curvature of 20 centimeters has refractive index of 1.60 at 400 nm and 1.50 at 700 nm. What is the difference between the lens' focal lengths at those two wavelengths?
   a. 400 nm focal length is 3.33 cm shorter
   b. 400 nm focal length is 0.33 cm shorter
   c. 700 nm focal length is 0.033 cm shorter
   d. 700 nm focal length is 1 cm shorter
   e. No difference
2. Silicon has a refractive index of 3.42 at 6 μm in the infrared. What is the reflective loss for 6-μm light incident from air normal to the surface?
   a. 5%
   b. 10%
   c. 20%
   d. 30%
   e. 40%

3. For the silicon sample in problem 2, what is the reflective loss if a coating with refractive index of 2 is applied on the surface? (Remember that you cannot just add the reflective losses. You must multiply the fractions of transmitted light to get the total transmitted light to assess overall reflective loss.)
   a. 5%
   b. 10%
   c. 15.4%
   d. 17.2%
   e. 20.5%
4. Which of the following optical materials is transparent in the entire optical spectrum between 0.4 and 0.7 μm? (Use Figure 5-8.)
   a. Gallium arsenide
   b. Silicon
   c. Zinc selenide
   d. Magnesium fluoride
   e. Silver chloride
5. You need an optical material that is transparent between 1 and 2 μm in the infrared. Which of the following materials is not suitable?
   a. Sodium chloride
   b. Silicon
   c. Zinc selenide
   d. Magnesium fluoride
   e. Silver chloride
6. What type of filter would you use to block light from a narrow-line laser, but transmit other light?
   a. Neutral-density filter
   b. Interference filter
   c. Color filter
   d. Spatial filter
   e. Any of the above
7. What is the wavelength of the fourth harmonic of the ruby laser (694 nm)?
   a. 2776 nm
   b. 1060 nm
   c. 698 nm
   d. 347 nm
   e. 173.5 nm
8. What is the wavelength of the difference frequency generated

by mixing light from a ruby laser at 694 nm with an neodymium laser emitting at 1064 nm?
   a. 370 nm
   b. 532 nm
   c. 694 nm
   d. 1758 nm
   e. 1996 nm
9. Raman shifting does what to input light?
   a. Modulates its intensity
   b. Doubles its frequency
   c. Changes its wavelength by a modest amount
   d. A and B
   e. Nothing
10. Direct modulation by changing the drive current works best for which type of laser?
   a. Semiconductor
   b. Ruby
   c. Helium–neon
   d. Neodymium–YAG
   e. All are equally difficult
11. What is the length of a Q-switched pulse from a 10-cm long neodymium–glass laser with a 90% reflective output mirror. Assume that the refractive index of glass is 1.5.
   a. 1 nanosecond
   b. 5 nanoseconds
   c. 10 nanoseconds
   d. 20 nanoseconds
   e. 1 microsecond
12. A laser has a bandwidth of 4 gigahertz (4×10^9 Hz). What is the shortest mode-locked pulse it can generate, according to the transform limit?
   a. 1 picosecond
   b. 10 picoseconds
   c. 30 picoseconds
   d. 100 picoseconds
   e. 250 picoseconds

CHAPTER 6

TYPES OF LASERS

ABOUT THIS CHAPTER

Many different types of lasers have been developed. They often are classified by the type of active medium, but that is not the whole story. In this chapter, we will look at lasers in a variety of ways, including not just the nature of the medium but the way in which they are operated. We will start by comparing the fundamental nature of oscillators and amplifiers, then move on to media type, gain levels within the laser, and laser bandwidth. The chapter closes with a section on laser-like light sources. Chapters 7 through 10 will examine major types of lasers in more detail.

6.1 LASER OSCILLATORS AND OPTICAL AMPLIFIERS

The distinction between amplification and oscillation is a fundamental one in the laser world. Although the acronym "laser" specifies only light *amplification* by the stimulated emission of radiation, the real definition of "laser" includes the requirement for internal feedback that sustains oscillation. A laser acts as an oscillator in that it generates its own beam, at a wavelength that depends on the laser's physical characteristics. In contrast, an *optical amplifier* merely amplifies light from an external source, and lacks mirrors that form a resonant cavity

This difference leads to a fundamental difference in behavior. A laser oscillator sets its own wavelength, but the wavelength of an optical amplifier depends on the input signal that stimulates emission. Laser oscillation is relatively easy to measure because the output power is low below threshold, but increases rapidly in power and decreases in linewidth once the device exceeds threshold. Amplification is more difficult to measure, particularly if the gain is low.

Early developers focused their attention on laser oscillators because they wanted devices that generated beams internally. Optical amplifiers (sometimes called laser amplifiers) have found some applications in boosting the power available from pulsed laser oscillators, but today the main application of optical amplifiers is for boosting the signal strength in fiber-optic communication systems.

6.1.1 Laser Oscillators

As you learned in Chapter 3, a conventional laser is an oscillator, with mirrors on two ends forming a resonant cavity. A laser oscillator generates a signal on its own, starting when a spontaneously emitted photon stimulates emission from another excited atom. If the original photon and the stimulated emission are directed along the axis of the cavity (or within the narrow range of angles reflected by the cavity mirrors), they bounce back and forth between the mirrors, with some light emerging from the output mirror as the laser beam. The increase of power within the laser cavity must at least equal the power lost within the cavity and the power exiting through the output mirror. A laser oscillator is the optical counterpart of a radio oscillator as a signal source.

A laser oscillator can produce pulses or a continuous beam. In either case, the laser output builds quickly as light oscillates back and forth between the mirrors, stimulating emission from excited atoms or molecules. From an electronic standpoint, the mirrors provide feedback that stabilizes the laser wavelength. In many pulsed lasers, light emission depletes the population inversion, terminating the pulse. In other pulsed lasers, internal components such as the Q switches described in Section 5.5.2 modulate the round-trip gain of the laser cavity, switching oscillation on and off. In a continuous laser, an external energy source continually excites the laser medium, and the number of photons produced stabilizes to generate a steady beam.

Light circulates back and forth between the mirrors in a laser oscillator, sustaining the oscillation. Individual photons bounce back and forth until they are absorbed or emitted through the output mirror. The oscillation wavelength is selected by the optics of the resonant cavity, and by the nature of the laser medium.

6.1.2 Optical Amplifiers

An optical amplifier is a rather different device, as shown in Figure 6-1. It's a single-pass device with windows on the ends rather than mirrors. (You do not really have to have windows, but it helps you see the difference between the amplifier and oscillator.)

The input in the figure arrives as a single pulse, which is amplified as it stimulates emission inside the amplifier. The pulse is shown growing larger as it passes through the amplifier, eventually reaching a maximum before passing through the output window to the external world. The output window nominally provides no feedback, although in practice it would reflect a tiny fraction of light back through the amplifying medium. For most applications, this back-reflection should be reduced to a very low level.

This means that an optical amplifier and a laser oscillator serve fundamentally different functions. A laser oscillator generates a signal at a desired wavelength. An optical amplifier amplifies light from an external source, which makes only a single pass through the amplifier stage. The amplifier may be a tube containing gas, a slab or cylinder of a solid-state laser material, a semiconductor chip, or an optical fiber doped to amplify light passing through its core.

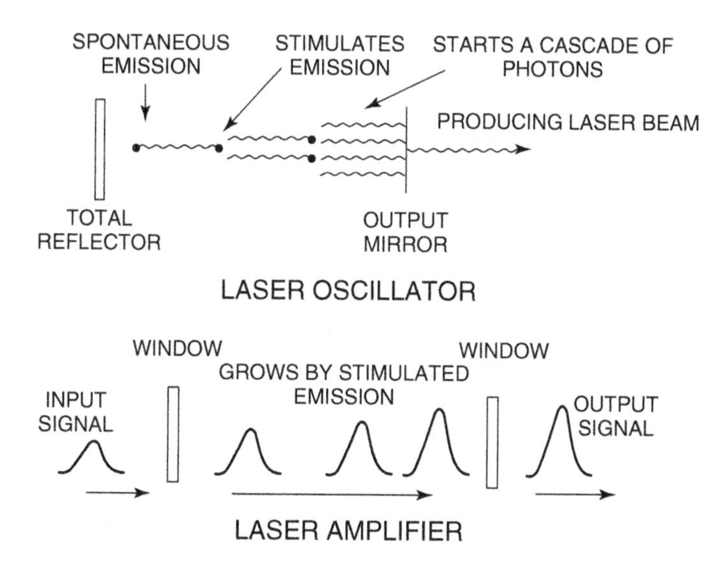

Figure 6-1. Light oscillates back and forth in the laser oscillator at top. Light makes only a single pass through the optical amplifier at bottom.

Optical amplifiers are used in two distinct configurations, one for producing powerful pulses and one for signal amplification.

6.1.3 Master Oscillator Power Amplifier (MOPA) Configurations

Internal dynamics limit how much power can be extracted from a laser oscillator in a single pulse, but that pulse can be amplified in external amplifier stages. This arrangement is called a *master oscillator power amplifier (MOPA),* and essentially separates the task of generating a pulse from amplifying it.

The master oscillator produces the input pulse, which then makes a single pass through each amplifier in a chain. The master oscillator output should be at a wavelength at which the amplifier stages have high gain.

Most MOPAs are solid-state lasers operated at relatively low repetition rates. To generate very high pulse powers, the amplifiers are arranged in stages, so the pulse passes first through amplifiers with successively larger gain volume, which can produce higher powers. The beam may be spread over larger areas in later amplifier stages. Thus, the pulse may pass through a small rod, then be expanded to pass through a larger rod, then be expanded again to pass through a slab or disk, as shown in Figure 6-2.

The amplifier shapes are chosen to spread out power so it does not reach a level high enough damage the surface of the laser material, which is the most vulnerable area. Disks or slabs are used at high powers to aid in dissipating excess heat.

The MOPA design is widely used in lasers with very high pulse power. The highest powers require long cooling times be-

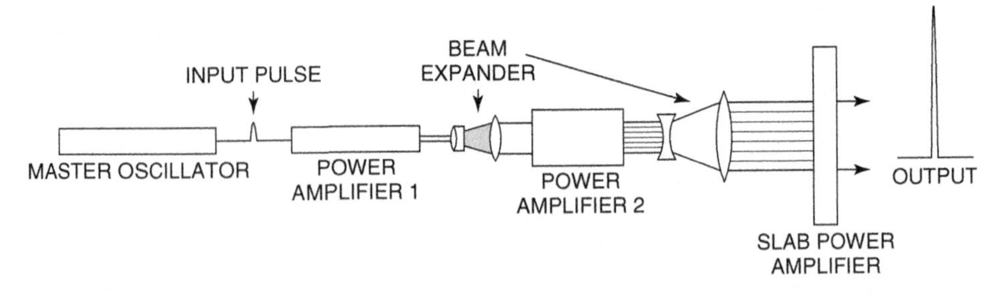

Figure 6-2. Master oscillator power amplifier puts a pulse through a series of larger amplifiers to increase pulse power.

tween pulses or active cooling of the final amplification stages, such as those used in laser fusion research.

6.1.4 Chirped Pulse Amplification

Another important variation on optical amplifiers is *chirped-pulse amplification.* The idea arose from efforts to overcome a fundamental problem in amplifying short and intense laser pulses; extremely high peak powers can damage optical materials and surfaces. The limit is an inherent one because it is the peak power that causes the damage, not imperfections in the material.

To overcome that limitation, chirped-pulse amplification stretches short pulses over longer periods of time, then recombines them. This is done by inserting dispersive optics that delay the input light by an amount that depends on its wavelength, spreading the spectrum out in time. This stretches the duration of the input pulse entering the amplifier, reducing the peak power to manageable levels that will not cause optical damage.

After the light passes through the amplifiers, the longer-duration amplified pulse is passed through another set of optics that compress it by delaying the wavelengths that led the dispersed pulse until the other wavelengths catch up. This squeezes the pulse to a much shorter duration with a much higher power. In this way, pulse duration can be increased by a factor of a thousand or more for amplification, then decreased by the same factor (or more) to multiply peak output power by a corresponding amount. This technique is at the heart of lasers that have reached peak powers greater than a petawatt (10^{15} watts) in ultrashort pulses.

6.1.5 Optical Signal Amplifiers

Low-power optical amplifiers are used to boost the strength of weak optical signals, primarily in fiber-optic communication systems. In this case, the input signal makes a single pass through the length of an optical amplifier.

One major attraction of optical amplifiers is that they can be designed to amplify signals across a range of wavelengths within the gain band of the active material. Optical fibers can simultaneously transmit signals at many different wavelengths, which optical amplifiers simultaneously amplify. Thus, if the input contained separate signals at 1551, 1552, 1553, and 1554 nm, an optical amplifier would increase the strength of all four signals.

This allows optical amplifiers to serve as signal boosters on long-distance fiber-optic communication systems. The amplifier takes the weak signals delivered by an input fiber, increases their power, then transmits them through the next span of fiber to the next amplifier. Optical amplifiers also can boost the strength of an optical signal before a receiver, or raise the strength of the signal from a transmitter at the start of a long length of fiber.

Three types of optical amplifiers are used in communications systems. One is an optical fiber with erbium or another element doped into its light-carrying core, which functions much like a solid-state laser and is covered in Section 8-8.

Another is an optical fiber that relies on a process called stimulated Raman scattering to amplify a weak signal, as described in Sections 6.2.9 and 8-10.

A third is a semiconductor optical amplifier, which will be covered in Section 9.6.4.

6.2 LASER MEDIA

Lasers generally are classified by the light-emitting medium or material. For purposes of keeping this book manageable, we divide the following four chapters on laser types into four broad categories: gas, solid-state, semiconductor, and the inevitable "other." But it is more complicated than that.

Population inversions and laser action can occur in a wide variety of materials, although by no means everything you can think of. Arthur Schawlow once made a "Jell-O" laser by adding a fluorescent dye to gelatin. But despite jokes by other laser physicists that you could make a telephone pole "lase" by dumping enough energy into it, opaque objects make very poor lasers, and the pole probably would start burning first.

Candidate laser media should be able to sustain a population inversion. You need a way to get the pump energy in, and a way to get the laser energy out, so the material must transmit the laser wavelength when a population inversion exists. The materials do not have to be stable, but they have to last long enough to produce stimulated emission. This adds up to quite a range of materials, which are described briefly here, and covered in more detail in Chapters 7 through 10.

6.2.1 Atomic Gases and Ions

Atomic gases are simple systems for physicists to work with, so they were among the first systems used to make lasers. Each atom can be considered as isolated except when it collides with its neighbors, or when photons or electrons transfer energy to or from it. Transitions can occur in neutral atoms, or in ions that have lost one or more electrons, although the more electrons that are removed, the more energy is needed for excitation.

Often, there are two gases involved, one to collect the excitation energy and the other to provide the laser transition. Energy is transferred between the two gases, as in the helium–neon laser, where the helium absorbs energy from an electric discharge in the gas and excited helium atoms transfer energy to the neon atoms.

A number of important visible transitions occur in ions of various elements, including the rare gases argon and krypton, and the metal cadmium in vapor form. Some atomic gases are metal vapors, which must be heated for the laser to operate. Transitions of outer-shell electrons generally are near the visible range; ions tend to be somewhat higher in energy, and provide many ultraviolet lines. Most practical atomic gas lasers are excited by an electric discharge passing through the gas, but the chemical oxygen–iodine laser is an exception.

6.2.2 Molecular Gases

Most molecular gas lasers emit light when the molecules make transitions between vibrational and rotational energy levels. Generally, vibrational energy levels are at higher frequencies, corresponding to wavelengths between about 3 and 12 μm. Molecules usually make rotational transitions at the same time, so vibrational–rotational transitions span a range of wavelengths. Longer-wavelength transitions usually are between purely rotational energy levels.

Molecular laser transitions tend to involve energy levels relatively close to the ground state, so more of the pump energy is converted into light than in shorter-wavelength atomic gas lasers. This higher efficiency produces less waste heat, and allows higher operating power, notably in the carbon dioxide laser. Molecular gas lasers can be excited electronically, but population inversions also can be produced in other ways, including by chemical reactions that produce excited molecules such as hydrogen fluoride.

6.2.3 Excited Dimers (Excimers)

Strong pulsed electrical discharges in certain gas mixtures can produce short-lived molecules in electronically excited states. The most important such lasers are called *excimers,* for the excited dimers (diatomic molecules), which emit the light. The discharge flows through a mixture of gases including a rare gas such as xenon or krypton, and a halogen such as fluorine or chlorine. It produces rare-gas-halide molecules such as krypton fluoride, which are bound together in an excited state, the upper laser level, but not in the lower laser level. As the excimers release energy by stimulated emission on an electronic transition, the molecules drop to a state in which the two atoms are not bound together, so the two atoms go their separate ways. This depopulates the lower laser level, assuring that the population inversion remains until the excimer population is thoroughly depleted. Excimer lasers have high gain.

Excimer lasers can be cumbersome because they use toxic gases, but they are valuable because they produce high-power pulses at ultraviolet wavelengths at which high power is otherwise hard to produce. A few other lasers operate similarly, including the molecular fluorine laser.

6.2.4 Highly Ionized Plasmas and Extreme Ultraviolet Emission

Laser action can occur on transitions in highly ionized plasmas. In this case, an intense pulse of energy vaporizes the laser material, stripping multiple electrons from the outer shells, and leaving vacancies in inner shells. The resulting ions can capture free electrons, which release energy when dropping back to vacancies in the inner electron shells. The transition energy is large, with emission lines for highly ionized plasmas in the extreme ultraviolet, at wavelengths of 10 to 100 nm.

Figure 6-3 shows the process schematically. An intense burst of energy, from an electric discharge or a powerful laser pulse, strips 20 electrons from a cadmium atom. One of the electrons

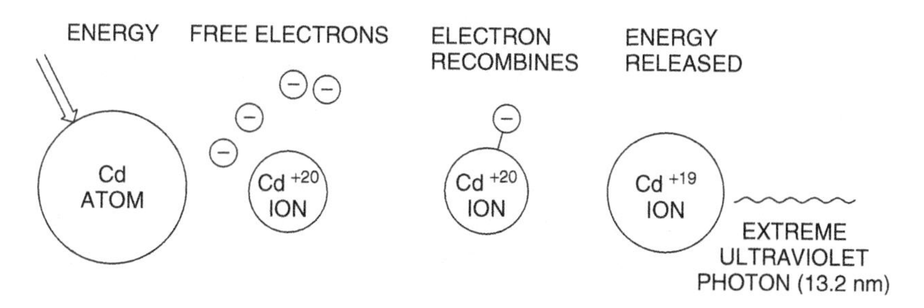

Figure 6-3. Recombination of a free electron with highly ionized cadmium (Cd^{+20}) produces an extreme ultraviolet photon.

then recombines with the cadmium ion. When it drops back into one of the empty orbits, it emits a photon of extreme ultraviolet light, in this case at 13.2 nm.

Such high-energy transitions generally have short spontaneous emission lifetimes, so ions do not stay in the upper laser level long. Spontaneous emission occurs quickly, and quickly stimulates emission, so gain is very high and pulses are short. These lasers are described in more detail in Section 10.3.

6.2.5 Dielectric Solids

The materials used in solid-state lasers are *dielectric solids.* The term *dielectric* means electrically nonconductive or insulating, a property of glass and many types of crystal. (In laser jargon, a solid-state laser is different from a semiconductor, because solid-state and semiconductor lasers operate in different ways.) Solid-state lasers are widely used and share some common features.

First, all are pumped optically by light from an external source. That pump light passes through the solid itself, which must be at least somewhat transparent at the pump wavelength. The pump light excites certain atoms within the solid, producing a population inversion that leads to laser emission.

Normally, the atoms that absorb light and generate stimulated emission make up no more than about 1% of the mass of the solid. They are essentially dopants added to a glass or crystal, like chromium atoms are added to sapphire (Al_2O_3) to make ruby. The chromium atoms give ruby its red color, absorb pump light, and produce red stimulated emission. The sapphire is transparent at the pump and laser wavelength and from an optical standpoint is merely a lattice that holds the light-emitting and -absorbing chromium atoms. In practice, the host material is chosen for its strength, manufacturability, heat dissipation, and transparency in the wavelength range of laser pumping and emission.

The pump wavelength is chosen to match an absorption band of the laser species, as shown in Figure 6.4. The pump source may be a lamp emitting a wide range of wavelengths overlapping absorption bands, or a laser emitting a narrow band of wavelengths that are strongly absorbed by the laser species. The excited atoms then drop to an upper laser level, producing a population inversion, and eventually emit laser light. The laser species may absorb light on the laser transition when there is no population inversion.

Although we show the laser transition levels as narrow lines, atoms in many solids have broader *vibronic energy levels,* because transitions can include changes in vibrational energy in the host as well as changes in electron energy level. As you will learn in Chapter 8, vibronic transitions can be tuned in wavelength. The deexcitation that removes atoms from the lower laser level generally transfers energy from the laser species to atomic vibration or heat in the host material.

There is a reason for emphasizing that the host materials in solid-state lasers should be insulators. Insulating solids do not con-

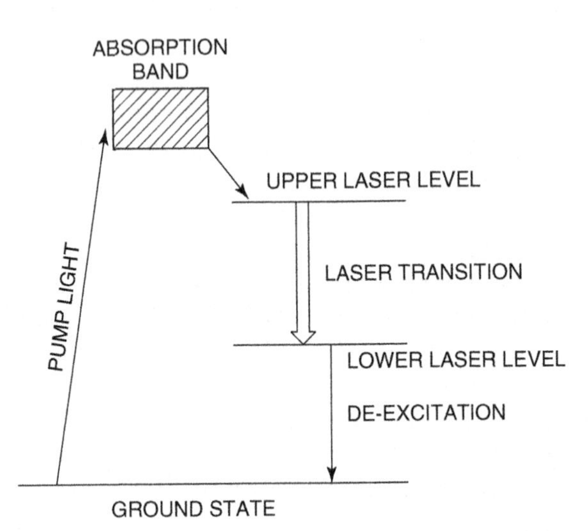

Figure 6-4. Energy levels of the light-emitting species in a generic solid-state laser. The solid host material is transparent at these wavelengths.

duct electricity because there is a wide gap between the energies of electrons forming valance bonds between atoms and the energies that atoms need to move freely in the solid. Pure crystals with such a large gap generally absorb little light at visible wavelengths.

6.2.6 Fiber Lasers and Amplifiers

Fiber lasers and amplifiers have become prominent in recent years. Fiber lasers can generate high powers from small packages, and fiber amplifiers are vital in telecommunications networks. However, fiber lasers and amplifiers are really just solid-state lasers in a different geometry. The light-emitting atoms are concentrated in the light-guiding core of the fiber, which from a physical standpoint is merely a very thin and very flexible rod of solid-state laser material. They are covered in Section 8.8.

6.2.7 Semiconductors

Semiconductor lasers work quite differently than the solid-state lasers described above, which is why laser specialists consider them an entirely separate category, although in electronics semiconductors are solid-state devices.

Semiconductor lasers are two-terminal electronic devices called diodes, which convert part of the current flowing through them to light. They are often called *diode lasers* or *laser diodes.* As you will learn in Chapter 9, parts of the semiconductor are doped with different materials, so electrons carry current in one region and holes (actually, vacancies in the electron shells of the semiconductor atoms) carry current in the other. When a semiconductor laser operates, both electrons and holes flow toward the *junction* of the two regions, where the two form an excited state called an *exciton,* which is the upper laser level. Spontaneous emission triggers a cascade of stimulated emission from these excitons, which is confined within a cavity fabricated in the semiconductor material.

The best semiconductor lasers are made from compound semiconductors, including gallium arsenide, indium phosphide, and gallium nitride. These are called III–V compounds because one element comes from Group III and the other comes from Group V of the periodic table. The precise formulation of the compound determines the wavelength.

Ironically, it is extremely difficult to make lasers from silicon, the semiconductor most widely used in electronics. Section 10.5 described the slow progress being made on silicon lasers.

6.2.8 Organic Dyes and Liquids

So far we have described gas and solid lasers, but what about liquids? Laser action has been demonstrated in the laboratory in a number of liquids, but only one type of liquid laser has ever found practical use, the organic dye laser.

Like the solid-state laser, the dye laser consists of a light-emitting species dissolved in a host material. In this case, the light emitter is an organic dye and the host usually is a liquid solvent in which the dye is dissolved. (Solid plastics also can serve as a dye host.) The dyes are fluorescent materials that absorb light, store the energy briefly, then drop to a lower energy level and emit light at a longer wavelength. The dyes are also complex molecules that vibrate as well as undergo electronic transitions, so their transitions are spread over a range of wavelengths, which can be tuned by adjusting the laser cavity. This has made them particularly valuable for scientific research, although they are cumbersome to operate. You will learn more about them in Section 10.1.

6.2.9 Stimulated Scattering

The concept of laser action can be extended to some fundamentally nonlinear processes in which light waves interact in odd ways with atoms.

One important process is stimulated Raman scattering, in which a light wave is absorbed and then reemitted by an atom at a different wavelength. The process transfers a bit of energy from the photon to the atom or, less often, when the atom has extra energy to start with, from the atom to the photon.

In the usual case, the atom transfers part of the photon energy to a vibrational mode within the material that contains it. These modes have characteristic frequencies for various materials, which cause the frequency of the emitted photon to shift by that frequency. For silicon atoms in silica (SiO_2), the most likely frequency shift is 13 GHz, so Raman scattering of light at 1480 nm reduces the energy of the emitted photon, shifting its wavelength to about 1580 nm.

Raman scattering is a nonlinear process, so the more light illuminates the material, the more likely it is to produce spontaneous Raman scattering. But at higher powers, the spontaneous Raman scattering can stimulate some of the other illuminated atoms to emit Raman-shifted light—stimulated Raman scattering. This process is relatively weak, but it accumulates through long lengths of material, such as in an optical fiber, where it can be used to make both Raman lasers and Raman amplifiers, which you'll learn about in Section 8.10.

Other types of lasers and laser-like devices work on similar principles.

6.2.10 Free-Electron Lasers

Stimulated emission can also extract energy from unbound or *free electrons* in an electron beam passing through an array of magnets with their fields oriented in alternating directions. The interaction of the electrons with the magnetic field causes them to emit electromagnetic energy, which can stimulate the emission of more electromagnetic radiation from other electrons. The emitted wavelength can be tuned by adjusting the electron velocity or the structure of the magnetic field, making the laser tunable.

The free-electron laser is large and complex, but it is tunable across an exceptionally broad range of wavelengths, making it attractive for research. It is described in Section 10.4.

6.3 THE IMPORTANCE OF GAIN

Gain can be a crucial factor in the performance and design of lasers and optical amplifiers, particularly at the extremes of low or high gain. The best way to explore the possibility is to consider three extreme cases: low-gain laser oscillators, high-gain pulsed lasers, and amplified spontaneous emission.

6.3.1 Low-Gain Laser Oscillators

When laser gain is low, the laser cavity must be designed to confine light with the minimum possible loss. Photons are unlikely to stimulate emission if they travel a short distance in a low-gain medium, so they must be reflected back and forth through the cavity many times, and only a tiny fraction of the light circulating within the cavity can be coupled through the output mirror into the beam.

Repeated oscillation through the laser cavity makes the beam highly coherent, monochromatic, and of high optical quality, although output power may be low. High-reflectivity optics are essential for successful operation of low-gain lasers.

6.3.2 High-Gain Lasers

Cavities used with high-gain lasers are designed to couple a large fraction of the incident light through the output mirror. The higher the gain, the fewer round trips of the cavity are needed.

High-gain laser media usually have high spontaneous emission rates, so there is a continual influx of new photons stimulating emission. These photons inevitably are emitted at a range of wavelengths, and the low-gain cavity does not strongly select the highest-gain wavelengths, so the output has relatively broad bandwidth. Likewise, beam divergence tends to be large, and beam quality and coherence tend to be limited.

High gain in a laser medium is essential to achieve laser action at wavelengths at which highly reflective optics are not available.

6.3.3 Amplified Spontaneous Emission

Stimulated emission begins as the amplification of spontaneous emission, but as the process continues in a low-gain laser oscillator most of the new stimulated emission is stimulated by photons that were themselves produced by stimulated emission. That is, stimulated emission does most of the stimulating, so most photons emitted are descendents of a small number of spontaneously emitted photons.

This is not the case in high-gain materials. The higher the gain, the more emission is stimulated by spontaneously emitted photons. That is important because spontaneously emitted photons are not coherent with each other; each has its own phase and may have a slightly different wavelength. The more important the contribution of spontaneous emission, the less coherent the beam.

If spontaneous emission dominates, the result is *amplified spontaneous emission* or *superradiance*. It can produce plenty of light, but the light is not very coherent. This can occur in very high-gain laser materials, and does not require mirrors or a resonant cavity.

Amplified spontaneous emission is desirable in some applications for which low coherence is desired. However, it also can create background noise in optical signal amplification in fiber-optic systems. Figure 6-5 shows the effect in an erbium-doped fiber amplifier, which receives as input signals at four wavelengths, shown as narrow lines. If the amplifier is operated with high gain, it will not only amplify the signals, but pick up and amplify spontaneous emission by erbium atoms in the fiber. That spontaneous emission noise occurs across the entire erbium bandwidth, not just at the input wavelengths, producing a noisy background of amplified spontaneous emission.

6.4 BROADBAND AND WAVELENGTH-TUNABLE LASERS

Although lasers are well known for their narrow-line emission, an important class of lasers can emit across a broad range of wavelengths, with the laser design selecting the particular emission wavelengths. This capability allows construction of lasers tunable across a range of wavelengths, which is important for scientific research and for other applications for which wavelength selection is particularly important. Because the shortest possible laser pulse duration depends inversely on the range of wavelengths, lasers with broad linewidth are important for producing ultrashort wavelengths. Ultrashort-pulse lasers, in turn, have proven able to generate light at many precisely spaced wavelengths, which is important for cutting-edge research and measurement.

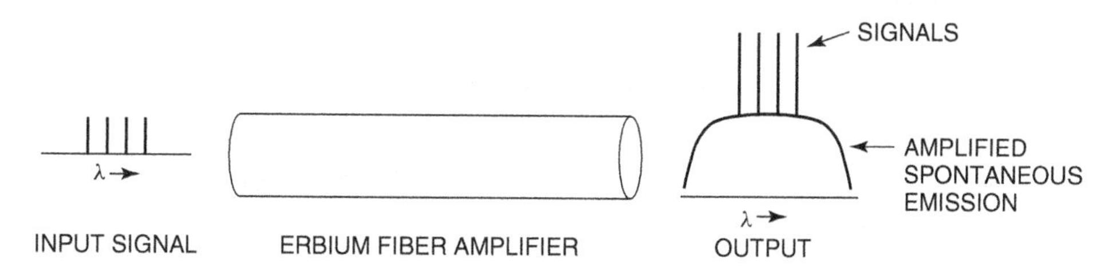

Figure 6-5. Amplified spontaneous emission in an erbium-doped fiber amplifier creates noise.

The two most important types of broadband lasers are the family of "vibronic" solid-state lasers, including titanium–sapphire laser described in Section 8.6, and the organic dye lasers described in Section 10.1.

6.4.1 Wavelength-Tunable Lasers

Tunable-wavelength lasers require both a laser medium capable of emitting across a range of wavelengths and an optical cavity that can be adjusted to select the oscillation wavelength.

A laser resonator oscillates at the wavelength at which gain is the highest. The trick in building a tunable laser cavity is restricting oscillation to a wavelength selected by the cavity, which is not at the peak of the laser medium's gain curve.

One approach is to insert a prism or diffraction grating that spreads light out into a spectrum, then align the cavity mirrors in a way that picks the desired wavelength. Figure 6-6 shows the basic idea, using a prism to disperse light generated by optically pumping dye molecules dissolved in solution. Stimulated emission from the dye passes through the prism, which bends different colors in different directions. The rear cavity mirror is moved so it reflects one wavelength back through the prism and dye cell to the output mirror. In the figure, the cavity mirror is lined up to reflect green light, in the middle of the spectrum, so it oscillates back and forth in the laser cavity.

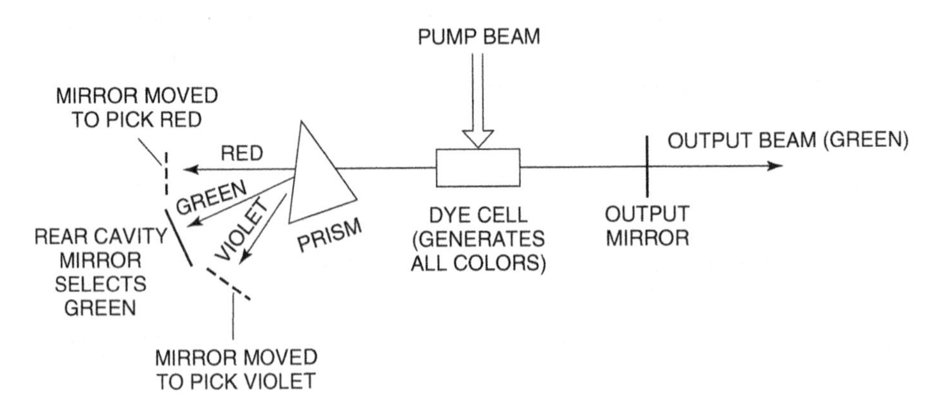

Figure 6-6. Prism in a tunable laser cavity disperses wavelengths; moving the rear cavity mirror selects one wavelength to oscillate.

Light waves at the red and violet ends of the spectrum are bent to the sides of the mirror so they are not reflected back through the dye cell to the output mirror. The dashed lines show where the rear cavity would have to be moved to select those wavelengths. Only the green light is passed back and forth through the laser cavity, so stimulated emission amplifies the green light to high power, overwhelming the red and violet light, which do not oscillate in the laser cavity. We should warn you that no real dye emits over such a broad range of wavelengths; a typical dye bandwidth is 10 to 70 nm, only a fraction of the visible spectrum. Real tunable laser cavities may be more complex.

In this simple example, moving the mirror picks the wavelength based on how the prism refracts the light. Other designs are possible. The prism could be turned to aim the desired wavelength toward a fixed mirror. Or a reflective diffraction grating could replace both the rear cavity mirror and prism, and the grating could be turned to select the wavelength reflected back through the dye cell and the output mirror.

Other types of optical elements could also be used to tune the wavelength, or to limit oscillation to a very narrow range of wavelengths for precision measurements. For example, a device called a Fabry–Perot etalon can be inserted into a tunable laser cavity to limit oscillation to a narrow range of wavelengths. The etalon is a pair of closely spaced, partly reflecting surfaces that bounce light back and forth between them. Interference effects allow the etalon to transmit only wavelengths that resonate between the two reflective surfaces. If the etalon is placed inside the cavity of a laser, tilting changes the distance the beam has to travel through it, changing the resonant wavelength transmitted. Combining an etalon with a movable mirror cavity can select a very narrow range of wavelengths.

High-performance tunable lasers have far more elaborate designs, such arrangements of several mirrors that form the cavity into a ring. Using such advanced techniques, the spectral bandwidth of a continuous-wave tunable laser can be limited to well below one part per billion.

6.4.2 Producing Short-Duration Pulses

As you learned in Section 4.6.2, a broad range of laser wavelengths is needed to generate short pulses because the shortest possible pulse duration Δt (in seconds) is a function of the laser bandwidth $\Delta\nu$ (in hertz):

$$\Delta t = \frac{0.441}{\Delta\nu} \tag{6-1}$$

Thus, concentrating laser energy into ultrashort pulses lasting only 10 femtoseconds (10×10^{-15} second) requires laser pulses spanning about 44 terahertz (44×10^{12} Hz). That much bandwidth is very difficult to get from a laser, so researchers developed new tools to stretch the spectrum of laser pulses.

Ultrashort pulses provide an extremely powerful tool for probing matter on a very fast time scale, and are becoming important in many areas. They also can be used for extremely precise frequency measurements, using ingenious techniques described in Section 13.2.3 that earned John Hall and Theodor Hänsch the 2005 Nobel Prize in Physics.

6.5 LASER-LIKE LIGHT SOURCES

A number of optical techniques using lasers or laser-related concepts can produce light that resembles laser light in its coherence, intensity, or monochromacity. We will briefly describe the most important ones, and explain their relationship to laser technology. See also Section 6.3.3 on amplified spontaneous emission.

6.5.1 Harmonic Generation

As you learned in Section 5.4.1, nonlinear processes can generate harmonics of light frequencies. The simplest of these processes is *second-harmonic generation* or frequency doubling, shown schematically in Figure 6-7. Passing light waves through a nonlinear material generates waves at twice the input frequency or, equivalently, half the wavelength. Not all the energy is converted to the second harmonic, but if the input power is high, and the light is concentrated into a suitable material, a reasonable fraction of the energy is upconverted to the second harmonic.

In practice, second-harmonic generators are often packaged within the laser, so it is not obvious that it is a separate function. A green laser pointer is a good example. The solid-state neodymi-

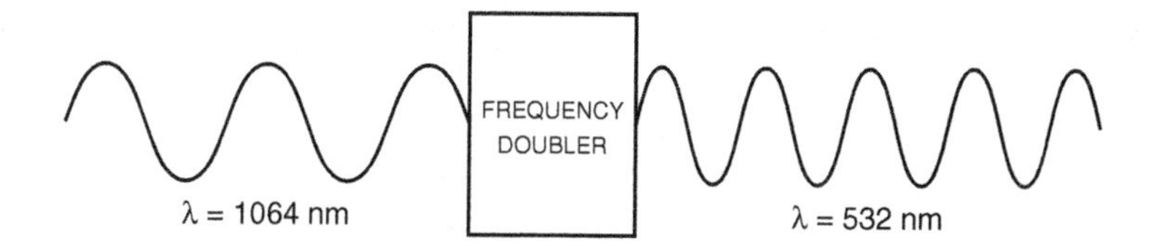

Figure 6-7. Second harmonic generation from 1064 nm infrared output of neodymium laser produces 532 nm green light.

um laser in the package emits at 1064 nm, and part of the energy is frequency-doubled to give a visible green beam at 532 nm, but the user does not see the harmonic generator as separate.

The third and fourth harmonics can be generated in superficially similar ways, although they require different steps.

Extremely intense laser pulses can produce higher-order harmonics in gas, generating much shorter wavelengths in the extreme ultraviolet, as described in Section 10.2.2.

6.5.2 Optical Parametric Oscillators

Nonlinear optics also are used in a tunable laser-like source called an *optical parametric oscillator* or *OPO*. Although it is pumped by a laser source and oscillates in a resonant cavity, the OPO is not a true laser because it does not rely on stimulated emission.

Operation of an optical parametric oscillator depends on a nonlinear process called *three-wave mixing* to convert light from a shorter wavelength to a longer one or, equivalently, from a higher frequency to a lower one. It starts with a strong pump beam at a frequency ν_{pump} directed into a nonlinear crystal that is placed inside an optical cavity resonant at one or both of the lower frequencies ν_{signal} and ν_{idler}, where $\nu_{\text{pump}} = \nu_{\text{signal}} + \nu_{\text{idler}}$. That is, the nonlinear interaction generates light at two wavelengths, one called the *signal* and the other the *idler*.

The only input to the OPO is the pump beam, but background waves are always present at the two other wavelengths, so the high-power pump mixes with the weak signal and idler beams, converting some of the pump beam to the signal and idler wavelengths, which is functionally equivalent to gain at those wavelengths. As long as the cavity loss at those wavelengths is higher than the gain, essentially nothing happens. However, if the gain at one (or both) wavelengths exceeds the cavity loss, the

OPO begins to oscillate. This occurs at a threshold power, as in a laser.

The oscillation wavelength depends on the cavity optics, the orientation and temperature of the nonlinear material, and other factors. Generally, only one wavelength oscillates at a time and the other wavelength and the remaining pump beam are dumped. OPOs can emit pulsed or continuous beams. Their output is coherent, and the spectral width of a continuous-output OPO can be very narrow, making it attractive for research.

An OPO also can be used for downconversion if the signal and idler waves are tuned to the same wavelength, so $\nu_{pump} = 2\nu_{output}$. This converts one input photon into two output photons at twice the wavelength. The two photons have correlated quantum properties, so they can be used in quantum-mechanical experiments.

Another logical extension of the optical parametric oscillator is the *optical parametric amplifier,* which is arranged so the pump beam transfers energy to a weaker input beam at the signal (or idler) wavelength as the two pass through a nonlinear material. (A similar process has been demonstrated in optical fibers, although its nature is slightly different.)

6.6 WHAT HAVE WE LEARNED?

- A laser oscillator has feedback from the resonant cavity and generates a beam on its own at a wavelength depending on the laser medium and the cavity. It is the optical equivalent of a radio-frequency oscillator.
- An optical amplifier lacks a resonant cavity and feedback; it amplifies light from an external source.
- A master oscillator power amplifier (MOPA) is a multistage pulsed laser, with a laser oscillator followed by one or more optical amplifier stages.
- The MOPA design can produce higher energy pulses than a single-stage oscillator.
- Chirped pulse amplification starts by stretching a short-duration input pulse into a longer pulse with lower peak power, then amplifying it, before compressing the amplified pulse to short duration. In this way, it can generate extremely high peak powers without damaging the amplifier.

- Optical signal amplifiers boost the strength of weak signals in fiber-optic communication systems. They can simultaneously amplify signals on multiple wavelengths within their gain bands.
- Many types of lasers have been developed based on emission on transitions of neutral or ionized atomic gases.
- Molecular gases typically emit on infrared transitions. Vibrational transitions usually occur between about 3 and 12 μm and normally are accompanied by rotational transitions. Rotational transitions alone occur at longer wavelengths.
- Some molecular gas lasers can be quite efficient, particularly carbon dioxide.
- Excimer lasers are based on molecules consisting of a rare gas and a halogen that only exist in the excited state; when they drop to the lower energy level, the molecules fall apart.
- Solid-state lasers are based on nonconductive solids such as glasses and crystals, which typically contain 1% or less of the laser species. They are optically pumped with a light source matched to absorption band of the laser species.
- Semiconductor lasers are electrical diodes, in which recombination of electrons at holes in a thin junction layer produces a population inversion.
- Dye lasers are optically pumped lasers in which the light is emitted by an organic dye dissolved in a liquid solvent or, sometimes, contained in a solid plastic.
- Stimulated emission can be produced by Raman scattering from atoms in a gas or solid.
- Highly efficient resonant cavities are needed for low-gain lasers. Few spontaneous emissions generate the stimulated emission.
- Light in high-gain lasers makes a few round-trips in the laser cavity; many spontaneously emitted photons contribute to the stimulated emission.
- Amplified spontaneous emission is produced by amplifying spontaneous rather than stimulated emission; this corresponds to a high-gain material without a resonant cavity.
- Tunable laser emission requires a laser medium with a wide linewidth and a cavity that can be adjusted to resonate across a range of wavelengths.
- Wide laser bandwidth is needed to generate very short laser pulses.

- Harmonic generation produces laser-like light by doubling the frequency of laser light.
- An optical parametric oscillator is an oscillator incorporating a nonlinear crystal that produces two new frequencies from an input laser beam; the sum of the two new frequencies equals the frequency of the laser light. The output can be tuned by adjusting the cavity optics.

WHAT'S NEXT?

In Chapter 7 you will learn about gas lasers, the most diverse family of lasers.

QUIZ FOR CHAPTER 6

1. What is required for a laser to operate as an oscillator?
   a. Stimulated emission must occur within a resonant cavity.
   b. Stimulated emission must occur without a resonant cavity.
   c. The gain within the cavity must equal the loss within the cavity.
   d. The gain within the cavity must equal or exceed the loss plus the power leaving through the output mirror.
   e. Spontaneous emission must occur.
2. What is needed for an optical amplifier?
   a. Stimulated emission must occur within a resonant cavity.
   b. Stimulated emission must occur without a resonant cavity.
   c. The gain within a resonant cavity must equal the loss within the cavity.
   d. The gain within the cavity must equal or exceed the loss of light in the cavity and through the output mirror.
   e. Spontaneous emission must occur
3. How many oscillators and amplifiers are in a MOPA?
   a. One oscillator followed by only one amplifier.
   b. One amplifier preceded by one oscillator, then a second amplifier.
   c. One oscillator followed by one or more amplifiers.
   d. One amplifier followed by one or more oscillators, then a second amplifier.
   e. Multiple oscillators with amplifiers to couple light between them.

4. How many signals can an optical amplifier amplify at separate wavelengths?
   a. One only
   b. Two
   c. One per amplifier stage
   d. Four
   e. As many as can fit within the amplifier's gain bandwidth
5. What item in the following list cannot be used as the active material in a laser?
   a. A telephone pole
   b. Gelatin doped with fluorescent dye
   c. An atomic gas
   d. Plastic doped with a fluorescent dye
   e. An ionized gas
6. At what wavelengths do molecular gas lasers typically emit on vibrational transitions?
   a. No wavelengths are typical
   b. Red wavelengths
   c. Infrared wavelengths shorter than 3 μm
   d. Infrared wavelengths of 3–12 μm
   e. Infrared wavelengths longer than 12 μm
7. What produces the population inversion in a solid-state laser?
   a. Light from an external source excites atoms contained in a host material.
   b. Light from an external source excites all atoms in the solid.
   c. Electrons in current passing through the host material transfer energy to the laser species.
   d. The host material emits light that is absorbed by the laser species.
   e. Heat melts the laser material so atoms can move freely.
8. What does a dispersive element do in a tunable laser?
   a. Convert a narrow-line laser into a broad-line laser.
   b. Absorbs excess light.
   c. Spreads out wavelengths so the cavity can select one.
   d. Directs pump light to the right atoms.
   e. Converts pump light to the absorption band of the laser species.
9. You have a laser that emits at 800 nm with a linewidth of 40 nm. Assuming that you can use all that linewidth, approximately what is the shortest pulse you can produce?
   a. 2.4 femtoseconds
   b. 24 femtoseconds

c. 39 femtoseconds
d. 240 femtoseconds
e. 390 femtoseconds

10. You have a narrow-line laser with bandwidth of only one megahertz that emits at 800 nm. What is the shortest pulse you could produce with that laser?
    a. 441 femtoseconds
    b. 441 picoseconds
    c. 4.41 nanoseconds
    d. 44.1 nanoseconds
    e. 441 nanoseconds

CHAPTER 7

GAS LASERS

ABOUT THIS CHAPTER

Gas lasers are one of the three most important laser families, and are in many ways the most varied. In this chapter, we will explore the basics of gas lasers, then learn about the most important types.

7.1 THE GAS LASER FAMILY

The family of gas lasers has grown tremendously since Ali Javan, William R. Bennett Jr., and Donald R. Herriott demonstrated the first gas laser at Bell Laboratories in December 1960. Their helium–neon laser was the first laser to generate a continuous beam, and versions modified to emit visible beams remain in widespread use. Since then, laser action has been demonstrated at literally thousands of wavelengths in a wide variety of gases, including other rare gases, metal vapors, and many different molecules.

Gas lasers vary widely in their characteristics. The weakest commercial lasers emit under a thousandth of a watt, but the most powerful you can buy emit thousands of watts in a continuous beam. Experimental high-energy lasers used to assess the potential of laser weapons can generate up to a couple of million watts, but only for seconds at a time. Some gas lasers can emit continuous beams for years; others emit pulses lasting a few billionths of a second. Their outputs range from deep in the vacuum ultraviolet—at wavelengths so short they are blocked completely by air—through the visible and infrared to the borderland of millimeter waves and microwaves.

What makes the gas-laser family so diverse? Gases are easy to study, so that extensive data have been collected on their spectra and energy levels. Experiments also are straightforward. In the heyday of early laser research, some modest university laboratories ran up impressive lists of discoveries by pumping gas into a glass tube, testing it, then replacing with other gases and testing them.

Today, gas lasers are fading, their sales are declining, and semiconductor and solid-state lasers have replaced them for many purposes. However, gas lasers remain at work in some applications. Red semiconductor lasers have become common, but gas lasers offer higher powers in some parts of the spectrum, and better beam quality. Over a billion dollars worth of gas lasers were sold around the world in 2006, according to *Laser Focus World* magazine. The biggest sellers in dollars were carbon dioxide lasers used in industry and medicine, and excimer lasers used largely in medicine and industry. The biggest sellers in numbers were comparatively inexpensive helium–neon lasers, well over 30,000 of which were sold in 2006, more than any other type of gas laser. Several thousand rare-gas ion lasers were sold, and over 1000 helium–cadmium lasers.

7.2 GAS-LASER BASICS

Most gas lasers share common features, as illustrated by the generic gas laser in Figure 7-1. The laser gas is contained in a tube with cavity mirrors at each end, one totally reflecting and one transmitting some light to form the output beam. Most gas lasers are excited by passing an electric current through the gas; the discharge usually is *longitudinal,* along the length of the tube, as shown in Figure 7-1. However, some gas lasers are excited *transversely* across the width of the tube. Electrons in the discharge transfer energy to the laser gas, generally with another step needed to excite atoms or molecules to the upper laser level and generate a population inversion. Stimulated emission then resonates within the laser cavity and produces the laser beam.

Each laser gas has its own distinct characteristics, but this general picture is useful. A bit more elaboration will prepare you to study individual types of gas lasers.

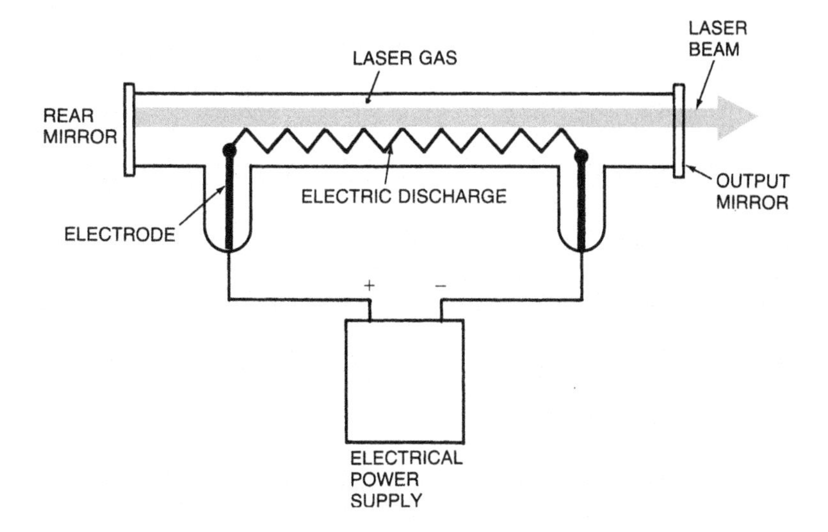

Figure 7-1. Generic gas laser

7.2.1 Gas Laser Media

Gas laser media can be divided into three main types.

Atomic gases include both neutral and ionized atoms, which emit light on transitions between electronic energy levels. Typically, the output of atomic gas or ion lasers is in the near ultraviolet, visible, or near infrared. Most can emit a continuous beam. Examples are the helium–neon and argon-ion lasers.

Excimers are diatomic molecules, most of which are stable in an electronically excited state, but dissociate in the lower laser level. They emit pulses of ultraviolet light on electronic transitions. Examples are the krypton fluoride and xenon chloride lasers.

Molecular gases are gases that usually emit on transitions between vibrational and/or rotational energy levels, although some such as nitrogen can operate on electronic transitions. Vibrational and rotational transitions can generate pulses or continuous output in the near-, mid- and far-infrared wavelengths. The most powerful commercial gas laser is the carbon dioxide laser.

Only a single type of atom or molecule emits light in almost all gas lasers; the only significant exception is the mixed argon–krypton ion laser. However, many gas lasers contain additional species that serve other purposes, such as transferring energy to the laser species, helping to depopulate the lower laser level, or helping to remove waste heat. The additional gases used depend on the type of laser. In the helium–neon laser, helium atoms capture energy from electrons passing through the gas, and the excited helium atoms then excite neon atoms to the upper laser level. Carbon dioxide gas lasers contain nitrogen molecules that absorb energy from an electric discharge and transfer it to CO_2 molecules, as well as helium atoms that help CO_2 drop from the lower laser level (maintaining the population inversion), and assist in heat transfer. The optimum gas mixture for laser operation depends on the operating conditions and power level as well as on the light-emitting species.

Total gas pressure is an important variable because it affects energy transfer and how well the gas conducts electricity. The stable discharge required for continuous laser operation is easiest to sustain at pressures only a small fraction of one atmosphere. Pulsed lasers can operate at much higher pressures because they do not require sustaining a stable discharge for a long time.

Metal vapors are the light-emitting species in a few atomic gas lasers, notably the helium–cadmium and copper-vapor lasers. These metals are solids at room temperature, so the laser tube must be heated to vaporize some of the metal contained inside it before the laser can emit light.

7.2.2 Gas Replacement, Flow, and Cooling

Most gas lasers normally operate inside sealed tubes that isolate the pure gas from the atmosphere and allow operation at the optimum pressure for laser emission. Early gas lasers needed periodic gas replacement because of gas leakage, particularly of tiny helium atoms. Great strides have been made in glass-sealing technology, and helium–neon lasers now are rated to operate for 20,000 hours or more.

Some sealed gas lasers require periodic replacement of the laser gas because contaminants accumulate and gradually degrade laser action. This is a particular problem with excimer lasers, so their tubes are designed for periodic purging and refilling with fresh laser gas. Other types of laser tubes, such as argon-ion, can be refurbished by adding new gas, cleaning the tube, and replacing some internal elements.

Gas flows through the tubes of many higher-power lasers, both to keep the gas at optimum operating temperature and to remove contaminants produced by electric discharges. Many flowing-gas lasers operate in a closed cycle, but some have open cycles that exhaust spent gas. Innocuous gases such as the CO_2, N_2, and helium in CO_2 lasers can be exhausted to the atmosphere, but hazardous gases, such as those in chemical lasers, are collected in cartridges.

Gas lasers emitting a few milliwatts often do not require active cooling, but higher-power gas lasers typically require some form of active cooling, with fans, closed-cycle refrigeration, or flowing water. Because CO_2 lasers are more efficient than most other gas lasers, they can be operated at higher powers without cooling than other gas lasers.

7.2.3 Gas Laser Excitation

The most common way to excite gas lasers is by passing a *longitudinal* electric discharge along the length of the laser tube, as shown in Figure 7-1. As in a fluorescent tube, an initial high-voltage pulse ionizes the gas so it conducts electricity. After the gas is ionized, the voltage is reduced to a lower level that can sustain the modest direct current needed to excite the laser medium. Operating at low gas pressure allows the stable discharge needed for a continuous laser.

Moderate-energy gas lasers also can be excited by a strong microwave field that excites atoms or molecules in the gas.

High-power and/or pulsed gas lasers often are excited *transversely,* by an electric discharge perpendicular to the length of the tube, as shown in Figure 7-2. This approach can pump more electrical energy into the laser medium faster than a transverse discharge, and can be used at gas pressures too high for a stable longitudinal discharge.

As we will see in Section 7.9.3, a few gas lasers are pumped optically, usually by an electrically excited gas laser with shorter wavelength. Although the overall efficiency is limited, this approach can generate otherwise unobtainable wavelengths.

7.2.4 Tube and Resonator Types

So far, our discussion of laser tubes and resonators has been vague on details such as placement of mirrors and the nature of the laser tubes. That is because things are a bit more complex than they might seem.

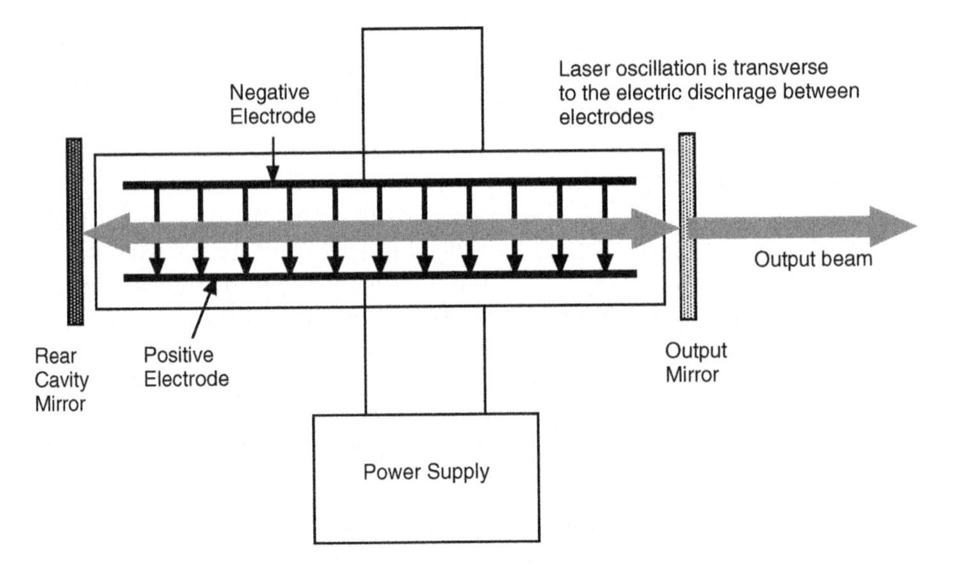

Figure 7-2. Transverse excitation of a gas laser

Gas-laser tubes need windows on their ends that let light through but isolate the laser gas from the atmosphere. They also need mirrors to form a reflective cavity. Applying a reflective coating to a window at the end of the laser tube can cut manufacturing costs, but it exposes the mirror coating to the discharge inside the laser tube, shortening its life. Using separate windows and mirrors improves performance, but adds to cost.

A common way to separate windows and mirrors is to mount the windows at the ends of the tube at Brewster's angle, shown in Figure 4-12. The cavity mirrors are mounted outside the Brewster window, as shown in Figure 7.3. Brewster windows polarize the beam because they have higher losses for light polarized perpendicular to the plane of incidence than parallel to the plane of incidence, and amplification by stimulated emission eliminates the weaker polarization. Figure 7-3 shows curved cavity mirrors defining a stable confocal resonator, which produces a good-quality, diffraction-limited beam with low divergence and is widely used in continuous-wave gas lasers at visible wavelengths.

Different types of cavities are used in high-gain, pulsed lasers, notably the rare-gas halide excimer lasers described later in this chapter. Such lasers do not need as careful optimization to oscillate, and they use cavities in which only the rear mirror is highly reflective. (The few percent reflection from an uncoated glass sur-

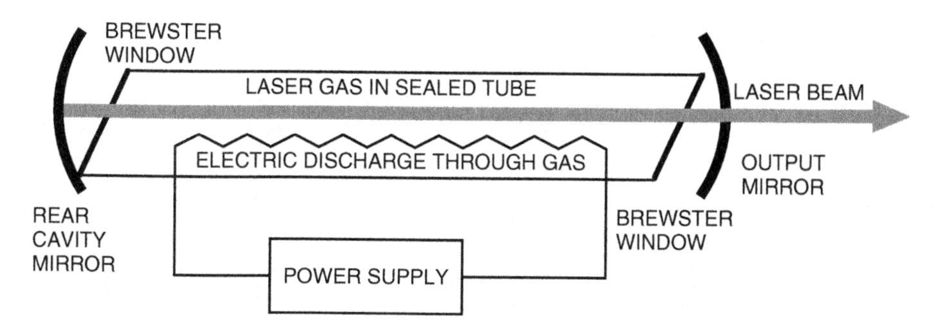

Figure 7-3. Laser tube with Brewster-angle window and confocal resonator.

face provides enough feedback to serve as an output mirror.) This design gives high-gain gas lasers comparatively large beam divergence and diameter.

Most tubes are glass or ceramic, and their prime role is to confine the laser gas and control excitation and gas flow. Some CO_2 lasers are operated in hollow waveguides that confine the laser light optically.

7.2.5 Wavelength and Bandwidth

Table 7-1 lists the nominal wavelengths and maximum power of important gas laser transitions. The electronic transitions have distinct nominal wavelengths; where ranges are given, they span a series of distinct lines, not a continuous range. Vibrational and rotational transitions are nominally distinct at low pressure, but they may be closely spaced.

The continual motion of atoms and molecules inevitably spreads gas-laser emission across a range of wavelengths. The velocity of the molecules along the length of the laser cavity causes a Doppler shift that increases or decreases wavelength, depending on the speed and direction of motion. The resulting *Doppler broadening* can spread laser output across a range of wavelengths proportional to the average speed of gas molecules, which depends on atomic mass M and the gas temperature T. The average velocity in this case is measured as the *root mean square* (the square root of the sum of the squares) $\langle v \rangle$:

$$\langle v \rangle = \left(\frac{3kT}{M} \right)^{1/2} \tag{7-1}$$

Table 7-1. Major gas lasers, by wavelength and power level, grouped under type of transition involved

Type	Approximate wavelength (nm)	Approximate power range (W)‡	Normal operation
Electronic Transitions			
Molecular fluorine (F_2)	157	1–5 (avg.)	Pulsed
Argon–fluoride excimer	193	0.5–50 (avg.)	Pulsed
Krypton–fluoride excimer	249	1–100 (avg.)	Pulsed
Argon-ion (UV lines)	275–305	0.001–1.6	Continuous
Xenon–chloride excimer	308	1–100 (avg.)	Pulsed
Helium–cadmium (UV line)	325	0.002–0.1	Continuous
Nitrogen (N_2)	337	0.001–0.01 (avg.)	Pulsed
Argon-ion (UV lines)	333–364	0.001–7	Continuous
Krypton-ion (UV lines)	335–360	0.001–2	Continuous
Xenon–fluoride excimer	351	0.5–30 (avg.)	Pulsed
Helium–cadmium (UV line)	354	0.001–0.02	Continuous
Krypton-ion	406–416	0.001–3	Continuous
Helium–cadmium	442	0.001–0.10	Continuous
Argon-ion	488–514.5	0.002–25	Continuous
Copper-vapor	510 and 578 nm	1–50 (avg.)	Pulsed
Helium–neon	543	0.0001–0.002	Continuous
Helium–neon	594	0.0001–0.002	Continuous
Helium–neon	612	0.0001–0.002	Continuous
Gold-vapor	628	1–10	Pulsed
Helium–neon	632.8	0.0001–0.05	Continuous
Krypton-ion	647*	0.001–7	Continuous
Helium–neon	1153	0.001–0.015	Continuous
Iodine and oxygen–iodine	1315	High-energy§	Pulsed or CW
Helium–neon	1523	0.001	Continuous
Vibrational Transitions			
Hydrogen fluoride (chemical)	2600–3000†	0.01–150, high-energy§	Pulsed or CW
Deuterium fluoride (chemical)	3600–4000†	0.01–100, high-energy§	Pulsed or CW
Carbon monoxide	5000–6500†	0.1–40	Pulsed or CW
Carbon dioxide	9000–11000†	0.1–20,000	Pulsed or CW
Vibrational or rotational transitions			
Far infrared	30,000–1,000,000†	<0.001–0.1	Pulsed or CW

*Other wavelengths also available.
†Many lines in this wavelength range.
‡For typical commercial lasers.
§High energy" in the power column denotes types that have been developed to emit higher power for military laser-weapon programs.

where k is the Boltzmann constant. This velocity at normal operating temperatures is large enough to make Doppler broadening the main factor determining the spectral width of most gas lasers. For example, the Doppler width (defined as full-width at half maximum) of a typical helium–neon gas laser is about 1.4 gigahertz, or about 0.0019 nm. Although this is small compared to the 4.738×10^{14} hertz frequency of the laser's 632.8-nm transition, it is much larger than the 1-megahertz bandwidth of one longitudinal mode of a typical helium–neon laser cavity, and large enough to include several longitudinal modes, separated by about 500 MHz.

Now that we have covered these basics, we will move on to important types of gas lasers.

7.3 HELIUM–NEON LASERS

Until the 1990s, the laser people were most likely to have seen was the low-power red helium–neon laser, which was used for supermarket checkout, construction alignment, and educational demonstrations. The least expensive of all gas lasers, the helium–neon laser was sometimes called a "He–Ne." Its coherence and its visible beam long made the He–Ne the standard laser for educational demonstrations and recording holograms. It remains widely used in scientific and medical instruments, and is still used for holography, but red semiconductor lasers have replaced the helium–neon laser in most other applications because they are much cheaper and smaller.

Sales of helium–neon lasers dropped from about 430,000 in 1991 to 39,000 in 2006, according to *Laser Focus World,* which forecasts a continuing decline as semiconductor and solid-state lasers crowd the venerable helium–neon laser out of markets it once dominated. Some two-thirds of the helium–neon lasers sold in 2006 were used in scientific and biomedical instruments.

7.3.1 Physical Principles

Helium–neon lasers can emit on several transitions among the energy levels shown in Figure 7-4. Electrons passing through a mixture of five parts helium and one part neon excite both species to high energy states, but the more abundant helium atoms collect more energy. Excited helium atoms readily transfer their excess

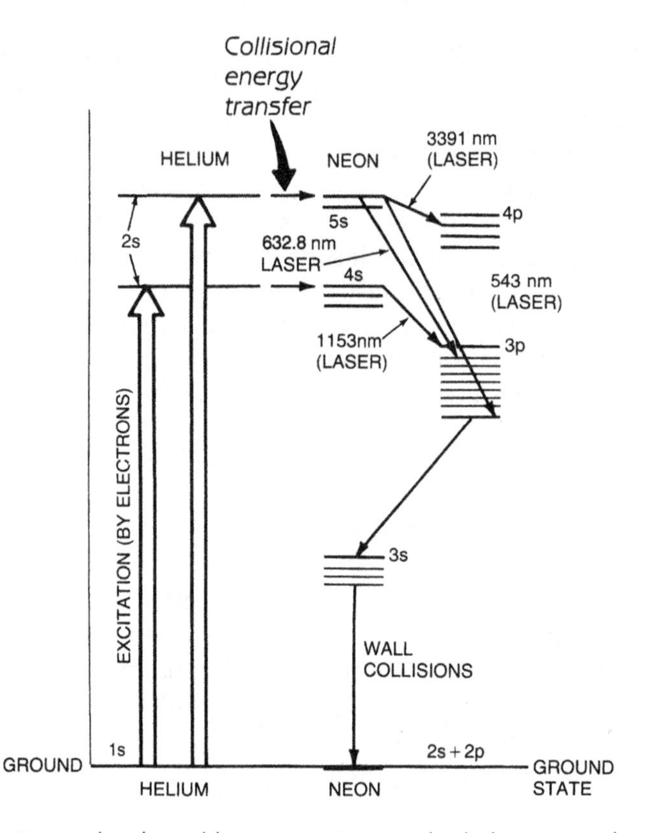

Figure 7-4. Energy levels and laser transitions in the helium–neon laser shown on a relative scale.

energy to neon atoms when the two collide, raising the neon atoms to the 5s and 4s energy levels. (The energy levels are spectroscopic labels convenient to use as identification.) The 5s and 4s energy levels of neon are metastable, so atoms stay in those states for a comparatively long time, producing a population inversion.

The transitions shown in Figure 7-4 are the strongest and most useful of the several possible in helium–neon mixtures. Note that some transitions descend from the 5s level and others from the 4s level. The first helium–neon laser operated at 1153 nm in the infrared, but the demand for visible beams made the 632.8-nm red line the standard for helium–neon. He–Ne lasers are now available on several weaker lines.

Once the neon atoms drop to the lower laser level, they quickly drop through a series of lower energy levels to the ground state. Energy transfer from helium atoms can then raise them again to the upper laser level. The overall efficiency of the helium–neon is low—typically 0.01% to 0.1%—because the laser transitions are far above the ground state. (Figure 7-4 is not to scale in energy.)

Overall gain of the helium–neon laser is very low, so care is needed to minimize laser cavity losses, as we will see below. However, the helium–neon laser is simple, practical, and inexpensive; mass produced sealed-tube versions can operate continuously for tens of thousands of hours. Ionizing the gas takes about 10,000 volts, but a couple thousand volts can maintain the current of a few milliamperes needed to sustain laser operation.

7.3.2 He–Ne Laser Construction

Figure 7-5 shows the internal structure of a typical mass-produced helium–neon laser. The discharge passing between electrodes at opposite ends of the tube is concentrated in a narrow bore, one to a few millimeters in diameter. This raises laser excitation efficiency, and also helps control beam quality. The bulk of the tube volume is a gas reservoir containing extra helium and neon. Gas pressure within the tube typically is a few tenths of a percent of atmospheric pressure.

Mirrors are bonded directly to mass-produced helium–neon tubes by a high-temperature process that produces what is called a

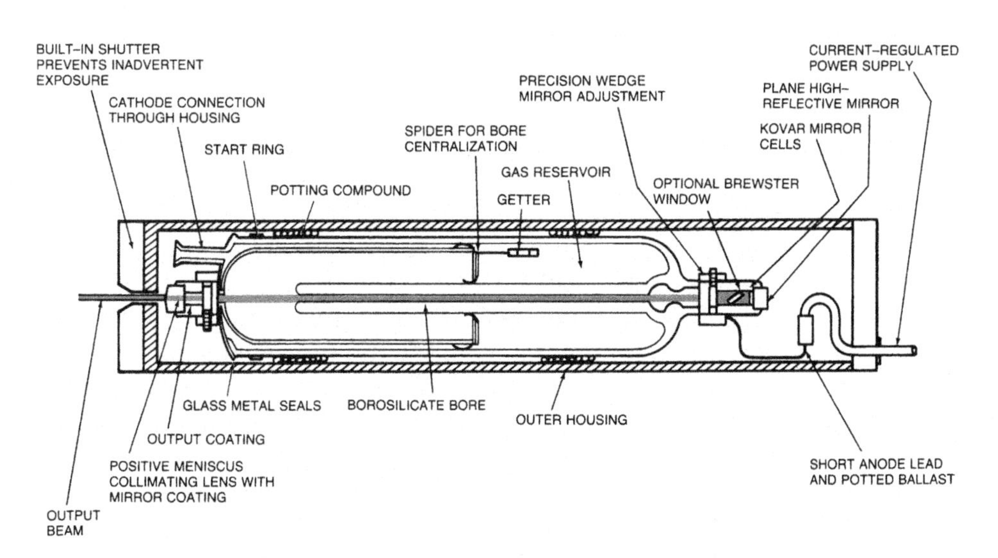

Figure 7-5. A mass-produced helium-neon laser (courtesy of Melles Griot).

"hard seal," which slows the helium leakage that otherwise might limit laser lifetime. The mirrors must have low loss because of the laser's low gain. The rear cavity mirror is totally reflective. The output mirror transmits only a few percent of the intracavity power to produce the laser beam. One or both mirrors have concave curvature to focus the beam within the laser cavity, which is important for good beam quality. An option is to seal the rear of the laser cavity with a Brewster window, and mount the rear mirror separate from the laser tube, to give polarized output.

The output power available from helium–neon lasers depends on tube length, gas pressure, and diameter of the discharge bore. Researchers have found that output power for a given tube length is highest when the product of gas pressure (in torr; 760 torr equals one atmosphere) times bore diameter (in millimeters) is 3.6 to 4. Extending the tube can raise output power somewhat, but the improvements are limited, and 50 mW is about the maximum practical output.

The helium–neon laser also can be operated in a ring cavity to sense rotation. The ring laser is actually a square or triangular array of tubes, with mirrors at each corner reflecting the beam into the next arm. A ring laser "gyroscope" can detect rotation about an axis perpendicular to the ring plane by measuring phase differences in light traveling in different directions around the ring.

7.3.3 Practical Helium–Neon Lasers

Mass-produced helium–neon lasers can deliver 0.5 to about 10 milliwatts of red light. They range in size from little bigger than fat pens, 10 centimeters long and 1.6 cm in diameter (4 by ⅝ inch), to 30 cm (a foot) or more long and 2–5 cm (1–2 inches) in diameter. A few special-purpose helium–neon lasers are considerably larger and can produce up to 50 mW in the red.

Although other helium–neon wavelengths listed in Table 7-1 have been available commercially since the mid-1980s, most helium–neon lasers emit the 632.8-nm red line. Output is much weaker, typically no more than a couple of milliwatts, at other visible wavelengths: 543 nm in the green, 594 nm in the yellow–orange, and 612 nm in the red–orange. Milliwatt power is available on infrared lines at 1.153, 1.523, and 3.39 micrometers. Although a few models can be made to emit at different wavelengths by switching their optics, most helium–neon lasers emit only a single wavelength.

Surplus red helium–neon laser tubes may sell for as little as $20 without a power supply or safety equipment required by Federal regulations. Complete new red helium–neon lasers ready for laboratory or other use start at a few hundred dollars; other wavelengths are more expensive.

Helium–neon lasers typically emit TEM_{00} beams, with diameter about a millimeter and divergence about a milliradian. The typical Doppler broadened bandwidth of a red helium–neon laser is 1.4 GHz, which corresponds to a coherence length of 20 to 30 centimeters, adequate for holography of small objects. To get a longer coherence length, you must buy a laser limited to a single longitudinal mode with 1-MHz bandwidth, with 200- to 300-m coherence length.

7.4 ARGON- AND KRYPTON-ION LASERS

Like helium–neon lasers, argon- and krypton-ion lasers are powered by electric discharges passing through elements of the rare-gas (Group VIII) column of the periodic table. All three emit continuous beams. However, there also are crucial differences. Argon and krypton lasers have laser transitions in ions rather than the neutral atoms in the helium–neon laser. For that reason, they are often called simply *ion lasers,* although ions are the light emitting species in some other gas lasers, notably helium–cadmium. Argon and krypton generate more powerful beams and emit at shorter wavelengths than helium–neon, with argon more powerful and more important commercially.

Historically, argon- and krypton-ion lasers long were the highest-power visible lasers emitting continuous beams. Improvements in solid-state lasers have changed that, and argon and krypton lasers have suffered because of their cost, complexity, and low efficiency. Their sales have dropped, but not as sharply as those of helium–neon lasers. *Laser Focus World* estimated that over 20,000 argon and krypton lasers were sold in 1991, a total that had dropped to just under 10,000 in 2006, and is expected to decline further. Like helium–neon lasers, the biggest current application of argon and krypton lasers is in scientific and biomedical instruments.

7.4.1 Properties of Argon and Krypton Lasers

The active medium in rare-gas ion lasers is argon or krypton at a pressure of roughly 0.001 atmosphere. Tubes containing mixtures of the two gases can emit on lines of both elements across the visible spectrum, a capability used mostly in laser light shows. Table 7-2 lists important lines of the two elements in the near-ultraviolet, visible, and near-infrared parts of the spectrum.

The laser lines of argon and krypton are in ions with one or two electrons stripped from their outer shells. Ultraviolet wavelengths shorter than 400 nm come from atoms with two electrons removed (Ar^{+2} or Kr^{+2}). Visible output comes from singly ionized atoms (Ar^{+} or Kr^{+}). These lines are very high above the ground state, limiting efficiency. Argon is the more efficient laser gas and is used for most applications, but krypton lines are spread across more of the visible spectrum.

As in helium–neon lasers, an initial high-voltage pulse ionizes the gas, then lower voltage produces a sustained discharge. The current ionizes the atoms and excites them to high energy states and the upper laser level, as shown for singly ionized argon in Figure 7-6. Laser transitions occur between many pairs of upper and lower levels; there isn't room to show them all in the figure. If

Table 7-2. Major wavelengths of ion lasers

Argon	Krypton
275.4 nm	337.4 nm
300.3 nm	350.7 nm
302.4 nm	356.4 nm
305.5 nm	406.7 nm
334.0 nm	413.1 nm
351.1 nm	415.4 nm
363.8 nm	468.0 nm
454.6 nm	476.2 nm
457.9 nm	482.5 nm
465.8 nm	520.8 nm
472.7 nm	530.9 nm
476.5 nm	568.2 nm
488.0 nm (strong)	647.1 nm (strong)
496.5 nm	676.4 nm
501.7 nm	752.5 nm
514.5 nm (strong)	799.3 nm
528.7 nm	
1090.0 nm	

the optics permit, argon and krypton lasers can oscillate simultaneously on several different visible wavelengths, each produced by a transition between a different pair of levels.

The lower laser levels all have very short lifetimes. Argon ions quickly drop from the lower laser level (an excited state of the ion) to the ion ground state by emitting an extreme-ultraviolet photon at 74 nm, which does not leave the laser tube. The ground-state ion can recapture an electron or again be excited to the upper laser levels. Similar things happen in krypton.

The need to ionize atoms makes argon and krypton lasers energy-hungry. Visible-wavelength lasers draw discharge currents of 10 to 70 amperes, more than a thousand times the level in helium–neon lasers, although the low resistance of the ionized gas permits voltages of 90 to 400 volts. Ultraviolet lasers require even more energy because they must remove two electrons from the light-emitting atoms. Ionization heats the gas, so the Doppler bandwidth of a single ion-laser line is about 5 GHz.

Typical visible output powers range from a few milliwatts to 25 W for argon; ultraviolet output is lower. Most argon and krypton lasers sold today emit less than 1 W and are used in scientific and biomedical instruments. Only 285 of the nearly 10,000 ion

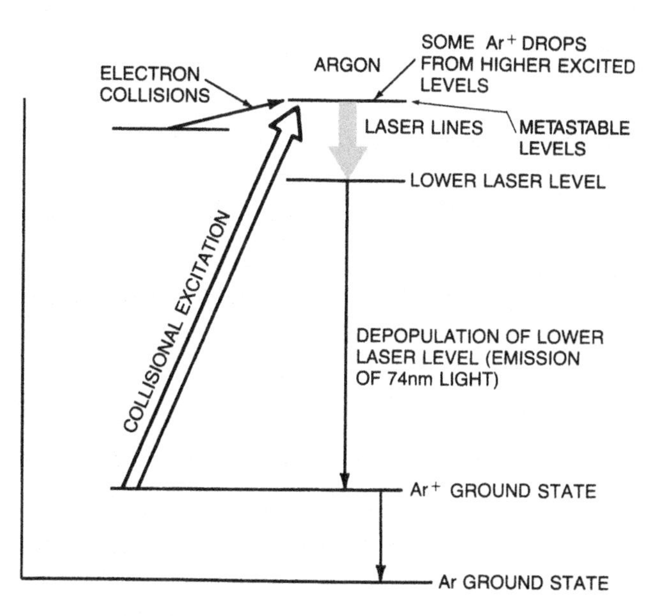

Figure 7-6. Energy levels that produce visible argon lines.

lasers sold in 2006 were more powerful than 1 W; most of them were used in scientific research and laser light shows.

7.4.2 Practical argon and krypton lasers

Argon and krypton lasers have low gain, so cavity losses must be minimized. Materials must withstand the intense extreme-ultraviolet generated when argon atoms drop from the lower laser level, so the tubes are generally ceramic. Cavity optics may allow oscillation on multiple lines, notably the 488 and 514.5 nm lines of argon, or oscillation can be restricted to a single wavelength. Generally, the optics produce TEM_{00} beams and include Brewster windows to select linear polarization.

As in helium–neon lasers, confining the discharge to a narrow region in the center of the tube enhances excitation efficiency. In newer ion lasers, a series of metal disks with central holes confine the discharge to the central region. Laser operation depletes the gas, so the tube must include a large gas reservoir. A return path is also needed for positive ions. With wall-plug efficiency only 0.01% to 0.001%, argon and krypton lasers need active cooling with forced air or, at powers above a few watts, flowing water.

Operating conditions inside argon and krypton laser tubes are extreme, so lifetimes are shorter than those of helium–neon lasers, typically ranging from 1000 hours for high-power tubes to 10,000 hours for low-power tubes.

7.4.3 Mixed-Gas Argon–Krypton Lasers

Argon and krypton can be mixed in a single laser tube to make a laser that oscillates simultaneously on many lines throughout the visible spectrum. The main use of such mixed-gas lasers are in light shows. Their lifetimes are relatively short because the two gases are depleted at different rates.

7.5 METAL-VAPOR LASERS

Gas lasers also include lasers in which the light-emitting species is a vaporized metal. The vapor may be ionized, as in the helium–cadmium laser, or neutral, as in copper-vapor lasers. The two families operate quite differently.

7.5.1 Helium–Cadmium Lasers

Helium–cadmium (He–Cd) lasers emit continuous beams at powers from under a milliwatt to tens of milliwatts, slightly more powerful than helium–neon, but less powerful than argon lasers. The strongest He–Cd wavelength is 441.6 nm in the blue part of the spectrum, but weaker 325 and 353.6 nm lines in the ultraviolet are also used. Their main applications are in industry, measurement, and instrumentation.

Figure 7-7 shows the energy levels involved in the He–Cd laser. An electric current flowing through a thin capillary bore excites helium atoms to high energy states, and collisions transfer the energy to cadmium vapor, ionizing the atoms and raising them to the upper laser levels of the three transitions. One metastable state is the upper level of the two ultraviolet transitions; the other is the upper level of the 441.6-nm blue line. Cavity optics select

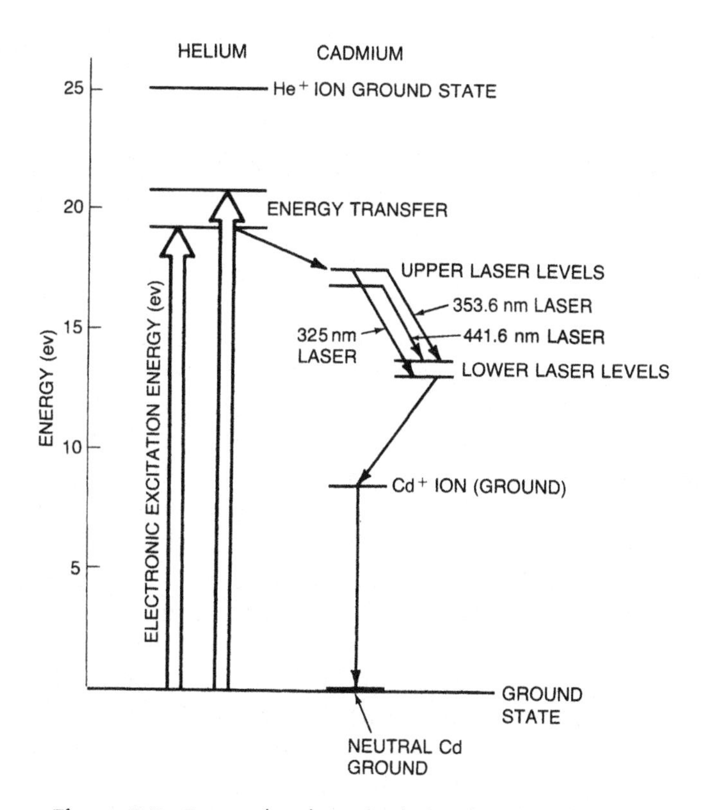

Figure 7-7. Energy levels in the helium–cadmium laser.

the oscillating wavelengths. Cadmium normally has only two electrons in an incompletely filled outer shell, so it is much easier to ionize than helium, argon, or krypton, which all have full outer electron shells.

Cadmium is a solid at room temperature, so chunks of metal in the laser tube must be heated to about 250°C to produce the several millitorr pressure of cadmium vapor needed for the laser to operate. Helium pressure is about a thousand times higher, several torr, but that is only about 1% of atmospheric pressure. Extra cadmium is put in the tube to replace metal vapor that condenses on cool parts of the tube during laser operation. Tubes normally also include a helium reservoir to replace gas that leaks out. Discharge voltages are around 1500 V.

He–Cd lasers continue to be used in some niche applications requiring high-quality blue or ultraviolet beams of modest power, but only 1100 were sold in 2006, according to *Laser Focus World,* and sales continue to decline. Other metal-vapor ion lasers have been demonstrated, but none are available commercially.

7.5.2 Copper-vapor lasers

The copper-vapor laser emits at 511 nm in the green and 578 nm in the yellow. The internal physics limits it to pulsed operation, but it can generate thousands of pulses per second and average powers of tens of watts.

To obtain the 0.1-torr vapor pressure needed for laser action, metallic copper in the laser tube must be heated to 1500°C, which takes about half an hour after the laser is switched on. Helium, neon, or argon may be added to the copper to improve discharge quality and energy transfer. Operation of the laser generates enough waste heat to keep the metal vaporized.

In operation, a pulsed discharge is fired along the laser tube, and collisions raise the vaporized neutral copper atoms to one of two excited states. A minimum pressure is needed to raise the excited-state lifetimes to the 10 milliseconds needed for laser operation. Stimulated emission drops the atoms to two lower levels, with emission of green or yellow light. Both of the lower laser levels are metastable, with lifetimes of tens to hundreds of microseconds, so the lower states quickly fill, terminating the population inversion and stopping the laser pulse in tens of nanoseconds. Depopulation of the lower levels leaves the laser ready to generate another pulse, allowing repetition rates of several thousand pulses per second.

Copper-vapor lasers are inherently powerful, with gain of 0.1 to 0.3 per centimeter. They do not require a high-gain cavity; commercial models have a totally reflective rear-cavity mirror, and an output mirror that reflects only about 10% of light back into the laser cavity. Average powers can reach tens of watts, with efficiency several tenths of a percent, attractively high for visible gas lasers.

Today, copper-vapor lasers have limited uses, primarily for high-speed imaging and research. Similar lasers have been operated with gold and lead vapor, but were never widely used.

7.6 CARBON DIOXIDE LASER

The carbon dioxide (CO_2) laser is exceptionally versatile and highly efficient; up to 20% of the input power can be converted into laser light. It operates under a wide variety of conditions, emitting a steady beam at low gas pressure or pulses at high pressures. Powers of continuous beams can range from milliwatts to more than 10 kW; output can be on a single narrow line, or spread across a series of lines in the 10-μm region.

These features have made the CO_2 laser very successful, with the highest dollar sales of any gas laser. In 2006, *Laser Focus World* counted $740 million in world-wide sales of carbon dioxide lasers. That covered 23,700 individual lasers, with most of them used for materials processing, 1660 used for surgery, and smaller numbers used for sensing and basic research. That's a huge jump over the 4600 CO_2 lasers the magazine counted as sold during 1991, mostly for materials working, with about 1000 for medicine.

The CO_2 laser emits on vibrational transitions of the CO_2 molecule in a band between 9 and 11 μm in the infrared. It can emit pulses or continuous beams, and it can produce the highest continuous power available from any commercial laser. The 10-μm output is transmitted reasonably well by the atmosphere, and absorbed strongly by a wide range of materials, allowing it to cut, drill, and weld both metals and nonmetals. The CO_2 wavelength also is strongly absorbed by the water in tissue, so it has long been used for some types of surgery.

7.6.1 CO_2 Laser Transitions

The CO_2 laser is excited by an electric discharge passing through the laser gas, which contains nitrogen and (usually) helium as well as carbon dioxide. The N_2 and CO_2 molecules absorb energy from electrons in the discharge. The lowest vibrational level of N_2 transfers energy easily to CO_2. Helium helps to maintain the population inversion by getting CO_2 molecules to drop from the lower laser levels to a lower level or the ground state.

The CO_2 molecule has three vibrational modes shown in Figure 7-8: the symmetric stretching mode ν_1, the bending mode ν_2, and the asymmetric stretching mode ν_3. The molecule can vibrate at different rates in each mode, so each has its own set of energy levels (0, 1, 2, 3, etc.). The laser transitions occur when CO_2 mole-

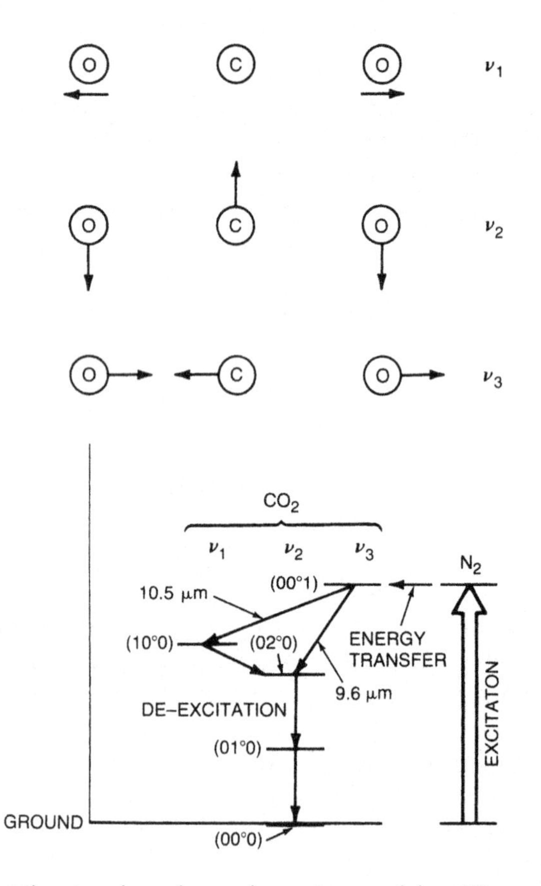

Figure 7-8. Vibrational modes and transitions of the CO_2 molecule.

cules drop from the higher-energy asymmetric stretching mode excited by the discharge to the lower-energy symmetric stretching or bending modes.

A CO_2 molecule in the asymmetric stretching mode would emit a 10.5 μm photon if it dropped directly to the symmetric stretching mode and a 9.6-μm photon if it dropped directly to the second excited level of the bending mode. However, the molecule changes its rotational state when it changes its vibrational state, so it doesn't emit those precise wavelengths. Speeding up its rotation takes some energy from the vibrational transition, so the emitted wavelength is longer than the nominal transition energy—for example, 10.6 μm rather than 10.5 μm. If the molecule slows down its rotation, the energy released adds to the energy from the vibrational transition, resulting in a shorter wavelength, such as 10.3 μm. This allows CO_2 lasers to emit on the range of closely spaced lines you saw earlier in Figure 4-5.

7.6.2 Sealed-Tube and Waveguide CO_2 lasers

The simplest CO_2 lasers operate in a sealed tube like other gas lasers described earlier. Some gas flow is possible by moving gas through the tube and the rest of a sealed system, which includes a gas reservoir. Typically, a high-voltage discharge passes between positive and negative electrodes at opposite ends of the tube, but the laser also can be excited by a radio-frequency induced discharge. Water or a catalyst must be added to regenerate CO_2 molecules which the discharge splits into oxygen and carbon monoxide.

Convenient and inexpensive, the sealed-tube design is widely used for CO_2 lasers with continuous powers under about 100 watts. The maximum power depends on length of the gain medium; a one-meter length can generate about 50 W. Mirrors can fold the laser beam inside the cavity so the gain medium is longer then the tube, but long paths are cumbersome, and as the power increases it becomes more difficult to remove waste heat, limiting the practical power level.

An alternative approach to making sealed-tube CO_2 lasers is to shrink the tube's cross section to a millimeter or two across, only about 100 times the 10-μm wavelength. A glass or ceramic tube of that size functions as an infrared waveguide, guiding the light waves through it (much like optical fibers guide visible light or hollow pipes guide microwaves). This structure avoids large diffraction losses that otherwise would occur with an output aperture that is so small compared to the wavelength. Gas flow is required in a waveguide laser, but it can be packaged as a sealed system with internal gas reservoir. Normally, waveguide CO_2 lasers emit continuous beams.

7.6.3 Longitudinal-Flow CO_2 Lasers

Higher power CO_2 output is possible if fresh gas flows the length of the laser cavity, in the same direction as the discharge current. This consumes gas, but pressures are low and the gases used in CO_2 lasers are not particularly hazardous or costly. Normally, some of the exhaust gas can be recycled by mixing it with fresh gas.

The straightforward design can produce output power per unit length somewhat higher than sealed-tube lasers, without the same limits on maximum total power. Continuous beams can deliver hundreds of watts. These lasers are more expensive to build and operate, so they are more costly than sealed CO_2 lasers.

7.6.4 Transverse-Flow CO_2 Lasers

Output power in a continuous CO_2 beam can reach about 10 kW per meter of tube length if the laser gas flow is *transverse* across the laser cavity rather than along its length. Transverse flow moves the gas much faster through the laser cavity than longitudinal flow along its length, removing waste heat and contaminants. The electric discharge that drives the laser also is applied perpendicular to tube axis, so it goes through a shorter length of gas, but pressure remains low to maintain a stable discharge.

Like longitudinal flow lasers, the gas pressure is low and the output beam is continuous. Typically, the gas is recycled, with some fresh gas added. Transverse flow normally is used only in very high-power lasers, with outputs of several kilowatts or more, and these lasers are big and expensive. (A diagram of one early model with 15-kW output labeled part of the flow loop as a "wind tunnel.")

7.6.5 Transversely Excited Atmospheric (TEA) CO_2 Lasers

Most sealed and flowing-gas CO_2 lasers operate at pressures below 0.1 atmosphere to stabilize the electric discharge needed for continuous operation, although the lasers can be pulsed. Operating at pressures near one atmosphere or higher requires pulsing electric discharges transverse to the laser cavity. Such *transversely excited atmospheric pressure* or TEA lasers are compact and can deliver powerful pulses lasting 40 nanoseconds to one microsecond, which are sought for some applications. The intense pulsed discharges break down CO_2, so gas replenishment or reconstitution is required.

7.6.6 Gas-Dynamic CO_2 Lasers

It is worth noting that electrical excitation is not the only way to produce a population inversion in carbon dioxide. Rapid expansion of hot, high-pressure CO_2 (typically mixed with other gases) through nozzles into a near-vacuum also can produce a population inversion. The rapid expansion cools the gas, but does not instantly drop all the molecules to low energy levels. The laser cavity is arranged perpendicular to the gas flow, which is transverse to the axis between the mirrors.

The gas-dynamic CO_2 laser was invented in the late 1960s, when it was a breakthrough that for the first time allowed laser power to reach the 100-kW range in a continuous beam. Its development stimulated a major military program to develop high-power laser weapons, but the gas-dynamic laser did not prove practical for that purpose, or for industrial use, and is not used commercially today.

7.6.7 Optics for CO_2 lasers

One thing that sets the CO_2 laser apart from other gas lasers described so far is its much longer 10-μm wavelength, far into the infrared range. Standard transparent materials used for visible optics are opaque at 10 μm, so other materials must be used, and reflective optics are often preferred where they can be used. That has some important consequences in designing and using CO_2 lasers.

CO_2 lasers have good but not extremely high gain, so laser cavities normally have a totally reflective rear mirror and a partly reflective output mirror, which are usually separate from the windows at the end of the laser tube. Metal mirrors are often used in the laser cavity, with output coupling through a hole in the mirror rather than through a partly transmissive coating. Windows on the laser tube are transparent to the 10 μm laser beam, but generally not to visible light.

Research and measurement applications often require limiting CO_2 emission to a single laser line, and these lasers are made with tuning optics that select one specific wavelength. These wavelengths are discrete lines at the low pressures used in most CO_2 lasers, but broaden to become a continuum at the high pressures used in TEA lasers.

Most other CO_2 laser applications are not sensitive to a specific wavelength and instead require delivery of the highest possible power for operations such as drilling, welding, or cutting. These lasers are made with cavity optics reflective throughout the 9 to 11 μm range, to extract as much energy as possible from the CO_2 laser cavity in a good-quality beam.

Refractive optics for 10-μm beams are made of different material than conventional glass and plastic optics for visible light. Important materials are shown in Figure 5-8. A few of these materials are transparent at visible wavelengths but not used for standard optics, including sodium chloride (which is transparent but must be protected from moisture), and zinc sulfide (which appears orange). However, most are opaque at visible wavelengths.

You cannot directly see 10-μm beams, and they can emerge from optics that look opaque to the eye. Dust particles burn when exposed to very high-power CO_2 beams, an effect called "fireflies." Many infrared viewers are sensitive to shorter bands near 1 μm or 3 to 5 μm, and do not show light in the 10-μm range. To see 10 μm, you need a "thermal" infrared viewer, so-called because the thermal (heat) radiation from room-temperature objects peaks near 10 μm. These may be available as military surplus.

7.7 EXCIMER LASERS

Excimer lasers are a family in which light is emitted by short-lived molecules made up of one rare gas atom (e.g., argon, krypton, or xenon) and one halogen (e.g., fluorine, chlorine, or bromine). They are often called rare-gas halides. Table 7-3 lists the most important excimer lasers and the closely related fluorine laser.

First demonstrated in the mid-1970s, excimer lasers have become important because of their unique ability to generate high-

Table 7-3. Major excimer lasers

Type	Wavelength
F_2*	157 nm
ArF	193 nm
KrCl	222 nm
KrF	249 nm
XeCl	308 nm
XeF	350 nm

*Not a rare-gas halide, but usually grouped with excimers.

power ultraviolet pulses. *Laser Focus World* estimates that about 1450 excimer lasers were sold in 2006, with more than half used for eye surgery, about 500 used in semiconductor fabrication, and the rest used in research. That number continues to grow. Although the number of units is small, excimer lasers are complex and expensive, with total sales about $450 million. The most costly lasers are used in semiconductor fabrication, with average price around $700,000. That is expensive for a laser, but only a small fraction of the cost of building a state-of-the-art chip plant.

7.7.1 Excimer Laser Physics

Excimers are peculiar diatomic molecules that are bound together only in an electronically excited state. The laser is excited by passing a short, intense electrical pulse through a blend of gases. Normally, 90% or more of the mixture is a buffer rare gas (typically helium or neon) not directly involved in the reaction. The rare gas that becomes part of the excimer (argon, krypton, or xenon) is a few percent of the mixture. A smaller fraction of the gas mixture is a gas that supplies the needed halogen atoms, either halogen molecules such as F_2, Cl_2, or Br_2, or molecules that contain halogens such as nitrogen trifluoride (NF_3).

The discharge splits halogen-containing molecules, freeing excited halogen atoms to react with the rare gas and forming electronically excited molecules like xenon fluoride (written XeF*, with the * meaning excited). The exact sequence of reactions are complex and depend on the gas mixture. The excited dimers remain in the electronically excited upper laser level for about 10 nanoseconds, then drop to the unbound lower laser level, so the molecule breaks up. That process rapidly empties the lower laser level, sustaining a population inversion. The duration of the electrical drive pulses and the molecular kinetics limit laser operation to pulses typically lasting several nanoseconds to about 100 ns.

Figure 7-9 shows the energy levels of a typical rare-gas halide as a function of the spacing between the two atoms in the molecule, *R* (the rare gas) and *H* (the halide). The dip in the excited-state curve shows the spacing where the two atoms have a minimum energy, binding them together to form a metastable molecule. The horizontal lines are vibrational levels which exist in the potential well. No such minimum exists in the ground state curve; the energy of the two molecules increases with their separation, so they are unbound and drift apart.

Excimer-laser repetition rates depend more on the power supply than on the gas. The principal limitation is speed of the high-voltage switches, which can operate at up to 4000 Hz in commercial lasers. Pulse energies of typical lasers range from millijoules to about a joule; they drop as repetition rate increases and vary somewhat among gases, with KrF and XeCl generally the most energetic. Average power—the product of pulse energy times repetition rate—can reach a couple hundred watts, although lower values are more common.

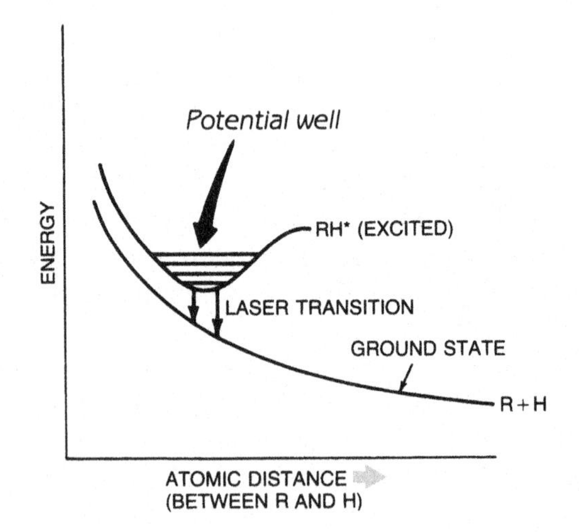

Figure 7-9. Internal energy of a rare-gas halide molecule in excited and ground states.

7.7.2 Excimer Laser Operation

Excimer lasers have such high gain that they almost do not need cavity mirrors. In practice, they have fully reflective rear mirrors and uncoated output windows that reflect a few percent of the beam back into the cavity and transmit the rest. The discharge is perpendicular to the length of the tube. Excimers are efficient for gas lasers, with wall-plug efficiency as high as 2%.

The corrosive halogens used in excimer lasers require special handling. The tubes must be made of halogen-resistant materials. After being filled with gas, the tubes are sealed and operated until the laser power declines to a point where the gas must be replaced. Tubes include a large volume of reserve gas to extend operating life, and often include a recycling system that helps regenerate the gas mixture. Eventually, the spent gas must be pumped out of the laser and replaced.

A gas fill can last many millions of shots for longer-lived mixtures such as xenon chloride, but at high repetition rates that does not add up to a long time. A 200-Hz laser in steady operation fires 720,000 pulses per hour. That means any excimer laser installation requires both gas supply and disposal.

7.7.3 Excimer Laser Configurations

Excimer lasers typically are designed for particular applications. More than half of all excimer lasers sold today are used in medicine, primarily for surgery that ablates tissue from the surface layer of the eye, the cornea, to correct refractive defects. The interaction of laser beam with the cornea depends strongly on the wavelength, and is best for the 193-nm ArF laser, which is the standard for refractive surgery. The lasers are built into refractive surgery systems used in physicians' offices or clinics.

About one-third of excimer lasers are used in the production of semiconductor integrated circuits. Short-wavelength ultraviolet lasers are needed to fabricate small details on the chips, with shorter wavelengths needed to fabricate details in smaller geometries. As semiconductor production lines have moved to finer geometries, the cutting edge has moved from the 249-nm KrF laser to the 193-nm ArF laser. These lasers are built for power and heavy use, and are optimized for a particular gas. The 157-nm F_2 laser would seem the next logical step for semiconductor lithography, but limited power levels and the difficulty of using a wavelength at which the atmosphere is opaque so far have stalled its use.

The remaining share of the excimer laser market is in laboratory research, where flexibility is more important than the number of pieces processed per minute. Many laboratory lasers are designed to handle several different gas mixtures rather than the single gas mixture used in most industrial and medical excimers.

7.8 CHEMICAL LASERS

Gas lasers powered by a chemical reaction are known as *chemical lasers.*They have come to occupy a peculiar niche in the laser world—high-power lasers used in military research aimed at producing high-energy laser weapons. Military developers hope to ultimately use chemical lasers as weapons to defend against long- or short-range missiles, and have shot down a few test targets. However, at this writing no high-energy lasers have been deployed for actual weapon use.

The primary attractions of chemical lasers are their high-power output and remote operability anywhere their fuel can be delivered. The high-power output comes from the relatively efficient conversion of chemical energy to laser light, and flowing-gas designs that remove waste heat. Chemical fuels are portable, so they could be stowed on an aircraft or in an armored ground vehicle, avoiding the impractical need to connect to the power grid. However, some military officials do not want to deploy weapons that are useless without fresh supplies of special chemical fuels.

A number of chemical lasers have been demonstrated, but military developers have focused on two families that offer high power and beams that propagate reasonably well in the air and/or in space. These are the hydrogen fluoride/deuterium fluoride (HF/DF) laser and the chemical oxygen/iodine laser (COIL). Both have been used in major military test beds.

7.8.1 Hydrogen Fluoride (HF/DF) Lasers

Vibrationally excited hydrogen fluoride (HF) molecules emit on many lines between about 2.6 and 3.0 μm. HF lasers can generate very high powers, but air absorbs those infrared wavelengths, so they do not propagate well enough in the atmosphere to be used effectively as weapons. Replacing the normal hydrogen-1 isotope with the heavier isotope deuterium (hydrogen-2) shifts the laser wavelength to a band between 3.6 and 4.0 μm, which propagates well in the atmosphere. Deuterium fluoride (DF) lasers have been used in some important laser-weapon demonstrations. The 2.2-megawatt Mid Infrared Advanced Chemical Laser (MIRACL) at the White Sands Missile Range is a DF laser. So is THEL, the Tactical High-Energy Laser, which shot down mortar shells and rockets in a series of tests conducted by the United States and Israel.

The initial stage of an HF/DF laser is something like a rocket engine. A fuel containing hydrogen is burned with an oxidizer containing fluorine to produce the excited HF molecules. A number of different compounds are used in practice for ease and safety of handling. Excited HF molecules emerge from the combustion zone and pass quickly through the optical cavity of the laser, where they release their extra energy to produce the laser beam. The exhaust is then pumped from the laser and collected. Although the concept may sound simple, the engineering is tricky, like making rockets go up rather than blow up.

The actual chemical process is a chain reaction:

$$
\begin{gathered}
H_2 + F \rightarrow HF^* + H \\
H + F_2 \rightarrow HF^* + F \\
H_2 + F \rightarrow HF^* + H \\
H + F_2 \rightarrow HF^* + F \\
\text{ad infinitum}
\end{gathered}
$$

Each step produces a vibrationally excited HF molecule (HF^*), which emits an infrared photon, and a free atom of either hydrogen or fluorine, which initiates the next step in the chain reaction. The process continues as long as hydrogen and fluorine atoms remain to keep it going. The chemical laser can operate continuously if the gases flow rapidly through the laser and the laser cavity, as shown in Figure 7-10. Although the population inversion does not last very long in any particular group of gas molecules, the gas flows through the laser so fast that it does not matter—fresh gas flows into the laser cavity between the resonator optics, maintaining a steady population inversion that can produce a continuous beam.

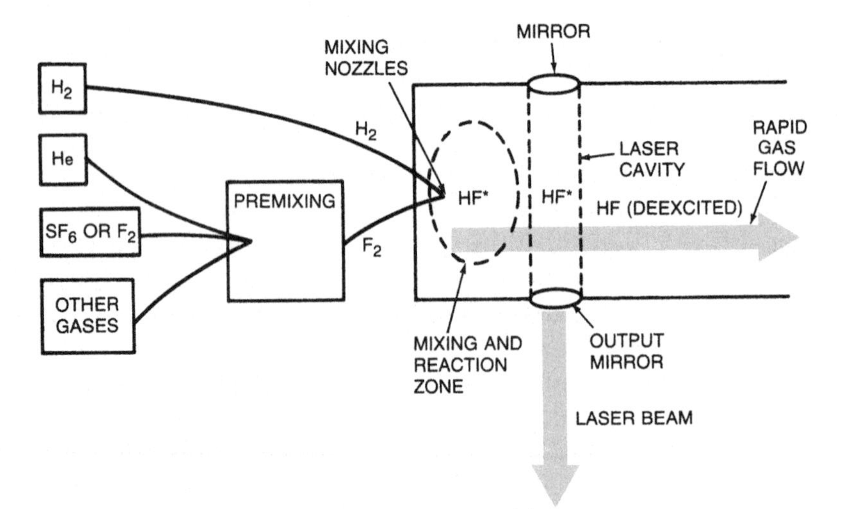

Figure 7-10. Laser beam is perpendicular to gas flow in an HF–DF chemical laser.

7.8.2 Chemical Oxygen–Iodine Lasers

The chemical oxygen–iodine laser (COIL) is at the heart of two major laser-weapon demonstrations: the megawatt-class Airborne Laser for defense against nuclear missiles, and the Advanced Tactical Laser to assess prospects for laser weapons on the battlefield. The crucial advantage of the COIL over a DF laser is a much shorter wavelength at 1.315 μm. The shorter wavelength can be focused to a small spot on the target using much smaller optics than required at the longer wavelength.

The COIL is a relatively complex system in which a chemical reaction generates excited oxygen molecules, which then excite iodine atoms to an excited electronic state that is the upper level of a 1.315-μm transition. In the Airborne Laser, a solution of hydrogen peroxide and potassium hydroxide dissolved in water is mixed with chlorine gas (Cl_2), causing a reaction that frees oxygen molecules excited to a long-lived high-energy state, and heats the gas. The expanding gas is then mixed with iodine molecules. Reactions with excited oxygen split the iodine molecules, freeing iodine atoms, which then are excited by other oxygen molecules, producing the population inversion that allows stimulated emission. The whole process occurs at low pressure, but the gas must flow quickly to remove waste energy.

High-power operation of the COIL built for the Airborne Laser has been demonstrated on the ground, but at this writing the laser has not been demonstrated in flight in the converted Boeing 747 aircraft that will carry it.

7.9 OTHER GAS LASERS

Many other gas lasers have been demonstrated in the laboratory, and some of them have been manufactured commercially. There is not enough room here to catalog them comprehensively, but a few types deserve brief mention.

7.9.1 Nitrogen Lasers

The 337-nm nitrogen laser resembles excimer lasers in having high gain and producing short ultraviolet pulses with high peak powers. It operates on a combined electronic–vibrational transition of molecular nitrogen (N_2), which is excited by a pulsed electric discharge at pressures from about 0.03 to one atmosphere. Excitation is efficient, and the gain is so high that it does not even require a rear cavity mirror (although adding one will double output power). However, the lower laser level has a 10-μs lifetime, so nitrogen accumulates in that state and quickly terminates the population inversion and laser pulse. This limits laser efficiency to 0.1% or less, pulse length to a few nanoseconds, and pulse energy to about 10 mJ.

Although these effects limit average power of nitrogen lasers, they are smaller, easier to use and very much less expensive than excimer lasers. Operation is also simpler because they use harmless nitrogen rather than the corrosive and hazardous halogens required for excimer lasers.

Another unique attraction of the nitrogen laser is that it is the easiest laser for an amateur scientist or student to build. Most gas lasers require very low pressure, hazardous gases, or precision mirror alignment. The nitrogen laser works best somewhat below atmospheric pressure, but it can operate at ambient pressure. It also requires a relatively simple power supply, and does not require glass-blowing skills. You can find plans on the Internet by searching for "home-built nitrogen laser" or "nitrogen laser kit."

7.9.2 Carbon Monoxide Lasers

The carbon monoxide (CO) laser is something of a less-successful cousin of the carbon dioxide laser. Emitting on vibrational–rotational transitions of CO, which are mostly between 5 and 6 μm, it can be even more efficient than the CO_2 laser. Like the CO_2 laser, the CO laser can emit powerful continuous beams. However, CO lasers have been plagued by serious practical problems, including strong absorption of some lines by the atmosphere, and the need to cool the gas below room temperature for efficient operation. It remains the most powerful laser source in its wavelength range.

7.9.3 Far-Infrared Lasers

The 10.6-μm CO_2 laser dominates the market for infrared lasers at wavelengths longer than about 5 μm. Many other molecular lasers have been demonstrated across the infrared spectrum, all the way to its long-wavelength end at 1000 μm, but none are as powerful, efficient, or practical as the CO_2.

This large laser gap is filled by a family often called *far-infrared lasers,* molecular gas lasers that are optically pumped, usually by light from CO_2 lasers, and emit laser beams between 30 and 1000 μm. The pump laser is tuned to a narrow band to raise the molecules to an excited vibrational level. The laser transition occurs between two rotational levels in the excited vibrational state. Typically, the pump beam enters at one end of the tube, and the far-infrared beam exits at the other end. The pump and output beams pass through holes in the cavity mirrors. These lasers are generally used in laboratory applications. If desired wavelengths cannot be produced by pumping with CO_2 lasers, other molecules can be pumped with less-efficient gas lasers emitting in the same region that are not readily available commercially, such as nitrous oxide (N_2O) lasers.

7.10 WHAT HAVE WE LEARNED?

- Gas lasers have been developed based on many different gases and gas mixtures. Each type has distinct characteristics.
- Sales of gas lasers are declining as semiconductor and solid-state lasers replace them for many purposes. However, annual sales remain over $1 billion.

- Most gas lasers can operate with sealed tubes, but some require flowing gas or regular replacement of the gas.
- An electric discharge passing the length of the laser tube excites most low-power lasers. Higher-power lasers are excited with a discharge transverse to the tube. Pulsed lasers can be powered by firing electrical pulses through the gas.
- Atomic gases emit on electronic transitions of neutral or ionized atoms.
- Molecular gases usually emit on vibrational or rotational transitions, but a few can emit on electronic transitions.
- Many gas lasers contain other gases in addition to the one that emits light.
- Gas flow through laser tubes removes contaminants and waste energy, and replenishes the laser gas.
- In many gas lasers, windows are mounted at Brewster's angle to minimize losses, and mirrors are separate from the tube. Brewster windows polarize the beam.
- Thermal motion of gas atoms causes Doppler broadening, which can dominate a gas laser's bandwidth.
- The red He–Ne laser remains the most common gas laser, but red semiconductor lasers have replaced it for many applications.
- In a He–Ne laser, the electric discharge excites helium atoms, which transfer energy to neon atoms, which emit the low-power laser beam. It has low gain.
- Coherence length is an important advantage of He–Ne lasers.
- Argon- and krypton-ion lasers emit shorter wavelengths and more powerful continuous beams then He–Nes. Both have many emission lines.
- He–Cd lasers emit milliwatts of blue and ultraviolet light. The tube must be heated to vaporize cadmium, which is solid at room temperature.
- Neutral metal atoms emit short repetitive pulses in copper-vapor lasers, which have very high gain.
- The CO_2 laser is the most versatile and powerful gas laser. emitting on vibrational transitions at 9 to 11 micrometers. Optics determine the oscillation wavelength.
- Sealed-tube CO_2 lasers can emit up to 100 W. Longitudinal-flow CO_2 lasers emit higher powers, and transverse-flow lasers produce powers to about 10 kW per meter of tube length. These lasers operate at low pressure.

- TEA CO_2 lasers emit high-power pulses when a transverse discharge excites gas at pressure near one atmosphere.
- CO_2 lasers are used in industry and medicine.
- Excimers are short-lived rare-gas halide molecules that dissociate when they emit ultraviolet photons and drop to a lower-energy state.
- Excimer lasers are the most powerful pulsed ultraviolet lasers; they are used in medicine and semiconductor fabrication.
- Chemical lasers get their energy from chemical reactions.
- The chemical oxygen–iodine laser (COIL) is being developed for use as a weapon.
- Nitrogen lasers have high gain and emit 337-nm pulses with high peak power, but have low average power and efficiency. They can be made to be compact and inexpensive.

WHAT'S NEXT?

In Chapter 8, we will learn about solid-state crystalline and glass lasers, which have different properties than gas lasers. We will cover semiconductor lasers in Chapter 9, and other lasers in Chapter 10.

QUIZ FOR CHAPTER 7

1. Which of the following general statements is not true about gas lasers?
   a. Most gas lasers are excited electrically.
   b. Only materials that are gaseous at room temperature can be used in gas lasers.
   c. The laser tube cannot contain atoms or molecules other than the one species that emits light.
   d. Gas lasers can emit continuous beams.
   e. Gas lasers can emit ultraviolet, visible, or infrared light.
2. Which of the following lasers can emit the shortest wavelength in a continuouswave beam?
   a. Helium–neon
   b. Helium–cadmium

c. Nitrogen
d. Argon–fluoride
e. Carbon dioxide

3. Which of the following lasers emits the shortest wavelength?
   a. Helium–cadmium
   b. Krypton–fluoride
   c. Nitrogen
   d. Argon–fluoride
   e. Argon-ion

4. Gas lasers have electronic transitions across approximately what wavelength range?
   a. 150–1500 nm
   b. 250–1500 nm
   c. 400–700 nm
   d. 400–5000 nm
   e. 150–15,000 nm

5. Why would an engineer select a red He–Ne laser over a red semiconductor laser?
   a. The He–Ne is smaller.
   b. The He–Ne uses power more efficiently.
   c. The He–Ne has better beam quality.
   d. The He–Ne costs less.
   e. Only if designing a retro optics lab.

6. The strongest emission lines of the argon-ion laser are at
   a. 275.4, 300.3, and 302.4 nm
   b. 488.0 and 514.5 nm
   c. 446 and 532 nm
   d. 632.8 and 1152 nm
   e. 647.1 and 799.3 nm

7. Application of a high voltage initially ionizes the gas in an argon laser. Then what happens?
   a. The laser emits intense light in the vacuum ultraviolet until the argon ions drop to the metastable upper laser level.
   b. Laser operation begins in the near ultraviolet, then voltage is reduced for laser operation at visible wavelengths.
   c. The voltage is increased further until the argon gas crosses the threshold for laser emission.

d. The voltage is decreased, and the resulting current produces laser emission.
e. The power supply burns out because argon-ion lasers do not require high voltage.

8. What is fundamentally different between a helium–cadmium laser and a He–Ne laser?
   a. He–Cd is pulsed; He–Ne emits continuous wave
   b. He–Cd emits a red beam; He–Ne emits a blue beam
   c. He–Cd oscillates on only one line; He–Ne can oscillate on several lines.
   d. He–Cd lasers are solid-state; He–Ne lasers are gas
   e. He–Cd lasers must be heated to generate light; He–Ne lasers do not require heating.

9. Carbon dioxide lasers are excited by
   a. Longitudinal electric discharge
   b. Transverse electric discharge at low pressure
   c. Transverse electric discharge at high pressure
   d. Transverse electric discharge in moving gas
   e. All of the above

10. Which gas laser can generate the highest power?
    a. Carbon dioxide
    b. Chemical oxygen–iodine
    c. Argon–fluoride
    d. Argon-ion
    e. Carbon monoxide

11. What happens when excimer molecules drop to the lower laser level?
    a. They emit light and dissociate because they are unstable in the ground state.
    b. They emit light and accumulate in the ground state, ending the population inversion so laser action stops.
    c. They absorb stimulated emission, reducing gain and terminating the laser pulse.
    d. They dissociate, then each of the two atoms that were in the molecule remains in an excited state to produce a population inversion.
    e. They react with each other and immediately return to the excited state.

12. Which of the following molecules is not used in an excimer laser?
    a. ArF
    b. HF
    c. KrF
    d. XeCl
    e. XeF

CHAPTER 8

SOLID-STATE AND FIBER LASERS

ABOUT THIS CHAPTER

In this chapter, you will learn about solid-state lasers, in which light is emitted by atoms embedded in a crystal, glass, or other transparent solid. After first explaining the general operation of solid-state lasers, this chapter describes the most important types, starting with the classic ruby laser, then covers neodymium and a family of related materials in which laser emission can be tuned across a range of wavelengths. The chapter also covers fiber lasers and amplifiers, in which the solid-state laser material is drawn into an optical fiber.

8.1 WHAT IS A SOLID-STATE LASER?

The first step in understanding solid-state lasers is to recognize that the laser community uses a different definition of "solid-state" than electronic engineers or physicists. Solid-state physics occurs in a solid. Solid-state circuits are semiconductor devices that conduct electricity and perform electronic operations. In electronics, semiconductors are considered solid-state devices in contrast to vacuum tubes, in which electrons move through a vacuum rather than through a solid. However, in the laser world semiconductor devices do not count as solid-state lasers because they operate in fundamentally different ways, as you will learn in Chapter 9.

The fundamental difference arises from the way the laser medium is excited. The solids in solid-state lasers transmit light that excites atoms contained within the solid to produce a population inversion on the laser transition. The light-absorbing atoms are dispersed in a solid that transmits the pump light that excites the atoms. In Maiman's first laser, the light-absorbing atoms were chromium dispersed in aluminum oxide (sapphire). Normally, the light-absorbing atoms are present only in small concentrations, added intentionally to an otherwise transparent solid as dopants. Solids transparent at optical wavelengths are insulators and conduct electrical current poorly. In contrast, semiconductor lasers are excited by an electrical current, as described in Chapter 9.

Chromium atoms both absorb and emit light in ruby, and give the crystal its red color. The simple design of Maiman's laser, shown in Figure 8-1, clearly illustrates the principles of solid-state laser operation. Light from the spring-shaped flashlamp illuminated the little ruby rod that it surrounded. Green and violet light passed through the red-colored ruby crystal and excited chromium atoms. Some chromium atoms then emitted red light, which stimulated other excited chromium atoms to release their excess energy as identical red photons. Thin metal films coated on the ends of the rod formed a reflective laser cavity, with the beam emerging through a small hole in the middle of the film on one end. The laser pulsed only during the flash of the flashlamp.

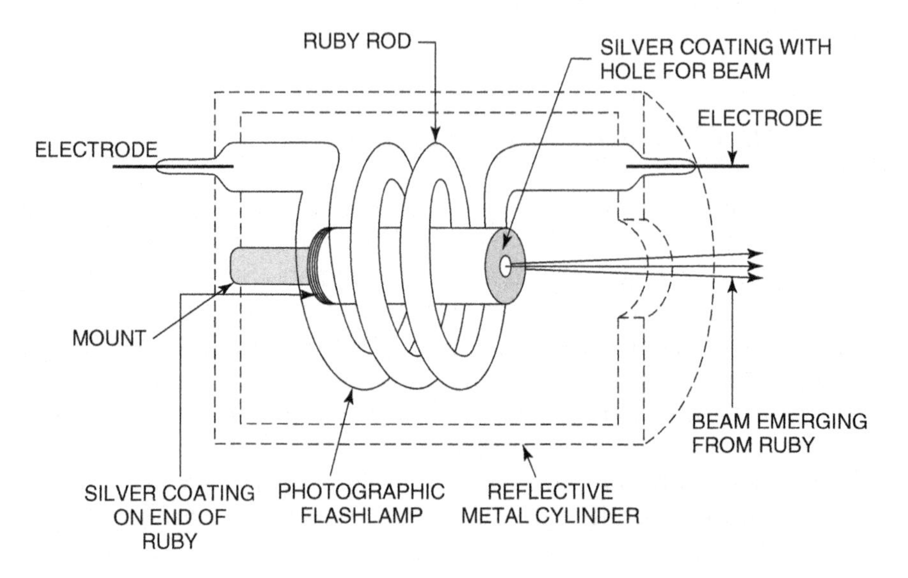

Figure 8-1. Structure of the first ruby laser.

Today's diverse family of solid-state lasers have been refined in many ways, but the same principles underlie their operation. Photons from an external source excite atoms dispersed in a solid host, producing a population inversion. Spontaneous emission triggers a cascade of stimulated emission, which oscillates between the mirrors in a laser cavity, and produces a beam.

Solid-state lasers can take various forms. Typically, the solid is shaped into a rod, but sometimes it may be a slab with mirrors arranged so light oscillates through the slab. The solid also may be in the form of an optical fiber, essentially a very long and very thin rod.

Solid-state lasers can be made from many different materials, although in practice most solid-state lasers are made of a few materials selected because they perform best as lasers. The lasers may be packaged with nonlinear devices that generate shorter-wavelength harmonics of the fundamental laser frequency. Solid-state laser materials can be used in laser amplifiers as well as in oscillators, particularly for optical amplification in fiber-optic communication systems, or to produce short high-energy pulses.

The technology has come a long way since Maiman's first laser. Sales of solid-state and fiber lasers were close to $1.2 billion dollars in 2006, according to *Laser Focus World*. That total is more than five times the dollar value of solid-states lasers sold in 1991, and that number is expected to continue rising.

Let us look first at some general issues of materials and optical pumping, then turn to specific types of solid-state lasers.

8.2 SOLID-STATE LASER MATERIALS

The two essential components of a solid-state laser material have specific names. The light-emitting atoms are the *active species* and the material in which they are embedded is the *host*. In the ruby laser, the chromium atoms that give the ruby its reddish color are the active species, and the host is the aluminum oxide or sapphire. Together, they are called ruby, but in most laser materials the active species is identified first, often by its chemical symbol, followed by the host material. For example, a Ti–sapphire (or Ti:sapphire) laser has titanium as the active species and sapphire as the host. Similarly, a laser in which neodymium is the active species and a crystal called yttrium aluminum garnet is the host is called Nd:YAG (or Nd–YAG).

Atoms of the active species are dispersed within the host, as shown in Figure 8-2. If the host is a crystal, the active species typically replaces one of the atoms in the host crystal with similar valence. Thus, chromium atoms occupy aluminum sites in ruby, where both elements have +3 valence. The active species is considered a *dopant,* because it is added in small quantities and replaces only a small fraction of atoms in the host.

Both the active species and the host are important in solid-state lasers. The active species determines the laser transition, but its interactions with the host may shift the wavelength slightly. Thermal and mechanical properties depend largely on the host material, which usually makes up 99% or more of the laser rod. Often, there are trade-offs in selecting hosts. For example, crystals such as yttrium–aluminum garnet have much better thermal properties than glass, but glass can be produced in larger sizes. Crystalline hosts are used most often, but Nd–glass is used in some special cases.

It takes considerable time and money to optimize the combination of an active species and a host for solid-state laser performance, so developers tend to stay with established combinations that are well characterized.

8.2.1 Active Species in Solid-State Lasers

The active species in a solid-state laser absorbs light that excites it to a higher energy level, which in turn populates the metastable upper laser level. Some solid-state laser materials include two dif-

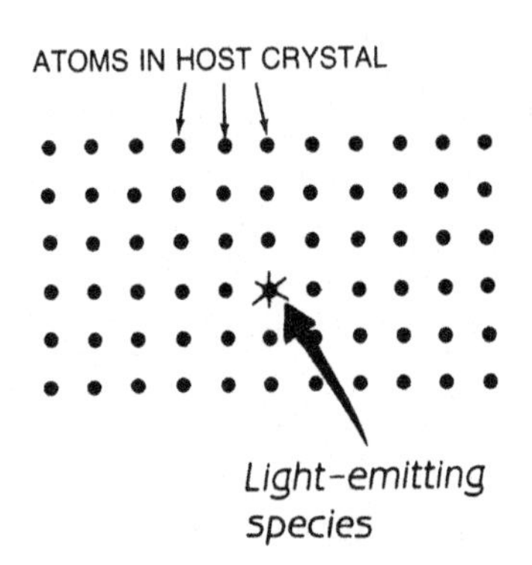

Figure 8-2. Light emitting atoms in a solid-state laser are embedded in a crystalline or glass host.

ferent species that play separate roles, with one absorbing pump light, then transferring energy to excite a second species to a metastable upper laser level. The most important active species are metals: chromium, neodymium, erbium, ytterbium, and titanium. Table 8-1 lists important commercial solid-state lasers using these materials.

These are far from the only metals used as active species in solid-state lasers. Solid-state cerium, cobalt, and holmium lasers also have been offered commercially, and solid-state laser action has been demonstrated in the laboratory from many other active species. However, the advantages of these other lasers have not overcome the high cost of developing new solid-state materials and optimizing their laser design.

The light-emitting species often are called "ions" because they have a nominal chemical valence in host crystals. The most common valence state is +3, including Cr^{+3}, Nd^{+3}, Er^{+3}, Yb^{+3}, and Ho^{+3}. These elements form covalent bonds in solid-state laser materials, so this ionization state is more nominal than real. The "ions" are fixed in the crystal, and the "missing" electrons are nearby, typically bonded with oxygen atoms.

Solid-state laser transitions are on electronic levels of the active species or ion. Valence bonds with the host material and interactions with adjacent atoms affect energy levels, changing laser

Table 8-1. Important commercial solid-state and fiber lasers

Output (nm)	Light emitter	Host	Pump source	Notes
675–1100	Titanium	Sapphire	Laser	Tunable
694	Chromium	Sapphire (ruby)	Flashlamp	Pulsed
701–826	Chromium	Alexandrite	Lamp	CW or pulsed
1064	Neodymium	YAG (yttrium–aluminum garnet)	Lamp, 808-nm laser	Pulsed or CW
1530–1570	Erbium	Glass, typically drawn into fiber	980 or 1480 nm laser	CW or amplifier
1047, 1054	Neodymium	YLF (yttrium lithium fluoride)	Lamp, 808-nm laser	Pulsed or CW
1064	Neodymium	YVO_4 (yttrium vanandate)	Laser	Pulsed or CW
1030–1100	Ytterbium	Glass rod or fiber	Laser	Pulsed or CW (fiber only)

wavelength. For example, neodymium emits at 1,054 nm when it is doped into phosphate-based glass, and at 1,064 nm in YAG (yttrium aluminum garnet).

Interactions between vibrational energy levels in the host and electronic energy levels in the active species can produce *vibronic* transitions, which change both electronic and vibrational energy levels. The electronic transition accounts for most of the energy, but the smaller vibrational transition adds or subtracts energy, and may allow wavelength tuning of the laser. The titanium–sapphire and alexandrite lasers are important vibronic lasers.

8.2.2 Host Materials

Host materials must meet stringent optical, thermal, and mechanical requirements. They must be reasonably transparent to the pump light, and absorb little light at the laser wavelength.

Thermal properties of the host are important because most of the pump energy that is not converted into laser light winds up as heat. Solids cannot dissipate heat as efficiently as gases, so heat removal is an important operating consideration.

Excessive heat can cause thermal and mechanical stress within a laser rod. Uneven distribution of heat can cause refractive-index differentials that degrade beam quality or reduce output power. In extreme cases, heat differentials can warp or crack the rod. Temperature increases also can affect energy-level populations, usually decreasing laser gain, efficiency, and output power.

Crystals with good thermal properties dissipate heat more efficiently than glass, so they can operate at higher powers and higher repetition rates than glass rods of equal size. However, there are trade-offs. Glass is easier to fabricate into optical fibers, which can dissipate heat efficiently because of their large surface area and relatively small volume. Glass also can be fabricated in much larger pieces than crystals, which must be grown under carefully controlled conditions to produce uniform crystals called *boules* without light-absorbing internal flaws.

A recently developed alternative is production of ceramic versions of laser crystals by heating and compressing powders until they solidify to make solid blocks of doped material, but this remains in development.

8.2.3 Laser Medium Configurations

The archetypical solid-state laser is a rod with the shape of a round pencil, but usually a little shorter and often thinner. Thin rods dissipate heat more readily than thick ones, but can generate an impressive amount of light.

One advantage of increasing rod size is that a larger volume of laser material can store a larger amount of energy that can be extracted in a single pulse. On the other hand, the amount of waste heat also increases with volume, but heat dissipation increases with surface area, so lasers with relatively large surface areas are better suited for high-power operation. Thus disks and slabs are better suited for high-power operation than fat rods, and fiber lasers dissipate heat quite efficiently.

Another way of increasing pulsed output is by shifting from a single laser oscillator to the master oscillator power amplifier (MOPA) configuration described in Section 6.1.3. By passing an oscillator pulse through a chain of amplifiers, as shown in Figure 8-3, a MOPA can build up powerful pulses. In the example, the amplifier volume increases along the line, allowing more energy storage. Power extraction efficiency can be increased by passing

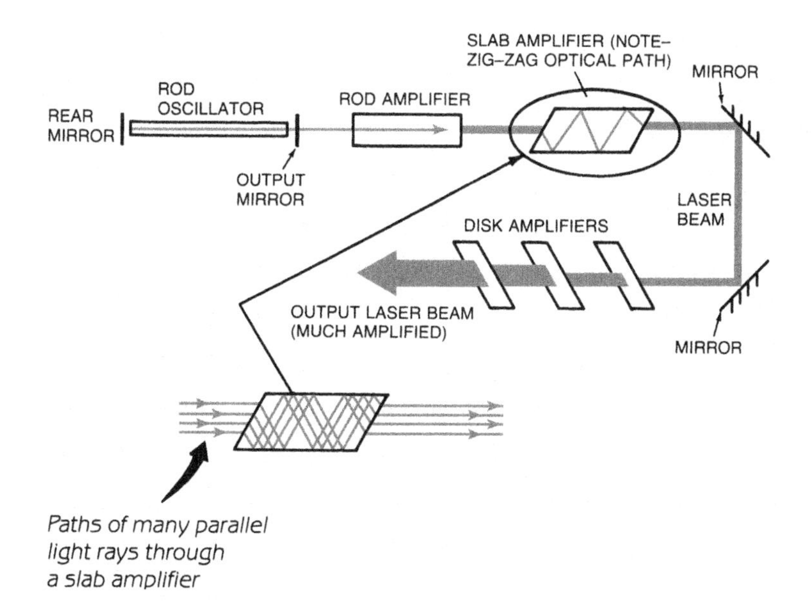

Figure 8-3. Laser oscillator with chain of slab and disk amplifiers.

the beam on a zigzag path through slab amplifiers to extract energy from a large fraction of the excited volume. Note that the path does not fold back on itself to create an oscillator. The final amplifier stages are large, flat disks that can generate large-diameter beams but still dissipate heat from its surface.

Reduce the diameter of a laser rod enough and you end up with an optical fiber. To make a fiber laser, you dope the active species into the light-guiding core of an optical fiber, which guides stimulated emission to the cavity mirrors at the ends of the fiber. The same type of fiber can be used as an optical amplifier by passing an optical input along the length of the fiber; such fiber amplifiers are widely used to amplify weak signals in long-distance telecommunication systems.

8.3 OPTICAL PUMPING

The only practical way to produce a population inversion in a solid that does not conduct electricity is to illuminate it with light, a technique called optical pumping. This section examines solid-state pump bands, and the light sources used to excite them.

8.3.1 Pump Bands and Absorption

Efficient optical pumping requires a light source that matches the absorption bands of the laser material. Figure 8-4 shows absorption in one of the most important solid-state lasers, neodymium–YAG. The sharp spikes are absorption on transitions between two sharply defined electronic energy levels. The broader bands show absorption on vibronic transitions.

Historically, the first optical pumping sources available for solid-state lasers were lamps. *Flashlamps* produce brief, intense flashes when an electrical pulse passes through a gas, and are widely used in photography. *Arc lamps* emit intense light continuously as a steady electric discharge flows through gas; applications include movie projectors. These lamps emit bright light at wavelengths absorbed by neodymium and other active species, but other wavelengths are wasted because they are not absorbed by the active species.

Alternatively, a laser can pump a solid-state laser at one of its strong absorption peaks, such as the 808-nm peak of Nd–YAG

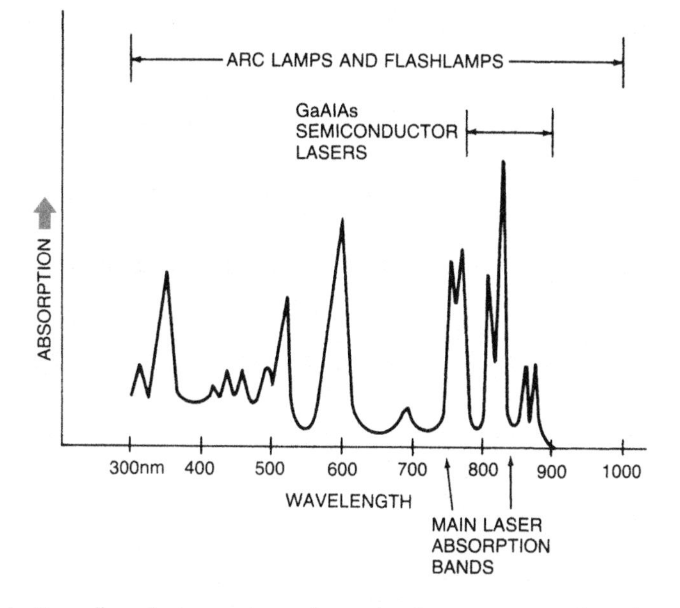

Figure 8-4. Broadband absorption of neodymium, compared with output of typical pump sources.

shown in Figure 8-4. Concentrating pump light on a strong absorption line makes laser pumping much more efficient than lamp pumping. Semiconductor diode laser pumps are ideal because they convert a large fraction of the input electrical energy into light. This means that a diode-pumped solid-state laser can have higher overall efficiency than a lamp-pumped laser, as long as the laser itself generates light efficiently. Some solid-state lasers also lack the broad absorption bands needed for lamp pumping, so they require laser pumping. One example is the titanium–sapphire laser described in Section 8.6.1. Let us look a bit closer at the options.

8.3.2 Flashlamp Pumping

Flashlamp pumping long dominated solid-state lasers, and remains common, but has been losing market share to diode-laser pumping. Its strength remains in producing powerful pulses.

The most common type of flashlamp used for optical pumping today is a long linear tube filled with a gas such as xenon. A brief high-voltage electrical pulse applied across the ends of the tube causes electrical breakdown of the gas, which conducts a current pulse that generates a brief flash. The whole process takes about a millisecond (0.001 second). The laser pulse produced inevitably is shorter than the lamp pulse because it takes time for the lamp power to exceed laser threshold, and is much shorter when the laser is equipped with pulse-shortening accessories such as the Q-switches described in Section 5.5.2. Pulse repetition rate is primarily limited by the switching electronics and the lamp.

Typically, the lamp is placed parallel to the laser rod inside a reflective cavity; Figure 8-5 shows a few variations. Some pump

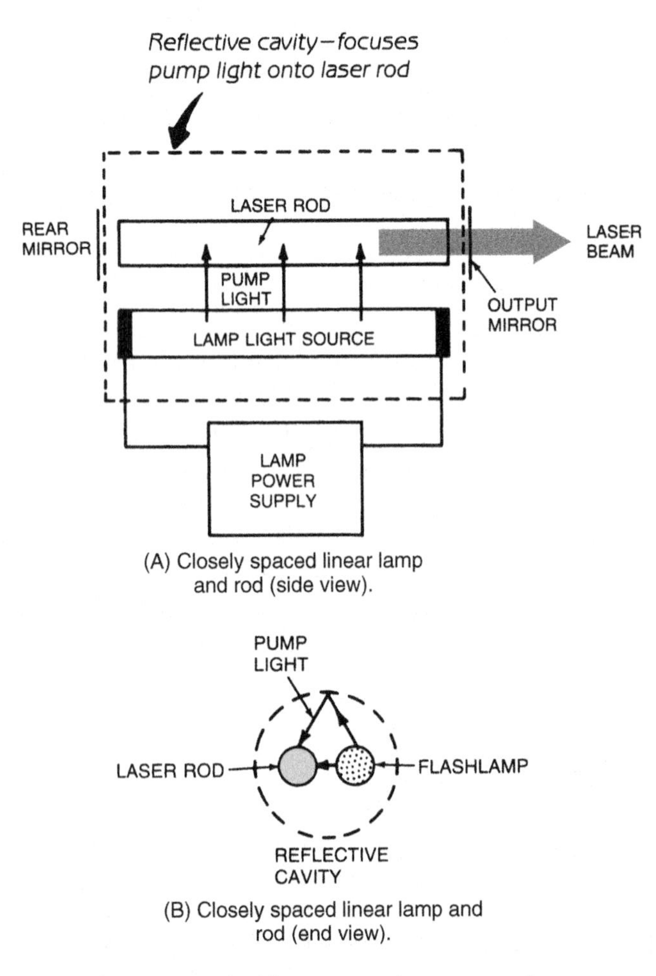

Figure 8-5. Flashlamp pumping geometries.

light travels directly from the lamp to the laser rod, and the remaining light is reflected from the inside of the reflective cavity toward the laser rod. Pumping is most efficient if the cavity is elliptical in cross section and the laser rod and the lamp are placed at the two foci, so a light ray from the lamp at one focus is reflected to the laser rod at the other focus. Alternatively, two lamps can be arranged on opposite sides of the laser rod in a dual elliptical cavity.

8.3.3 Arc Lamp Pumping

Continuous-wave solid-state lasers can be pumped with electric arc lamps, in which a steady electric current flows through a gas-filled tube, producing intense light. This light is bright enough to sustain a continuous population inversion in solid-state laser materials such as Nd–YAG, which are capable of continuous laser operation.

Solid-state lasers pumped with arc lamps can be pulsed with Q-switches, mode lockers, or cavity dumpers. This allows pulsing at higher repetition rates than is possible with flashlamps.

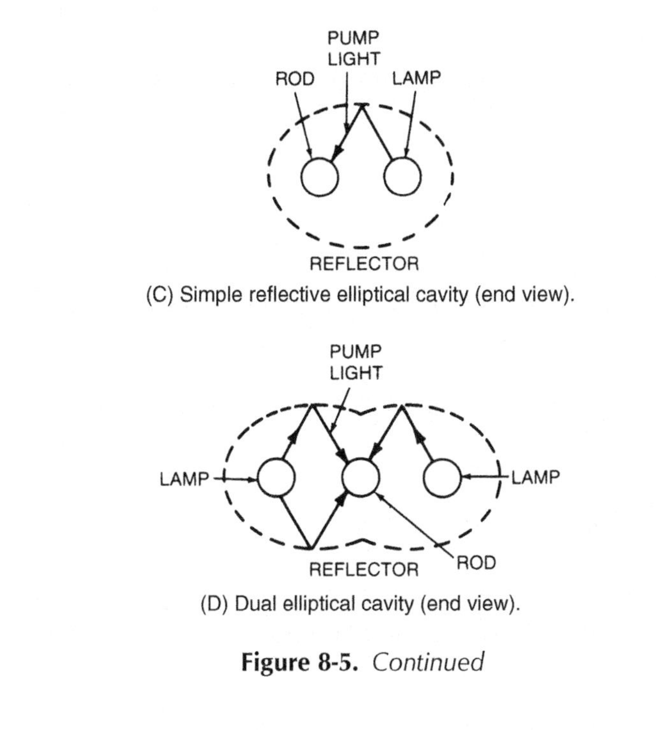

(C) Simple reflective elliptical cavity (end view).

(D) Dual elliptical cavity (end view).

Figure 8-5. *Continued*

8.3.4 Diode Laser Pumping

Just over half the 35,000 solid-state lasers sold in 2006 were pumped by semiconductor diode lasers, according to a *Laser Focus World* survey. Such *diode-pumped lasers* have gained that large market share because they are more compact and efficient than lamp-pumped solid-state lasers. However, diode-pumped lasers are expensive per watt, so diode pumping is less likely to be used at high powers.

Laser pumping is attractive when the wavelength of the pump laser matches an absorption peak of the active species in the laser rod, so a large fraction of the pump photons are absorbed. Diode pumping is particularly attractive because semiconductor lasers convert a large fraction of their electrical input into light. Combining the high efficiencies of diode emission and optical pumping means that the overall or wall-plug efficiency can be well over 10%, compared to 0.5 to 1% for flashlamp pumping.

An additional advantage of diode lasers is that their wavelength depends on the semiconductor composition, making it easier to match output to absorption lines. The standard pump line of neodymium lasers is 808 nm, readily available from gallium–aluminum arsenide (GaAlAs) lasers, as you will learn in Chapter 9. Pump diodes also can be matched to the absorption lines of erbium and ytterbium lasers.

Pump diodes typically illuminate laser rods along their sides, or are aimed along the length of the solid-state laser cavity. Figure 8-6 shows examples.

8.3.5 Other Laser Pumping

Other types of lasers can be used to pump solid-state lasers when no suitable semiconductor laser matches their pump bands. That is the case for titanium–sapphire lasers, where absorption near 500 nm in the green part of the spectrum, a wavelength available from frequency-doubled neodymium lasers and argon-ion lasers, but not from diodes.

8.4 RUBY LASERS

Ruby was the first laser material, but it is far from ideal. It is a three-level laser system, with the ground state also being the lower laser level, so efficiency normally is lower than in four-level lasers such

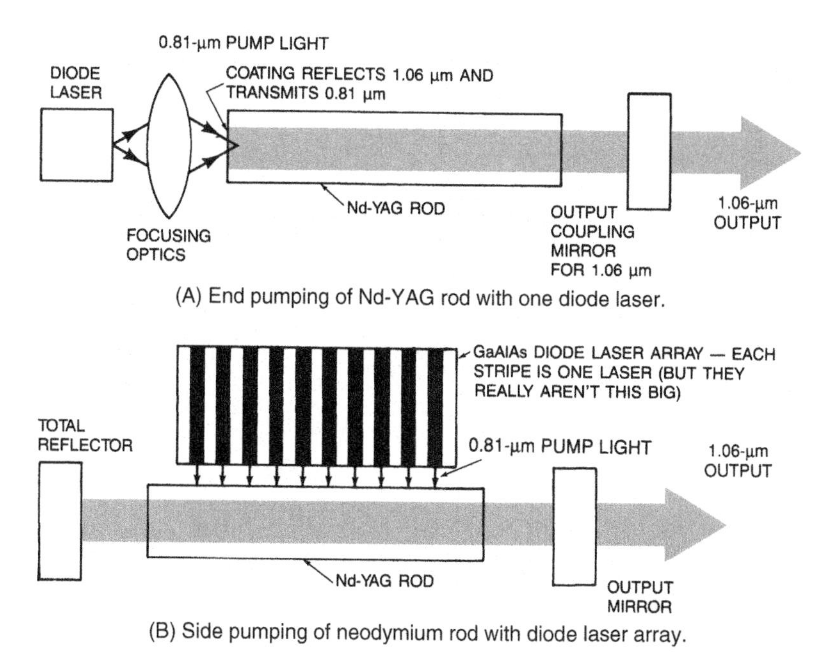

Figure 8-6. Diode-pumping of solid-state lasers.

as neodymium. Ruby is limited to pulsed operation at low repetition rates, but it is of historic importance and continues to be used in a few applications that require its high-power red pulses.

8.4.1 Ruby Laser Medium and Physics

Natural ruby is a gemstone, but ruby lasers use a synthetic ruby made by doping aluminum oxide with 0.01 to 0.5 percent chromium and growing crystals. The resulting ruby crystals are pink or reddish.

Figure 8-7 shows ruby-laser energy levels. Ground-state chromium absorbs light in bands centered near 400 and 550 nm, well matched to xenon flashlamps. Chromium atoms in the upper excited state drop to a pair of closely spaced metastable levels after about 100 nanoseconds, releasing energy to the crystalline lattice. The metastable state lifetimes are three milliseconds at room temperature, long enough for stimulated emission to depopulate them. Normally, emission is concentrated on the lower-energy laser line at 694.3 nm.

Chromium atoms in the ground state can absorb stimulated emission because they are at the lower level of the laser transition.

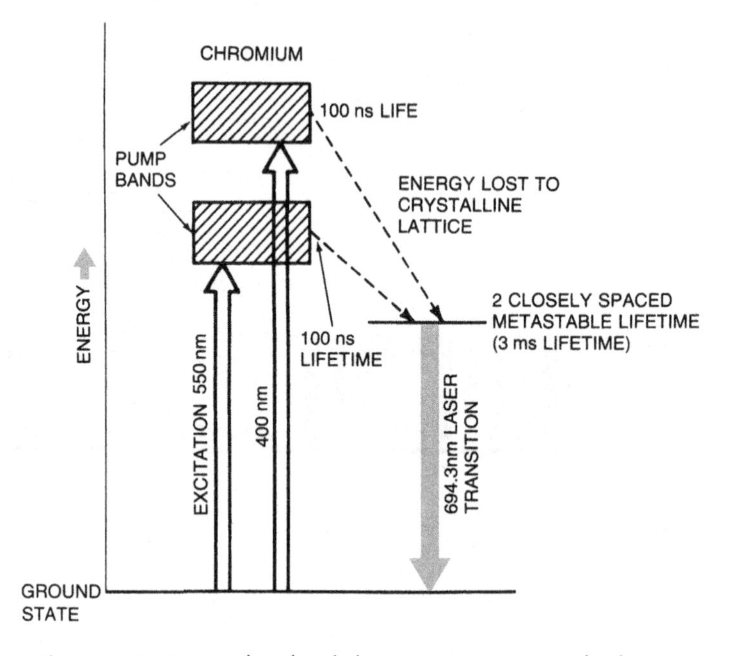

Figure 8-7. Energy levels of chromium atoms in ruby lase.

Atoms that drop to the ground state during the flashlamp pulse are quickly reexcited, maintaining the population inversion and laser action for around a millisecond. Once the flash stops, chromium atoms accumulate in the ground state and absorb stimulated emission, ending the laser pulse.

The long lifetime of the upper laser level allows a ruby rod to store energy effectively, and ruby lasers can be Q-switched to produce short pulses with energies of a few joules. Ruby can also be used in oscillator amplifiers.

Ruby is limited to a wall-plug efficiency below 1%, but the crystal conducts heat well, and resists damage from excess optical energy as long as its surface is kept clean of dirt that could absorb laser energy. The laser properties of a three-level system degrade rapidly as temperature increases, so ruby must be operated at no more than a few pulses per second to prevent heat build-up.

8.4.2 Ruby Laser Operation

Ruby lasers typically use rods 3 to 25 mm (⅛ to 1 inch) in diameter, and up to about 20 cm (8 inches) long. Energies of multimode pulses can reach 100 joules, but pulse energy typically is much lower, especially if oscillation is limited to TEM_{00} mode. Q-switching limits pulse duration to 10 to 35 nanoseconds and pulse energies to a few joules, but yields peak power in the 100-megawatt range.

Ruby lasers can be operated in a dual-pulse mode for holographic measurements by controlling a Q-switch so it produces two short pulses during a single long flashlamp pulse. The two pulses record a double exposure on the same holographic film, which can show small changes during the brief interval between pulses.

8.5 NEODYMIUM LASERS

The most common solid-state lasers use neodymium atoms in various solid hosts, based on the four-level energy-level structure shown in Figure 8-8. The most widely quoted value for the laser transition is 1064 nm, which is the value in Nd–YAG. The figure rounds the wavelength to 1.06 μm as a reminder that interactions between neodymium atoms and various hosts can change the

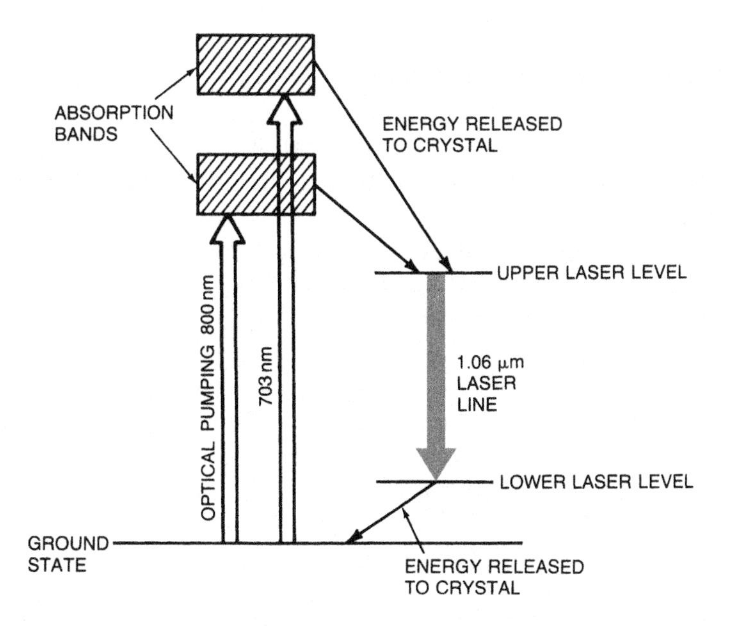

Figure 8-8. Major energy levels in 1.06-micrometer neodymium laser.

wavelength by about 1%. The figure does not show a number of secondary laser transitions that also occur in neodymium but are much weaker than the main line.

The two primary pump bands shown are in the 700–850 nm range, which raise neodymium atoms to one of two broad high-energy bands. The detailed absorption spectrum is more complex, as was shown in Figure 8-4, with a strong peak at 808 nm. Lamp pumping uses the broad absorption bands; diode-laser pumping uses the 808-nm line.

The optically excited neodymium atoms quickly drop to the metastable upper laser level, releasing energy to the crystalline lattice. When they drop from the metastable state, they emit on the strong 1.06-micrometer laser transition and drop to a short-lived lower laser level, then to the ground state.

8.5.1 Neodymium Hosts

The most common neodymium host is yttrium–aluminum garnet or YAG, a hard, brittle crystal with the chemical formula $Y_3Al_5O_{12}$. Dozens of other crystals have been tested as neodymium hosts; two that are often used are yttrium lithium fluoride (YLF) and yttrium vanadate (YVO_4). Silicate and phosphate glasses also are useful hosts for neodymium.

Neodymium atoms replace some yttrium atoms in YAG. Typical YAG rods are cylinders 6 to 9 mm (0.24–0.35 inches) in diameter and up to 10 cm (4 inches) long, cut from single-crystal boules. The crystal has good thermal, optical and mechanical properties, but its growth is slow and difficult, limiting sizes. Fusing powdered Nd–YAG under pressure makes a transparent ceramic with crystal-like properties used for making large slabs.

Nd–YAG emits most strongly at 1064 nm, and can emit about 20% as much power at 1318 nm and significant power at 946 nm. Packaging an Nd–YAG laser emitting on a strong line with harmonic generators produces the shorter wavelengths listed in Table 8-2. Continuous beams can generate the second harmonic, but producing the third, fourth, and fifth harmonics generally requires pulses with high peak power. The 532-nm green line and the 473 nm blue line are the most important visible lines.

The optical and thermal properties of Nd–YAG allow both continuous and pulsed laser operation. The small size of Nd–YAG rods and the optical properties of neodymium atoms limit energy

Table 8-2. Important Nd–YAG harmonics

Fundamental	1064 nm	946 nm
Second	532 nm	473 nm
Third	355 nm	
Fourth	266 nm	
Fifth	213 nm	

storage in a typical rod to about half a joule, far less than in a ruby rod. However, most of that energy can be removed from the rod in a Q-switched pulse, and the energy can be replenished quickly, in well under the millisecond duration of a flashlamp pulse, so a repetitively pulsed Nd–YAG rod can generate high average powers, as well as high peak power in Q-switched pulses lasting 10 to 20 nanoseconds.

Nd–YAG lasers can be quite powerful. The average power of repetitively pulsed lasers can exceed a kilowatt, although most operate at much lower powers. Power generally is lower in continuous lasers, and is lower at short-wavelength harmonics than at the fundamental wavelength.

Yttrium lithium fluoride ($YLiF_4$) does not conduct heat as well as YAG and is softer, but its refractive index changes less with temperature, and it suffers fewer heat-related problems. As in YAG, the neodymium atoms replace some yttrium atoms in the YLF crystal. Nd–YLF can store more energy than Nd–YAG, so it can generate higher-energy Q-switched pulses. The crystal is birefringent, so it generates two wavelengths, 1047 and 1053 nanometers, each with its own polarization orientation. Commercial versions operate pulsed or continuously.

Neodymium also can is doped into *yttrium vanadate* to form Nd–YVO_4, in which it also replaces some yttrium atoms. Nd–yttrium vanadate offers some advantages in energy storage, and is still in development.

8.5.2 Neodymium-Glass Lasers

Glass doped with neodymium also can serve as a solid-state laser medium. The wavelength emitted depends on the glass composition; it is 1,062 nm for silicate glass, 1,054 nm for phosphate glass, and 1,080 nm for fused silica. Its principal attraction is the well-developed technology for making glass with good optical quality in sizes much larger than available for crystalline laser hosts. Nd-glass lasers have a larger output bandwidth than crystalline lasers, allowing mode locking and generation of shorter optical pulses. Because it has lower gain than Nd–YAG, an equal volume of Nd–glass can store more energy than YAG, so the glass laser can generate higher-energy pulses.

The principal trade-off is that glass has poorer thermal characteristics than YAG, so glass lasers need more time to cool between pulses. Thus, glass laser oscillators cannot be pumped continuously, and normally operate at much lower repetition rates than Nd–YAG lasers. There is little market for glass oscillators.

However, the attractions of larger volumes and higher energy are compelling for use in giant research lasers, such as the National Ignition Facility, which is to be completed at the Lawrence Livermore National Laboratory in 2009. The system is a giant oscillator amplifier with 192 beam lines focused onto a single target to produce nuclear fusion. It contains more than 3000 slabs of neodymium glass, each weighing 42 kilograms and measuring 81 by 46 by 3.4 centimeters. Such big chunks of glass need a long time to cool down, so they can fire only a few shots a day, but they produce tremendous peak powers during their brief pulses.

8.5.3 Neodymium Laser Configurations

Neodymium laser rods are pumped with lamps or semiconductor diode lasers, as shown in Figures 8.5 and 8.6. Typically, lamp-pumped lasers have wall-plug efficiency in the 0.1–1% range, but diode-pumped lasers have much higher efficiency, sometimes exceeding 10%.

The high gain in neodymium lasers allows the use of stable or unstable resonators. A stable resonator can produce the standard Gaussian TEM_{00} beam with a bright central spot, but it does not extract laser energy from as much of the laser volume as an unstable resonator. The beams from unstable resonators differ from those of stable resonators in the near field, but in the far field both have bright central spots.

External amplifiers can boost the output power and energy produced by neodymium oscillators. Different stages do not need to have the same host material, but the amplification band must overlap the oscillator output. Thus, a Nd–YAG oscillator may be used with Nd-glass amplifier stages.

Neodymium lasers are versatile and can take a variety of forms. Some are made specifically for medical applications, materials working, or used in instrumentation or measurement. Others are made for general-purpose research. As a result, neodymium lasers can look strikingly different, ranging from a compact green laser pointer you can fit in your pocket to a massive drilling laser installed on a factory laser, yet Nd–YAG rods lie at the core of both.

The wide range of applications leads to different design choices. General-purpose laboratory lasers give the user many options, such as leaving room for the addition of accessories such as harmonic generators and Q-switches. Lasers designed for specific applications typically allow few modifications.

The smallest models can operate without active cooling, but larger types require either forced-air cooling (i.e., a fan), or flowing-water cooling to remove substantial amounts of waste heat. A 100-watt laser that is 1% efficient generates 9.9 kilowatts of waste heat!

8.5.4 Harmonic Generation and Wavelengths

The near-infrared wavelength of neodymium lasers is fine for some purposes, but visible or ultraviolet light is better for many others. Fortunately, neodymium lasers generate high enough powers that nonlinear harmonic generation can readily produce shorter wavelengths, as listed in Table 8-2 for commercial Nd–YAG lasers. YLF and YVO_4 lasers also can produce harmonics at similar wavelengths.

Frequency-doubling is the simplest form of harmonic generation, and can be done with continuous beams. Figure 8-9 compares the arrangement used to generate the 532-nm second-harmonic generation to the more complex configurations used to produce the third and fourth harmonics at 355 and 266 nm in the ultraviolet. Because the conversion is inevitably incomplete, some light remains at the fundamental 1064-nm wavelength and must be blocked or separated from the other wavelengths.

Some neodymium lasers are packaged with internal harmonic generators, so the packaged laser emits the harmonic wavelength rather than the fundamental, but descriptions may not clearly indicate the nature of the light source. The green laser pointers used by amateur astronomers to point out stars in the night sky are an

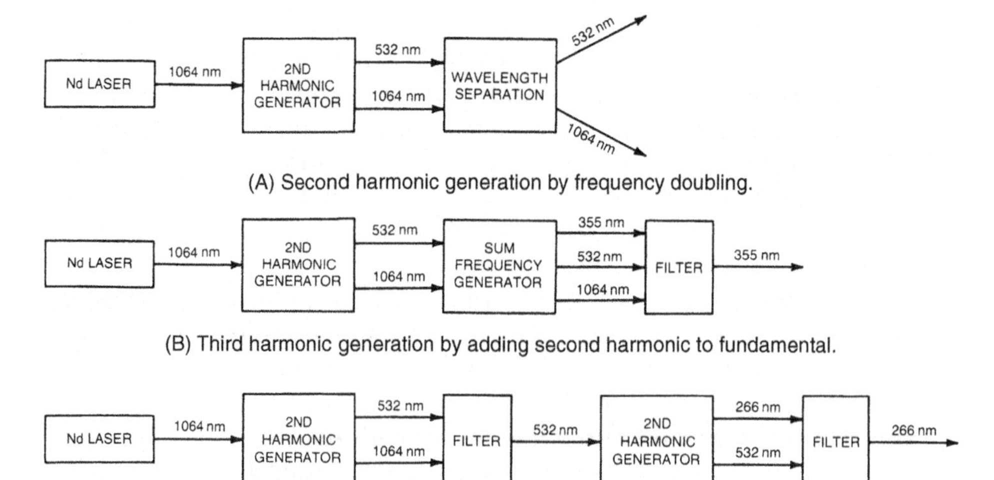

Figure 8-9. Generation of second, third, and fourth harmonics of a Nd-YAG laser.

example; they contain a neodymium laser and a harmonic generator that converts its 1064-nm output to 532 nm.

Sum-frequency generation also can convert near-infrared light from neodymium lasers into the visible and near ultraviolet. For example, combining the 1064 and 1342 nm Nd–YAG wavelengths in a suitable nonlinear material generates sum-frequency output at 593 nm. Recently, a variety of other nonlinear combinations have been used to generate visible beams from solid-state lasers. Often, the combinations used are not clearly identified, and some may actually be semiconductor diode lasers marketed as solid-state lasers.

8.5.5 Q-Switching and Pulse Length

Pulse duration of neodymium lasers depends on the pump source and pulse-control devices such as Q-switches. Without control devices, the pulse duration depends directly on the length of the pump pulses.

Q-switching of a flashlamp-pumped Nd–YAG laser generates pulses of 3 to 30 ns, with repetition rate equal that of the flashlamp. Q-switching of continuous Nd–YAG lasers generates much longer pulses, often hundreds of nanoseconds, because energy accumulates more slowly than in a pulsed laser. Cavity-dumped pulses also are shorter with flashlamp-pumped lasers than with continuous-wave models. (There is no difference for mode locking, which produces 30- to 200-picosecond pulses, depending on the range of wavelengths in the light.)

To provide this pulse control, Q-switches are often built into neodymium lasers, which are sold as short-pulse generators.

8.6 VIBRONIC AND TUNABLE SOLID-STATE LASERS

Vibronic lasers are a class of solid-state lasers in which the lower laser level is actually a band spanning a range of energy levels from atomic vibration in the solid-laser host, as shown in Figure 8-10. This makes it possible to tune their wavelength across a range determined by the width of the vibrational energy band. This is an important difference from solid-state lasers such as ruby and neodymium, in which laser transitions are between single upper and lower laser levels, as shown in Figures 8-7 and 8.8.

The name *vibronic* comes from vibrational–electronic, because the laser transition can involve changes in both vibrational and electronic states. In many cases, the upper laser level is itself the bottom of a vibronic band created by vibrational sublevels of a

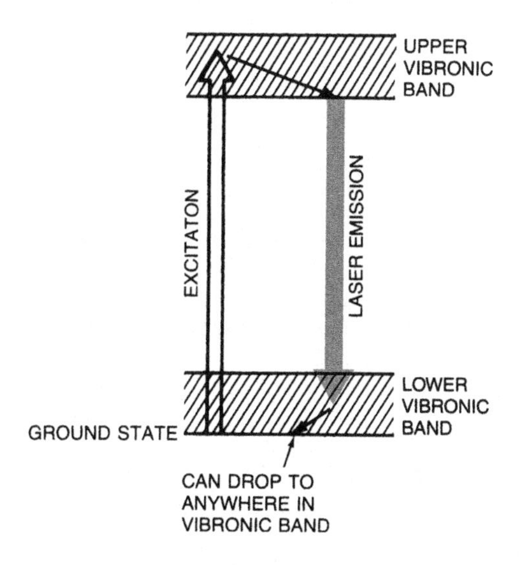

Figure 8-10. General energy level structure of a vibronic laser.

higher-energy electronic state. (Vibronic transitions also exist in gas molecules, although in that case the vibration is of atoms in the molecule relative to each other rather than of an atom relative to the adjacent atoms in a solid host.)

The excited atom can drop to anywhere within the lower vibronic band when it is stimulated to emit energy. However, the upper part of the lower level is less populated than the part of the lower level close to the ground state. With no wavelength-selective optics in the cavity, this causes gain to be stronger at the corresponding wavelengths. With suitable wavelength-tuning optics, vibronic laser emission can be tuned across a comparatively broad range of wavelengths, which can vary up to ± 20% from a central wavelength. Table 8-3 lists emission ranges and compositions of some vibronic lasers.

The broad gain bandwidth of vibronic lasers makes them suitable for both broadband and tunable output. As you learned in Section 6.4.2, broadband output is needed for short pulses, which have become a major application of titanium–sapphire lasers. Their broad gain bandwidth also gives vibronic lasers the ability to generate wavelengths not available from other solid-state lasers, and to tune the output wavelength for research and other applications.

Aside from their broad bandwidth and tunability, vibronic lasers work in much the same way as other solid-state lasers, and sometimes can operate in the same laser cavities.

Table 8-3. Representative broadband vibronic lasers

Name	Composition	Approximate range (nm)
Alexandrite	Chromium-doped $BeAl_2O_4$	701–858*
Co–MgF_2	Cobalt-doped MgF_2	1600–2400†
Cr–Emerald	Chromium-doped $Be_3Al_2(SiO_3)_6$	729–842
Cr–Forsterite	Chromium-doped Mg_2SiO_4	1167–1345
Cr–GSGG	Chromium-doped $Gd_3Sc_2Ga_3O_{12}$	740–850
Cr–YAG	Cr^{4+}-doped YAG	1350–1650
Cr–ZnSe	Cr^{2+}-doped ZnSe	2000–3100
LiCAF	Cr-doped $LiCaAlF_6$	720–840
LiSAF	Cr-doped $LiSrAlF_6$	780–1010
Thulium–YAG	Thulium-doped YAG	1870–2160
Ti–sapphire	Titanium-doped Al_2O_3	675–1100

*To 858 nm at elevated temperatures; only to 826 nm at room temperature.
†Operates at cryogenic temperatures.

8.6.1 Titanium–Sapphire Lasers

Titanium-doped sapphire has two major attractions for solid-state lasers: broad emission bandwidth and good material characteristics. The broad bandwidth makes Ti–sapphire suitable for two distinct types of applications described in Section 6.4: generating optical pulses shorter than one picosecond (10^{-12} s) and producing wavelength-tunable output. The characteristics and the convenience of solid-state lasers have made Ti–sapphire the laser of choice for these applications, and roughly 500 a year are sold, mostly for research.

The laser crystal contains about 0.1% titanium, which replaces aluminum in the Al_2O_3 crystal lattice. In that sense, it is similar to the ruby laser, in which chromium atoms replace aluminum in the sapphire lattice. The Ti^{+3} ion interacts strongly with the host crystal, and this combines with the structure of the titanium energy levels to make the bandwidth of the laser transition the broadest of any solid-state laser, from 660 nm in the red to 1100 nm in the near infrared. The peak output is at 700 to 900 nm.

No single laser emits over that entire range. Tunable emission usually is limited to a range of 100 to 300 nm with any particular set of cavity optics; changing cavity optics gives access to other parts of the wavelength range, as shown in Figure 8-11. Normally, tunable models are offered with multiple sets of optics with ranges that overlap at their edges, or with broadband optics centered near the 800-nm gain peak. Harmonic generation is possible, with pulsed Ti–sapphire lasers offering the second harmonic tunable at 350–470 nm, the third harmonic tunable at 235–300 nm, and the fourth harmonic tunable near 210 nm.

Picosecond and femtosecond Ti–sapphire lasers are designed to oscillate broadband, and are mode locked to concentrate their power into a series of short pulses spaced at regular intervals. As you learned in Section 5.5.4, mode locking produces short pulses by locking together all cavity modes, with the pulse width decreasing as the spectral bandwidth increases. The bandwidth of a Ti–sapphire laser is broad enough to allow generation of pulses in the 10-fs range. External optical devices can further extend the bandwidth to allow generation of pulses in the 5-fs range, but the arrangements are complex. Ti–sapphire pulses can be amplified in external amplifiers, and the chirped-pulse amplification technique described in Section 6.1.4 permits peak powers to reach very high levels, albeit for a very short time.

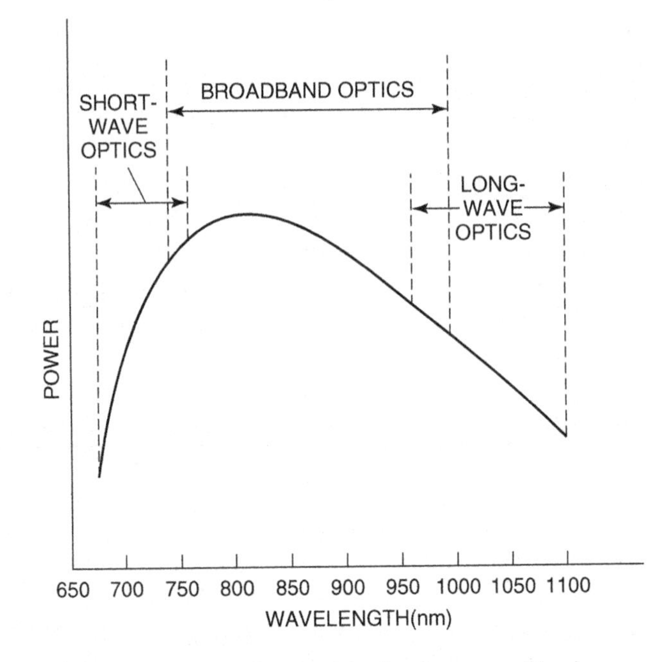

Figure 8-11. Output of a tunable titanium–sapphire laser.

Crystal growth is relatively easy, but Ti–sapphire has such high gain that most lasers use small crystals. The crystal's excellent optical and mechanical properties allow Ti–sapphire lasers to produce repetitive pulses or continuous beams.

One practical limitation of Ti–sapphire lasers is the difficulty of pumping the peak absorption band centered at 500 nm. The upper state has a 3.2-μs lifetime, too short for flashlamp pumping, and powerful diode lasers are not available in the pump band. Generally, pumping is with frequency-doubled neodymium lasers and argon-ion lasers, with wavelengths close to the 500-nm absorption peak.

8.6.2 Alexandrite Lasers

The first vibronic laser to be developed commercially was alexandrite. It is now used in medicine and cosmetic treatment as well as in research.

The light-emitting species is chromium, added to $BeAl_2O_4$ (a mineral known as alexandrite) in concentrations of about 0.01 to 0.4%. Its energy levels are similar to those of the ruby laser, as shown in Figure 8-12. The 380 to 630 nm pump bands in alexandrite are similar to those in ruby, but alexandrite has a vibronic band in its ground state that is absent in ruby.

Alexandrite can lase on a weak 680.4-nm fixed-wavelength transition to the ground state, but it is not as efficient as the equivalent 694.3-nm ruby transition. Its strongest room-temperature emission is on vibronic transitions at 700 to 830 nm. The pump bands of alexandrite extend far enough into the red for pumping by red diode lasers as well as lamps. However, red diode lasers do not generate light as efficiently as the near-infrared 808-nm diodes used to pump neodymium lasers.

Alexandrite has peculiar kinetics because two electronically excited states together function as the upper laser level. One is the bottom of a vibronic band of energy levels, the other is a fixed state with longer lifetime and only slightly less energy. This combination makes alexandrite gain increase with temperature, unlike most other lasers in which gain drops as temperature rises.

Temperature also affects wavelength. After stimulated emission, chromium atoms drop to one of the vibrationally excited substates of the ground state, then release the remaining energy by exciting a vibration in the crystal lattice. Rising temperature increases the steady-state population of the lowest vibrational sub-

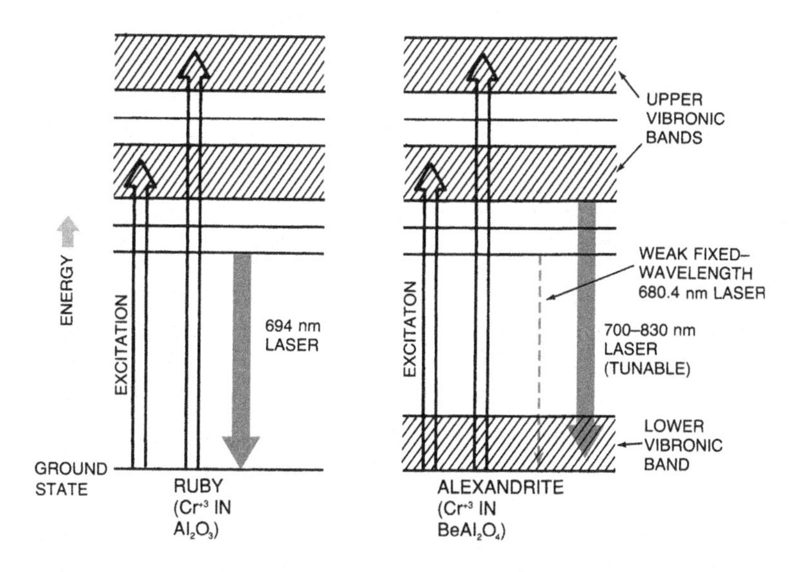

Figure 8-12. Comparison of ruby and alexandrite energy levels.

levels, making it harder to invert the population between these levels and the upper laser level. Because these low-lying levels correspond to the higher-energy end of alexandrite's tuning range, rising temperatures shift the gain toward longer wavelengths, so heated alexandrite lasers can operate at wavelengths to 858 nm.

Alexandrite has lower gain than neodymium lasers, making cavity design more difficult, but allowing alexandrite to store more energy. Like Nd–YAG, alexandrite can operate pulsed or continuous wave. In pulsed operation, average powers can reach 100 watts, lower than the most powerful Nd–YAG lasers, but still among the more powerful available lasers. Harmonic generation can produce tunable light at 360 to 400 nm, 240 to 270 nm, and 190 to 200 nm.

8.6.3 Other Vibronic Lasers

The most important of the other vibronic lasers listed in Table 8-3 is chromium-doped lithium-aluminum-strontium fluoride ($Cr^{3+}:LiSrAlF_6$), usually abbreviated Cr-LiSAF. It is tunable between 780 and 1010 nm, a broad enough range to produce pulses lasting tens of femtoseconds. The bandwidth does not quite match that of Ti–sapphire, but Cr-LiSAF has a broad absorption band in the red, so it can be pumped by 670-nm diode lasers instead of the green lasers required for Ti–sapphire. When pulsed, it can generate harmonics in the blue and ultraviolet.

Many vibronic lasers have been demonstrated in the laboratory, but have yet to find practical use. Those listed in Table 8-3 are only a sampling of the types demonstrated.

8.6.4 Tunable Color-Center Lasers

Another type of tunable solid-state laser is the color-center laser, in which the active media is crystal doped with impurities that introduce flaws in the crystalline lattice. The atoms at the flaw or "color center" absorb and emit light as they change position in the crystal. Some energy is converted into strain and vibration during the transition, allowing the color center to both absorb and emit light over a broad range of wavelengths. In practice, color-center lasers must be pumped with laser light and cooled to low temperatures, but they have gain over a broad range of wavelengths, so they can be made tunable. Their main applications have been for research and measurement in the near infrared, but they are little used because of the need for a costly pump laser and cryogenic cooling.

8.7 ERBIUM AND OTHER EYE-SAFE LASERS

Some solid-state lasers are grouped together because they emit at wavelengths longer than 1400 nm. Lasers in that range are less dangerous to the human eye than lasers emitting at 400 to 1400 nm because the fluid that fills the eye absorbs the longer wavelengths before they can reach the light-sensitive retina at the back of the eye.

Eye safety is an important consideration for military and civilian lasers used in the open air. The civilian systems are used largely for atmospheric measurements. Armies use laser range finders and target designators to locate and guide weapons to enemy targets, and need eye-safe versions of that equipment to use in training exercises with friendly troops. Eye-safe lasers are also used for some other applications, particularly medical treatment, which requires laser wavelengths longer than 1400 nm.

A few eye-safe lasers deserve particular mention:

- *Erbium–Glass Lasers.* Erbium-doped glass lasers emit pulses near 1540 nm, a wavelength for which reasonably good detectors are available. The lasers contain erbium-doped glass either in the form of rods or as optical fibers. The rod lasers are similar to neodymium lasers but lower in power; erbium fiber lasers and the 1540-nm erbium laser line are described in Section 8.8 below. Ytterbium may be added to improve energy absorption when the material is lamp-pumped.
- *Erbium–YAG Lasers.* Erbium-doped YAG lasers emit near 2900 nm, an eye-safe wavelength which hard tissue absorbs so strongly that the laser beams can cut bone and drill teeth.
- *Holmium Lasers.* Holmium is rare-earth element similar to erbium and neodymium that emits at a number of eye-safe wavelengths in crystals such as YAG and YLF. Its strongest wavelength is near 2100 nm.
- *Wavelength-Shifted Neodymium Lasers.* Another option is nonlinear Raman shifting of the wavelength of Nd–YAG lasers to longer eye-safe wavelengths. For example, passing Nd–YAG

pulses through a cell containing pure methane at high pressure can shift the 1064 nm Nd–YAG output to 1540 nm. This approach can produce dual-wavelength lasers with output at both the primary 1064-nm Nd–YAG line and the eye-safe 1540 nm line.

8.8 RARE-EARTH-DOPED FIBER LASERS

So far we have described solid-state lasers in which the laser medium is a bulk material, usually a rod, but sometimes a slab or parallelepiped. The fastest-growing part of the solid-state laser market in recent years has been fiber lasers, in which the active species is a rare earth element doped into the light-guiding core of an optical fiber. The laser is pumped optically from one or both ends or the side, almost always with diode lasers. The three most important rare-earth fiber lasers use ytterbium, erbium, and thulium.

The fiber configuration concentrates both pump light and stimulated emission in the volume that contains the laser species. By enhancing both excitation efficiency and the production of stimulated emission, this makes fibers attractive both as optical amplifiers and as laser oscillators. For lasers, the long, thin fiber offers a large surface area per unit volume, so it can dissipate heat more readily than a bulk rod, making it possible to sustain kilowatt powers. For communication systems, a fiber amplifier is highly efficient, can be low in noise, and is the right size and shape to couple energy to and from the transmission fibers that carry the signals. Section 8.9 covers fiber amplifiers; this section concentrates on fiber lasers.

8.8.1 Fiber Laser Structure

As you learned in Section 5.1.8, an optical fiber consists of an inner light-guiding core surrounded by a transparent cladding layer with lower refractive index. The refractive-index difference can be small—typically 1% and often less—but that small difference is enough to confine light directed within about 7 degrees of the axis of the fiber.

This simple structure can guide a pump beam directed into the fiber core as well as stimulated emission produced by the active species in the core. It is widely used in fiber amplifiers and can be used in low-power fiber lasers. However, a more elaborate dual-cladding structure is needed for higher-power fiber lasers. As shown in Figure 8-13, a small inner core contains the laser species and guides the stimulated emission. It is surrounded by a larger outer core with slightly lower refractive index, which collects the pump light at the end of the fiber and guides it along the fiber. Total internal reflection at the boundary between outer core and the surrounding cladding bounces the pump light back and forth through the inner core where it excites the active species. The drawing shows the outer core as having a simple shape but, in practice, the core has a more elaborate shape that reflects the pump light at varying angles so it passes repeatedly through the inner core as it travels along the fiber. Typically, the laser cavity is formed by cavity mirrors at the ends of the doped fiber, which transmit pump light into the laser but reflect the stimulated emission.

An alternative approach used at even higher powers is to connect a number of pump diode lasers to separate fibers that deliver pump light to an optical coupler at the end of the fiber, as shown in Figure 8-14. The coupler combines the inputs and directs the pump light into the outer core that guides pump light through the

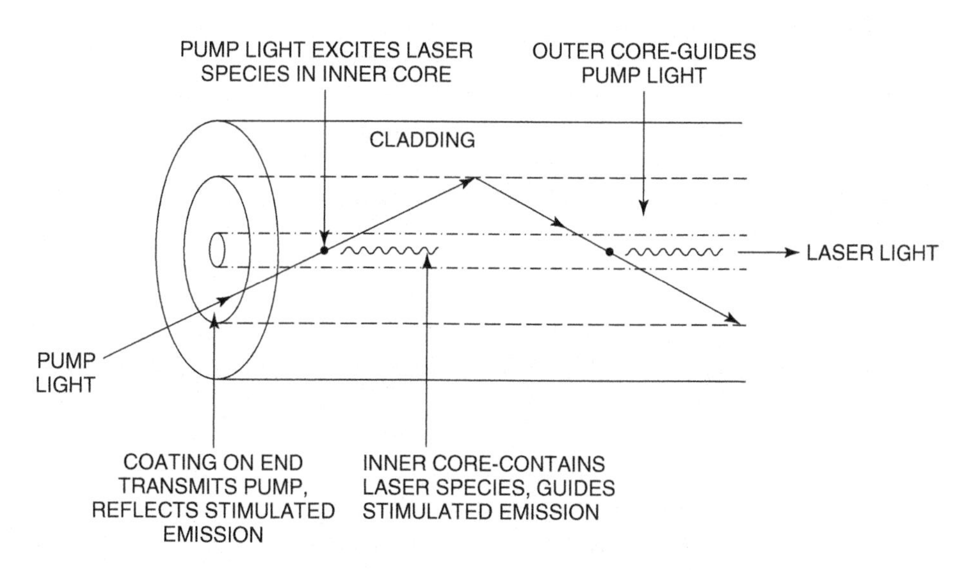

Figure 8-13. Internal structure of dual-core fiber laser.

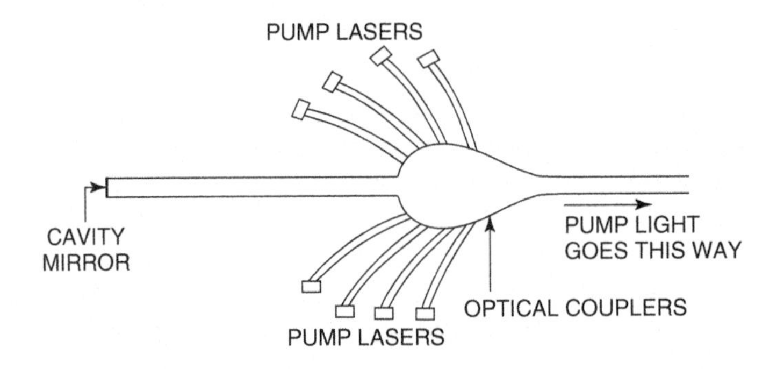

Figure 8-14. Pumping a fiber laser through an optical coupler spliced into the fiber.

laser section of the fiber. In the illustration, the pump light goes to the right; another set of pump lasers could be added on the other end of the fiber laser. In practice, the pump lasers actually are arrays of many separate semiconductor lasers with their emission collected and delivered through a single fiber to the fiber laser.

The relative size of the inner and outer core depend on the type of laser. The outer core should be large to collect and guide pump power; in some cases it approaches one millimeter in diameter. The inner core should be small to concentrate emission and produce a high-quality beam, but not so small that it concentrates power enough to cause optical damage. Ideally, the inner core should be smaller than 10 μm so it supports only a single transverse mode. However, optical tricks such as coiling the fiber can confine light in a single mode even if the inner core is as large as 40 μm. Continuous single-mode output has reached 2 kW in the laboratory, but commercial versions are now limited to about 1.5 kW in a single mode.

Operation at higher powers usually requires fibers with larger cores that emit multimode beams. Industrial fiber lasers can reach powers well above 10 kW by combining beams from several large-core multimode fibers. Their beam quality is adequate for materials working, for which beams are focused to a nearby target. Laboratory researchers are trying to combine outputs of multiple single-mode fiber lasers to produce a coherent multikilowatt beam that will retain good beam quality over longer distances for military applications, but this has yet to be demonstrated.

Pumping a fiber laser at a wavelength well matched to the active species absorption peak can produce extremely high efficiency. Eighty percent of the power from a 1-kW diode-laser pump can be converted into output energy in ytterbium-doped fiber, an incredible efficiency for a laser, leaving only 200 W of waste heat to be dissipated from the fiber. (The overall efficiency is much lower because the pump laser is much less efficient in converting electrical power to light energy.)

The principle of the fiber laser is quite general, and works particularly well when the inner core is doped with rare-earth elements such as ytterbium, erbium, and thulium. The host must be a glassy material that can be drawn into thin fibers, and typically is silicate glass. However, other glass-forming materials such as fluoride compounds are used in some fiber lasers and amplifiers. Operating characteristics depend largely on the choice of light-emitting species, and to a lesser extent on the composition of the glass fiber.

Like other solid-state lasers that emit high powers, fiber lasers can be packaged with harmonic generators or Raman shifters to generate additional wavelengths.

8.8.2 Fiber Laser Materials and Energy Levels

The rare-earth elements that are the active species in fiber lasers share a common type of energy-level structure, shown in Figure 8-15. An interaction called the Stark effect, which occurs between electrons and electric fields within the solid, splits atomic energy states into many closely spaced sublevels. Pump light ex-

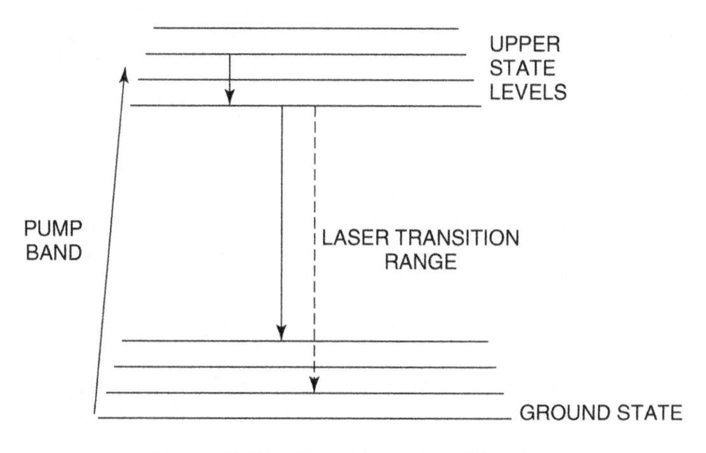

Figure 8-15. Transitions in a fiber laser.

cites the active species from the ground state into one of the sublevels in an electronically excited state; the presence of multiple sublevels allows pumping on different lines. Typically, the atom drops to a lower sublevel, and remains in that upper laser level until it is stimulated to emit light. Then the atom drops to a sublevel of the ground state, and ultimately returns to the ground state.

These interactions produce gain on transitions between sublevels of the upper state and sublevels of the ground state. If the transition drops to upper sublevels of the ground level, which are sparsely populated at normal temperatures, the system acts like a four-level laser. If the transition drops to a sublevel close to the ground state, which has a large population at normal temperatures, it acts like a three-level laser.

Such systems are called *quasi-three-level lasers* because they are not quite three-level or four-level lasers. Their gain is low when they act like three- or four-level lasers, but in between those extremes they have higher gain and attractive properties. Erbium and ytterbium are good examples.

One important consequence of this energy-level structure is that optical pumping can be quite efficient. If the pump excites the laser species directly to the top of the upper of the two bands shown in Figure 8-15, a large fraction of that energy will reappear in the stimulated emission photon. For example, erbium can be pumped at 1480 nm, and has peak gain around 1535 nm. Likewise, ytterbium can be pumped at 975 nm and oscillates around 1070 nm.

Another consequence is that gain is possible across a broad range of wavelengths covered by the spread among sublevels. Laser oscillation can be tuned to specific wavelengths in that range, or the laser cavity can be designed for broadband oscillation to produce short pulses.

8.8.3 Ytterbium-Doped Fiber Lasers

Ytterbium-doped fiber lasers are the most powerful type. They emit pulsed or continuously between 1030 and 1120 nm, with peak output at 1070 to 1080 nm. Typically, single semiconductor lasers or arrays pump the fiber from one or both ends. The main pump band is at 975 nm, but ytterbium also has other pump lines at 915 and 936 nm.

Ytterbium-doped fiber lasers can be used to produce tunable output or femtosecond pulses. However, their major application is for materials working. Ytterbium fiber lasers emit close to the 1064 nm line of neodymium lasers that were long standard for many such applications. They also are used in medical treatment, image recording, instrumentation, and research.

The main operational advantages of ytterbium-doped fiber lasers are high efficiency, high power, compact size, and ease of operation. Commercial versions are rated at wall-plug efficiency of 25%, and up to 75% of the pump energy can be converted to output light. That high efficiency reduces both power requirements and waste heat. Beam quality is quite good.

Continuous-wave powers range from 1 W to over 10 kW, with powers above 1 kW usually generated from one or more multimode fibers. Beam quality decreases at higher power levels, but is adequate for materials-working.

8.8.4 Erbium-Doped Fiber Lasers

Erbium-doped fiber lasers have the same type of laser transition as ytterbium, but at lower energy, corresponding to wavelengths of 1520 to 1620 nm. This range is particularly important for fiber-optic communications, in which erbium-doped fibers are used in optical amplifiers, described in Section 8.9.

Erbium has two primary pump lines, at 980 and 1480 nm. The 980-nm line excites the erbium atoms to a high energy state, which drops to the upper laser level. The 1480 nm line excites erbium directly to the upper laser level, but is not as widely used.

Erbium-fiber lasers cannot match the high power of ytterbium fiber lasers, but their broad bandwidth makes them attractive for generating ultrashort pulses. Their wavelength also makes them ideal for measuring the properties of fiber-optic systems. Their major applications are as repetitively pulsed tunable lasers, but they also are used in some industrial and atmospheric measurement applications because their wavelength is eye-safe.

8.8.5 Thulium-Doped Fiber Lasers

A third type of fiber laser offered commercially is the thulium-doped fiber laser, which emits from about 1750 to 2200 nm in the infrared. The power does not match ytterbium, but the wavelength is better for some applications. Examples include some medical procedures, measurements of pollution, and materials working on plastics that are transparent at shorter wavelengths.

8.9 RARE-EARTH-DOPED FIBER AMPLIFIERS

Rare-earth doped fibers are used in optical amplifiers as well as in lasers. Like the optical amplifiers described in Section 6.1.2, fiber amplifiers are gain media without resonant cavities, so stimulated emission amplifies light from an external source making a single pass through the fiber. The primary application of fiber amplifiers is to boost the strength of weak optical signals that have passed through tens of kilometers of optical fiber in a long-distance communication system. For this application, the fiber amplifier must amplify signals across a range of wavelengths accurately and uniformly, without adding noise and distortion. Erbium-doped fiber amplifiers turn out to work extremely well for this purpose, and their development launched a tremendous expansion of long-distance telephone and Internet networks. Understanding the design of these amplifiers requires a brief explanation of their application.

The first generation of fiber-optic communication networks built in the 1980s needed electrooptic repeaters to boost signal strength in order to span long distances. Before optical amplifiers were available, optical signals could only be amplified if they were converted into electronic form, amplified electronically, then converted back into light for transmission through the next length of fiber. The development of erbium-doped fiber amplifiers in the late 1980s led to dramatic advances in long-haul fiber-optic communications.

Erbium-doped fiber amplifiers can directly amplify optical signals at the 1550 nm wavelength where glass fibers are most transparent. Better yet, a single erbium fiber amplifier can simultaneously amplify light across a range of wavelengths from 1530 to 1570 nm, allowing separate signals to be sent at different wavelengths through the same fiber. The technique is called *wavelength division multiplexing* and works like transmitting radio signals through the air at frequencies far enough apart that they do not interfere with each other. Wavelength-division multiplexing multiplies the capacity of a single fiber; instead of transmitting one signal at 2.5 gigabits per second, the fiber could transmit 10 or 20, depending on how narrowly the spectrum could be sliced. The combination of fiber amplifiers and wavelength-division multiplexing spawned an explosive growth in long-haul fiber-optic transmission, the technology that today transmits telephone and Internet traffic around the globe.

Figure 8-16 shows operation of an erbium-doped fiber amplifier. Weak optical input enters one end of the erbium-doped fiber, and light from a pump laser at either 980 or 1480 nm is directed into the other end. The pump light produces a population inversion in erbium atoms, and the input signal stimulates emission from the excited erbium atoms. A coupler at the right end of the figure directs the output 1550-nm signal one way, and transmits pump light in the other direction. A filter on the left side blocks pump light from traveling backward through the fiber-optic system. Note that there are no mirrors. Erbium gain is not perfectly uniform across the 1530 to 1570 nm band, but optics can be added to maintain uniform signal strength at all wavelengths.

Erbium-doped fibers also can amplify light from 1570 to about 1620 nm, but because gain is weaker at those wavelengths, no single amplifier can provide the uniform gain across the whole erbium gain band from 1530 to 1620 nm.

Other types of doped fibers have been developed to amplify signals in other wavelength bands. However, so far there has been little demand for further expanding fiber transmission capacity.

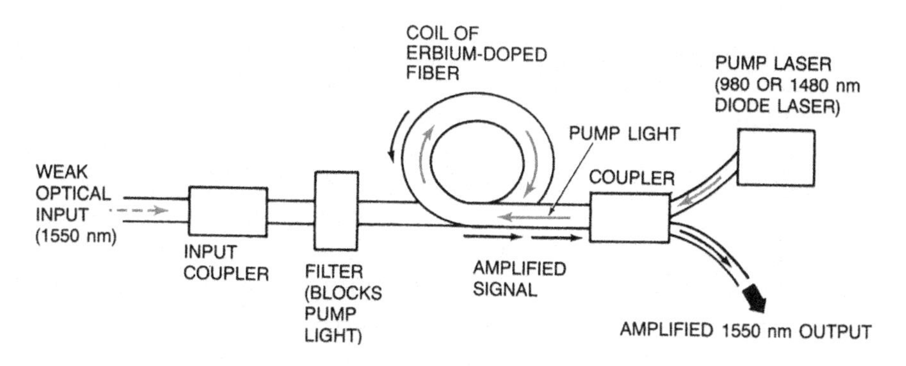

Figure 8-16. Erbium-doped fiber amplifier.

8.10 RAMAN FIBER LASERS AND AMPLIFIERS

A separate family of fiber lasers (and amplifiers) is based not on rare-earth doped fibers but on stimulated Raman scattering in ordinary fibers.

Raman scattering is a relatively weak process in which an atom absorbs a photon at one wavelength, then almost instantly emits a photon at a different wavelength, shifted by a change in vibrational energy. When a material is illuminated by intense light, many of its atoms are involved in this scattering process, during which they can be stimulated to emit a Raman-shifted photon by a photon at the Raman-shifted wavelength, a process called *stimulated Raman scattering* that was described in Section 6.2.9. Raman scattering is more likely to involve a loss of photon energy than a gain of photon energy so, in practice, stimulated Raman scattering produces light at a longer wavelength.

Stimulated Raman scattering can occur in any solid. It is normally unlikely, but the probability increases with the number of photons being scattered by the solid, that is, with the power density of the light being scattered, called the pump beam. The more photons are being scattered, the more likely atoms in the process of scattering light can be stimulated to emit light at the Raman-shifted wavelength.

The chance or *cross section* of stimulated Raman emission is very small in the silica (SiO_2) used in standard optical fibers. However, small-core single-mode fibers concentrate light in a small volume, and the light travels long distances through the fiber, increasing the chance for stimulated Raman scattering to occur. This allows amplification of light from an external source in the Raman gain band.

Raman scattering also can spontaneously generate light at a shifted wavelength. Put suitable mirrors on the ends of a fiber carrying high power at the pump wavelength, and the initial spontaneous emission can trigger a cascade of stimulated Raman emission, causing laser oscillation at the Raman-shifted wavelength.

Raman gain in silica peaks at a frequency 13 GHz lower than the pump laser frequency. Thus, a Raman fiber laser or amplifier effectively shifts the wavelength of the pump laser to a longer wavelength. Figure 8-17 shows how this works in a Raman fiber laser. Stimulated Raman scattering converts 1100 nm pump light to 1178 nm, which oscillates between the mirrors at the ends of the fiber.

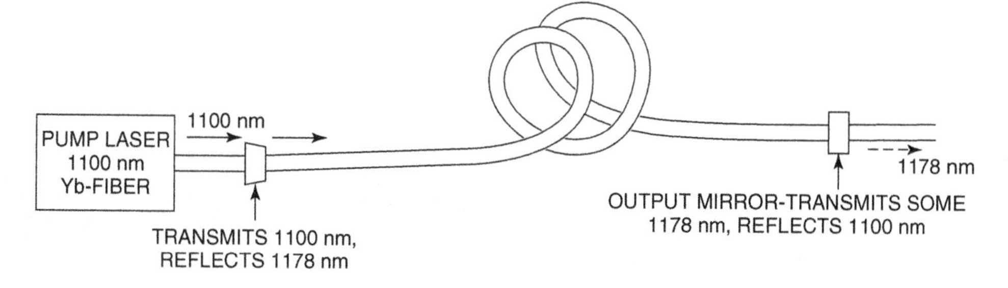

Figure 8-17. Raman laser converts the 1100 nm pump laser beam to the longer 1178 nm wavelength.

With enough pump power and hundreds of meters of fiber, more than half of the input energy can be converted to the shorter wavelength. Developers also are working to increase output power by making fibers from materials with stronger Raman scattering.

This ability to shift wavelength is important because efficient, high-power lasers are available only at a few wavelengths. Raman fiber lasers can be cascaded to produce more wavelengths. For example, the 1178-nm output of the Raman fiber laser in Figure 8-17 can pump a second Raman fiber laser to generate a longer wavelength. Although each step loses energy, the output is still more powerful than otherwise would be available at that wavelength. Likewise, the output of a Raman fiber laser can be frequency-doubled to produce visible wavelengths not otherwise available; the second harmonic of 1178 nm is 589 nm, for example.

Raman fiber amplifiers have been developed for communication systems that use stimulated Raman scattering to amplify a weak input signal by transferring energy from a strong pump beam. Like erbium amplifiers, Raman amplifiers have gain across a range of wavelengths, so they can amplify signals transmitted through the fiber at a range of wavelengths. The two types complement each other because Raman amplification tends to be stronger at longer wavelengths and erbium has higher gain at shorter wavelengths, so combining them makes gain even across a wider range of wavelengths.

8.11 WHAT WE HAVE LEARNED?

- Solid-state lasers are solids in which light-emitting atoms are contained in a solid that transmits both pump light and the stimulated emission that produces the laser beam. The active species is a dopant in the glass or crystalline host.

- Semiconductor lasers are not considered solid-state lasers.
- Solid-state lasers may be rod-shaped, a slab, or an optical fiber.
- The active species is identified first, then the host; Ti–sapphire means titanium is the active species and sapphire is the host.
- Neodymium, chromium, erbium, ytterbium, and titanium are important active species in solid-state lasers.
- Solid-state laser transitions are electronic transitions of the active species. The exact wavelength also depends on interactions with the host.
- Crystals have better thermal properties than glass, so they can operate at higher powers and repetition rates.
- Glass is easier to make in large pieces than crystals.
- Efficient optical pumping requires matching the pump source to absorption bands of the active species.
- Lamps pump broad absorption bands. Lasers pump narrow absorption lines and generate laser power more efficiently.
- Semiconductor diode lasers are preferred for laser pumping because they generate light more efficiently than other lasers.
- Ruby lasers are made from synthetic sapphire doped with chromium as the active species. They are three-level lasers that emit red pulses at 694 nm.
- Neodymium lasers are the most common solid-state lasers. Neodymium emits near 1.06 μm in several solid hosts.
- Common host crystals are YAG (yttrium–aluminum garnet), YLF (yttrium–lithium fluoride), and yttrium vanadate. Neodymium replaces some yttrium atoms in the crystals.
- The most important harmonics of neodymium lasers are the second at 532 nm, the third at 355 nm, and the fourth at 266 nm. The wavelengths differ slightly among host materials.
- Diode-pumped neodymium lasers can exceed 10% efficiency; lamp-pumped lasers are 0.1% to 1% efficient.
- Vibronic lasers can be tuned across a range of wavelengths. They include titanium–sapphire, alexandrite, and several chromium-doped crystals.
- Ti–sapphire lasers have a broad bandwidth from 660 to 1100 nm and are used to generate tunable output and ultrashort pulses.
- Alexandrite lasers are tunable from 700 to 850 nm; they use chromium as the active species and the host is a crystal called alexandrite.

- Eye-safe lasers emit at wavelengths longer than 1400 nm.
- Erbium–glass lasers emit pulses near 1540 nm. Erbium–YAG lasers emit near 2900 nm.
- Fiber lasers confine stimulated emission to a small core in the center of the fiber.
- Rare-earth elements used in fiber lasers include erbium, ytterbium and thulium.
- Many rare-earth fibers have dual cores: an inner core that contains the rare-earth active species and guides stimulated emission, and an outer core that collects pump light and guides it along the fiber.
- Ytterbium-doped fiber lasers are the most powerful type, and have emitted over 1 kW in a single-mode beam. Their output is at 1030 to 1120 nm.
- Rare-earth doped fibers are used as optical amplifiers in fiber-optic communication systems. The most important type is the erbium-doped fiber amplifier, which amplifies light in the 1530–1570 nm band where optical fibers have their lowest loss.
- Stimulated Raman scattering in optical fibers is the basis for both fiber lasers and fiber amplifiers. A strong pump beam can produce stimulated Raman scattering in many transparent solids.

WHAT'S NEXT?

In Chapter 9, we will learn about the fast-moving technology of semiconductor diode lasers.

QUIZ FOR CHAPTER 8

1. A host material for a solid-state laser must meet which of the following criteria?
   a. Must be transparent at the pump wavelength
   b. Must be transparent at the laser wavelength
   c. Must be able to conduct away waste heat
   d. A & B only
   e. A, B and C
2. Which of the following is the best pump source for a solid-state laser?

a. A flashlight
b. A flashlamp
c. A helium–neon laser
d. An electrical discharge
e. A fluorescent tube

3. What type of laser is the most efficient pump for neodymium lasers?
   a. GaAlAs semiconductor
   b. Argon-ion
   c. Helium–neon at 632.8 nm
   d. Ruby
   e. InGaAsP semiconductor
4. Which of the following elements is not used as an active species in solid-state lasers?
   a. Neodymium
   b. Titanium
   c. Chlorine
   d. Chromium
   e. Ytterbium
5. What is an important advantage of glass as a host for solid-state lasers?
   a. Glass can be drawn into optical fibers.
   b. Glass has better thermal characteristics than crystals.
   c. Glass is the only material transparent to all pump light.
   d. Glass is used in flashlamps.
   e. Glass melts at higher temperatures.
6. Which of the following lasers is a quasi-three-level system?
   a. Ruby
   b. Neodymium–YAG
   c. Titanium–sapphire
   d. Erbium-doped fiber
   e. There is no such thing.
7. What is the second harmonic of the ruby laser?
   a. 266 nm
   b. 347 nm
   c. 355 nm
   d. 532 nm
   e. 694 nm
8. Which of the following wavelengths cannot be readily generated from a Nd–YAG laser?
   a. 266 nm
   b. 355 nm

c. 477 nm
d. 532 nm
e. 1064 nm

9. What is the most powerful rare-earth doped fiber laser?
   a. Neodymium–YAG
   b. Erbium
   c. Thulium
   d. Yttrium
   e. Ytterbium
10. What is the function of an outer core in a fiber laser?
   a. It collects and guides pump light along the fiber.
   b. It collects and guides stimulated emission.
   c. It confines the active species so the dopant does not leak out of the fiber.
   d. It confines stimulated Raman emission.
   e. Coolant flows through it.
11. What differentiates vibronic lasers from other solid-state lasers?
   a. Can be pumped efficiently with semiconductor lasers
   b. Higher gain
   c. Shorter wavelengths
   d. Broader gain bandwidth and tunable output
   e. All of the above
12. What active species is used in fiber amplifiers?
   a. Neodymium
   b. Erbium
   c. Thulium
   d. Yttrium
   e. Ytterbium

CHAPTER 9

SEMICONDUCTOR DIODE LASERS

ABOUT THIS CHAPTER

The internal workings of semiconductor diode lasers are more complex than those of gas or solid-state lasers because they depend on the unique properties of semiconductors. This chapter starts by explaining the basics of semiconductor physics, then explains how semiconductor diodes can produce light, and how that light emission can be used to create a laser. Then it describes the diverse types of semiconductor diode lasers that are now in wide use in applications from playing CDs to powering high-power solid-state and fiber lasers.

9.1 BASICS OF SEMICONDUCTOR DIODE LASERS

Like the other lasers described so far, a semiconductor diode laser generates a beam when spontaneous emission triggers a cascade of stimulated emission from a population inversion inside a resonant optical cavity. As in a gas laser, the excitation energy comes from an electric current passing through the laser material, but in a semiconductor diode laser the laser material is a solid with a particular internal structure that produces and traps a population inversion.

The structure is called a *diode* and it is formed within the semiconductor by depositing layers with different compositions. We will get into the details later. It is possible to produce stimulated emission from semiconductors in other ways, such as stimulat-

Understanding Lasers: An Entry-Level Guide, Third Edition. By Jeff Hecht

ed Raman scattering in silicon (described in Chapter 10) but virtually all semiconductor lasers now in use rely on an internal diode to produce a population inversion. Thus, you really do not have to call them semiconductor diode lasers; you can call them semiconductor lasers, diode lasers, or laser diodes. Those terms are interchangeable. Sometimes, the devices are just called diodes in a context in which it is clear that they are lasers, such as a diode-pumped solid-state laser.

Diode lasers are close cousins of the *light-emitting diodes* or *LEDs*. LEDs came first, initially discovered in 1907 by Henry J. Round, and in the 1920s independently rediscovered and studied in much more detail by Oleg Losev in the Soviet Union. More LED demonstrations followed as physicists learned more about semiconductor physics, especially in the 1950s, but their emission was feeble. That changed in 1962 with the dramatic demonstration of bright infrared emission from gallium arsenide by Robert Rediker at the MIT Lincoln Laboratory. Within a few months, Robert Hall of General Electric in Schenectady, NY, had built on that to make the first diode lasers, and Nick Holonyak Jr., working for GE in Syracuse, NY, had made the first bright visible LEDs.

The fundamental difference between the LEDs and the diode lasers was that diode lasers were driven much harder with a much higher current. The first diode lasers were operated at the cryogenic temperature of liquid nitrogen, 77°K (–196° C or –321°F). It took years to develop versions that could generate a continuous beam at room temperature, and more years to improve them so they could operate for more than a few seconds or minutes at a time. But diode laser technology has followed the growth of other semiconductor technologies, and their technology has improved tremendously.

Diode lasers are now a pervasive if hidden technology. *Laser Focus World* estimated that in 2006 over 800 million diode lasers were sold around the globe, accounting for nearly $3 billion in sales. You use them when you listen to CDs, watch DVDs, or make long-distance calls that pass through fiber-optic cables. Diode lasers emitting violet light were a key advance behind the new generation of HD video players. People routinely use diode-laser pointers to highlight presentations and play with their cats. Diode laser scanners read price codes in stores and align construction equipment.

9.2 SEMICONDUCTOR BASICS

Semiconductor physics as a whole is beyond the scope of this book, but some fundamental aspects of semiconductors are crucial to the understanding of semiconductor lasers. This section will cover those aspects and semiconductor materials, before we get into the details of light emission and lasers.

9.2.1 Valence and Conduction Bands

Semiconductors get their name because they conduct electricity better than an insulator but not as well as a conductor. This arises from the nature of the bonds between atoms and their electrons in the solid. In an insulator like glass, the outer electrons are tightly bound to atoms, often in covalent bonds between pairs of atoms. In conductors like copper, the outer electrons are only loosely bound to atoms, so they can flow freely though the material if a voltage is applied across it. A semiconductor falls in between the two.

The difference can be explained by considering electrons to be in one of two levels: a *valence band* in which they are bound to atoms, and a *conduction band* in which they are free to move around in the solid. The valence band is at a higher energy level than the conduction band, but in a conductor the top of the valence band overlaps with the bottom of the conduction band, as shown in Figure 9-1. This reflects the low-energy bonding of electrons to met-

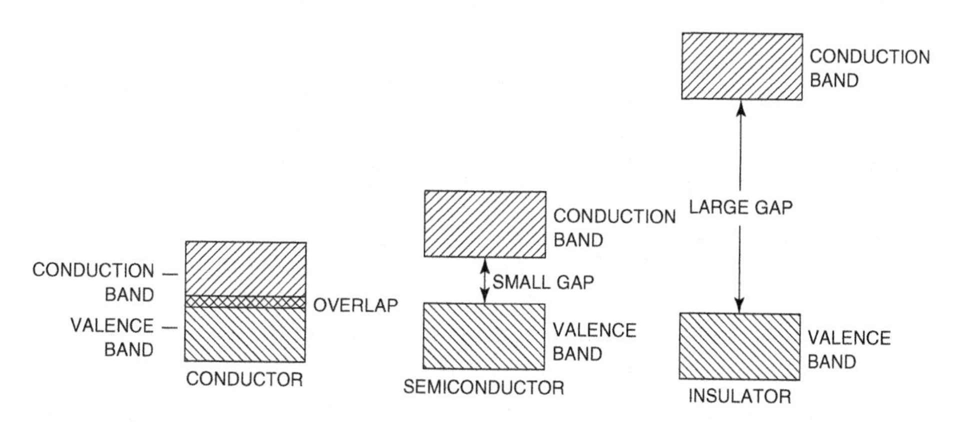

Figure 9-1. Valence and conduction bands in conductor, semiconductor, and insulator.

al atoms, which allows electrons to move and carry current easily in conductors. In a semiconductor, there is a small *band gap* between the lower valence band and the higher-energy conduction band, and a few valence electrons have enough energy to escape to the conduction band. In an insulator, the band gap is large, so a valence electron needs so much energy to escape that essentially no electrons are free to conduct current through the solid.

The number of electrons N in the valence and conduction bands depends on the bandgap energy ΔE and the temperature T:

$$\frac{N_{\text{conduction}}}{N_{\text{valence}}} = \exp^{\Delta E/kT} \tag{9-1}$$

where k is the Boltzmann constant. This is the same formula used to describe the relative proportions of atoms and molecules in a pair of different energy levels.

If you plug in numbers, it turns out that very few electrons are in the conduction band at room temperature. In 100% pure silicon, where the band gap is 1.1 electron volts, only 3×10^{-19} of the electrons are in the conduction band at room temperature. That is enough to carry a feeble current, but it gives pure silicon a high electrical resistance.

9.2.2 Electrons, Holes, and Doping

Electronic devices require higher conductivity. All it takes is a few impurities. To understand how that works, we will start by looking at silicon, the semiconductor most widely used in electronics, although silicon is a very poor material for making diode lasers.

Silicon atoms have four electrons in their outer shell, and in silicon crystals each of those four electrons forms a bond with an adjacent atom, as shown in Figure 9-2A. Very few of the valance electrons in such pure silicon crystals can escape to the conduction band at room temperature.

In reality, any silicon crystal inevitably includes a few impurity atoms, and some of them can fit into the crystal in positions that otherwise would be occupied by silicon atoms. Suppose the impurity atom had five electrons in its outer shell, such as phosphorus or arsenic. As shown in Figure 9-2B, four of the electrons would form bonds with adjacent silicon atoms, but the fifth would be left over and it easily could move about through the crystal.

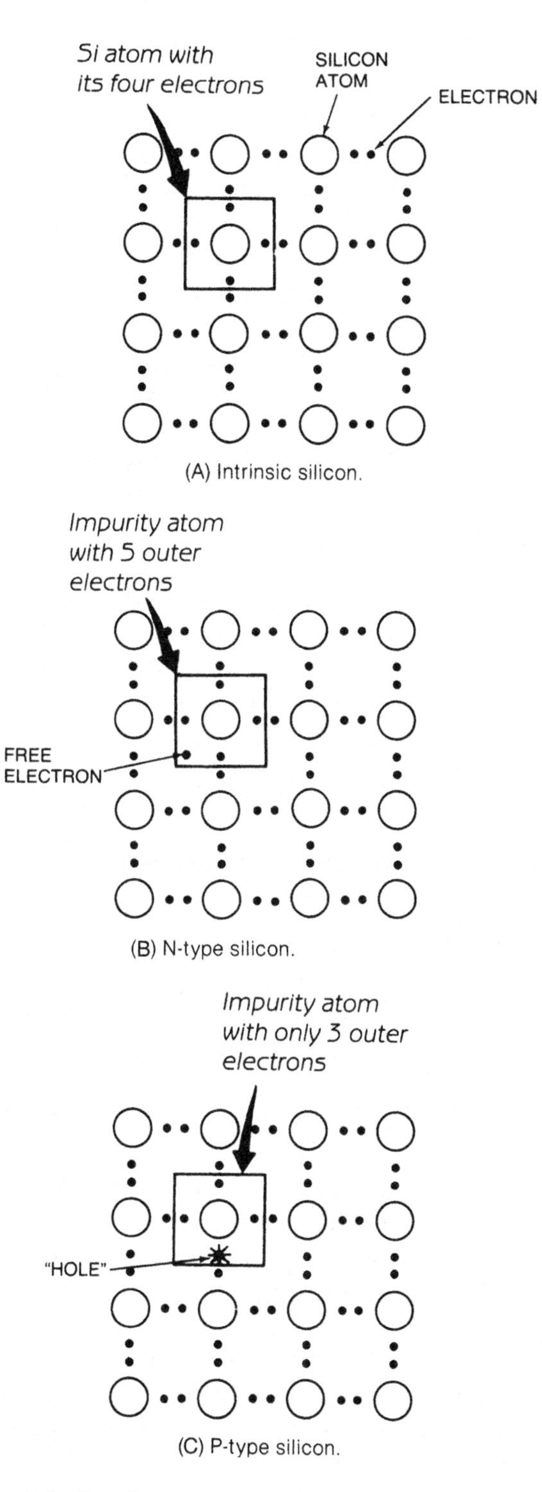

Figure 9-2. Bonding in pure and doped silicon crystals.

An atom with only three outer electrons such as aluminum or gallium also could fit into a space normally occupied by silicon, with its three outer electrons each bonding to adjacent silicon atoms. However, this would leave a *hole* at the place where the fourth outer electron of silicon normally would form a bond to an adjacent silicon atom, as shown in Figure 9-2C. The hole can "move" if a nearby electron moves to fill the hole, leaving behind a hole elsewhere in the crystal lattice. In that way, holes can move through the crystal as carriers of positive charge, to complement the negative charge carried by electrons.

Intentionally adding impurities can increase the conductivity of the semiconductor, making it more useful for electronic devices. A semiconductor doped with atoms that donate electrons to create free carriers, like phosphorus in silicon, is called *n*-type because it has negative current carriers. Semiconductors doped with atoms that produce holes (or electron acceptors) are called *p*-type because they contain extra holes that serve as positive current carriers. The degree of conductivity in *n*- and *p*-type materials depends on the impurity doping. Undoped semiconductors are called *intrinsic* or *i*-type.

9.2.3 Diodes, Junctions, and Recombination

The simplest semiconductor device is called a diode, a name meaning that it has two electrical terminals. The first electronic diodes were vacuum tubes that transmitted current in one direction but not the other. Semiconductor diodes, likewise, normally transmit current in only one direction, although there are exceptions.

A semiconductor diode consists of regions of *p*- and *n*-type material which meet at a *junction* layer. In practice, a diode often is made by diffusing an excess of one type of dopant into a slab of semiconductor doped with the other type. For example, an electron acceptor such as aluminum or gallium can be diffused into a slab of *n*-type silicon, forming a top layer of *p*-type material in which the holes outnumber the electrons. A junction region in which holes and electrons are equal in number separates the *p*-type region from the *n*-type region. The junction typically is only 0.1 to 1 μm thick, and it is the place where the current flow changes and important things happen. Exactly what happens depends on the voltage applied across the junction.

If there is no bias across the junction, charge carriers are distributed through the crystal in roughly the same way as impurities and not much happens. Near the junction, electrons from the *n*-type material can fall into holes in the *p*-type material, releasing energy by a process called *recombination* as the electron drops from the conduction band to fill the hole in the valence band. This process reaches an equilibrium, and no net current flows.

No current also flows if the diode is *reverse biased* by applying a positive voltage to the *n* side of a junction and a negative voltage to the *p* side. As shown in Figure 9-3, the positive electrode attracts electrons from the *n*-type material, and the negative electrode attracts holes from the *p*-type material. This draws carriers away from the junction, and virtually no current flows through the junction. The leakage current is slightly above zero, because the semiconductor's resistance is not infinite, and a much higher current can flow if a high voltage is applied across the diode, causing *breakdown*.

Current flows through the diode that is *forward biased* by applying a positive voltage to the *p* side and a negative voltage to the *n* side. As shown in Figure 9-4, this attracts the *p* carriers to the *n* side of the device and vice versa, making them *recombine* at the junction. The electron–hole pair is called an *exciton*, which exists briefly before releasing its extra energy as the electron falls into the hole in the valence band, releasing the band-gap energy at the

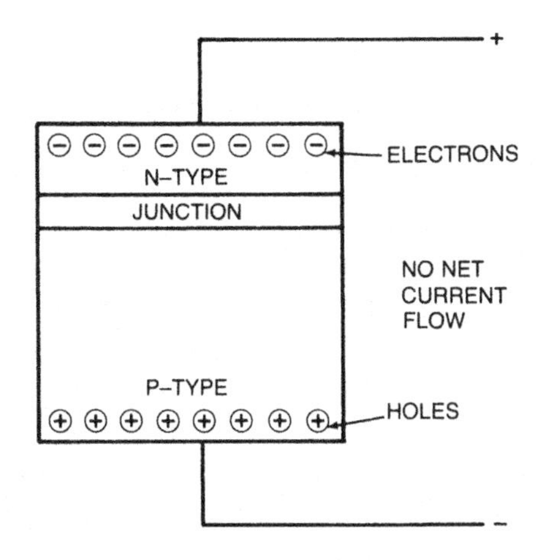

Figure 9-3. No current flows in a reverse-biased semiconductor junction.

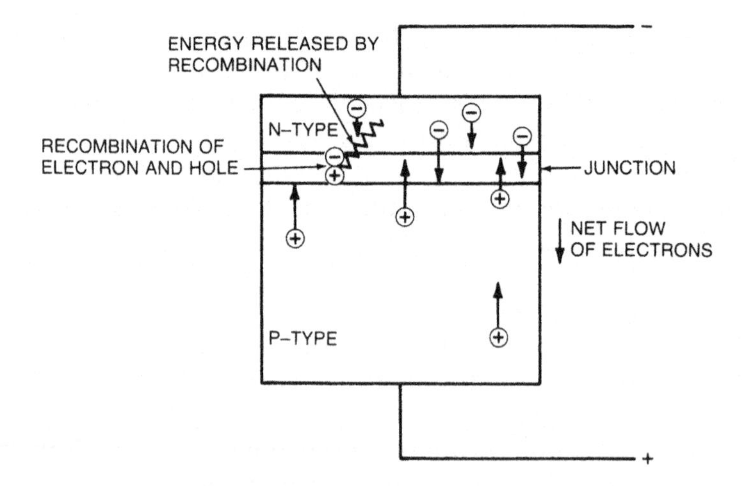

Figure 9-4. Forward-biasing causes electrons and holes to flow toward a junction in a semiconductor diode.

junction. Once the applied voltage exceeds the band-gap energy—typically 0.5 to 2 electronvolts—current flows through the diode. This produces a voltage drop at the junction equal to the band-gap energy.

LEDs are forward-biased diodes in which the recombination energy is released as light, described in more detail later. It is also possible for light shining on a semiconductor diode to release energy if it has enough energy to excite a valence-band electron to the conduction band, creating an electron–hole pair. This effect can be used to sense light in photodetectors (described in Section 5.8.1) or to generate electricity in solar cells.

9.2.4 Indirect and Direct Band Gaps

Silicon is the best-known semiconductor. It is the standard for most electronic applications, for light detection at visible and near-infrared wavelengths, and for solar cells. However, silicon is a very poor light emitter because it can only make an indirect transition from the conduction band to the lower-energy valence band. This condition, called an *indirect bandgap,* means that an electron in the conduction band must interact with something else in order to drop down to the valence band. Normally, that interaction takes a while, so it is milliseconds before the silicon can emit light, and by then other interactions with much shorter lifetimes have drained away the energy that could have been emitted as light. Germanium and some other semiconductors such as gallium phosphide also have indirect bandgaps that make them poor light emitters.

Silicon LEDs have been demonstrated in the laboratory, but they require special tricks to make the material behave differently. One example is fabricating nanostructures called quantum dots, which confine electrons and holes on the scale of a few nanometers. That quantum confinement changes the momentums of the electrons and holes to make it more likely that a conduction electron can drop directly into a hole in the valence band, making the silicon act more like a direct-bandgap material.

Direct-bandgap semiconductors can emit light efficiently because electrons can drop directly from the conduction band to the valence band without changing their momentum, which requires interactions that can drain away energy. The most important direct-bandgap semiconductors are compounds of elements from groups III and V in the periodic table, such as gallium arsenide and indium phosphide, known as *III–V compounds.* They are the type used for LEDs and diode lasers.

The distinctions between direct and indirect bandgap compounds are not always sharp or obvious. GaAs has a direct bandgap and GaP has an indirect bandgap. Mix a little phosphorous with GaAs, and it remains a direct-bandgap semiconductor until the phosphorous level crosses a threshold.

Because this book is about lasers, we will focus on the family of compound semiconductors with direct bandgaps that allow light emission, making them useful in LEDs and semiconductor lasers.

9.2.5 Compound Semiconductors

Compound semiconductors are inorganic compounds containing two or more elements with the electrical properties of semiconductors. In principle, all types of semiconductor devices can be made from compound semiconductors but, in practice, silicon dominates the market for electronic devices. Compound semiconductors fill specialized niches, and are particularly important for LEDs, lasers, and optoelectronic devices.

The most important compound semiconductors for laser applications are the III–V compounds, which contain equal amounts of elements from group IIIa and group Va of the periodic table. The important elements are listed below:

Group IIIa	**Group Va**
Aluminum (Al)	Nitrogen (N)
Gallium (Ga)	Phosphorus (P)
Indium (In)	Arsenic (As)
	Antimony (Sb)

The simplest of these materials are "binary" compounds containing two elements, such as gallium arsenide (GaAs), indium phosphide (InP), and gallium nitride (GaN). Each of these compounds has its own set of characteristics, including energy levels, band gap, and atomic spacing in the crystalline lattice.

Other elements can be added to the compound as long as they maintain the balance of equal numbers of atoms from group III and group V, which have different valence. This is desirable because it allows adjusting properties of the semiconductor, particularly the size of the band gap. For example, replacing some gallium in gallium arsenide with aluminum increases the band-gap energy in gallium–aluminum arsenide (GaAlAs). Such compounds containing three elements are called *ternary* and are written in the form $Ga_{1-x}Al_xAs$, where x is a number between 0 and 1. This format indicates what we said above, that the number of gallium atoms plus the number of aluminum atoms must equal the number of arsenic atoms.

Adding a fourth element to make a *quaternary* compound gives more flexibility and control over material properties. An example is indium–gallium arsenide–phosphide (InGaAsP), which is written $In_{1-x}Ga_xAs_{1-y}P_y$, where both x and y are numbers between 0 and 1. In this case, the total number of indium and gallium atoms must equal the number of arsenic and phosphorus atoms. It is also possible to have three elements in one group and only one in the second, such as indium–gallium–aluminum phosphide (InGaAlP), written as $In_{1-x-y}Ga_xAl_yP$, where both x and y are numbers between 0 and 1 which together add to less than 1. In this case, the total number of indium, gallium, and aluminum atoms must equal the number of phosphorous atoms to form a semiconductor crystal.

Ternary and quaternary compounds are harder to fabricate into good crystals, but they offer an ability to control the band gap and lattice constant, which is invaluable in fabricating devices such as diode lasers. Practical concerns impose some limits on material characteristics. It is difficult to grow ternary and quaternary compounds in bulk, so they normally are deposited on substrate wafers made of binary compounds, mostly GaAs and InP. Successful deposition requires careful matching of the lattice spacing of the deposited compound with the substrate, limiting the blends that are easy to fabricate. This affects the composition and wavelengths of semiconductor lasers, as described later in this chapter.

Light emission has been demonstrated from two other families of compound semiconductors. One is silicon carbide (SiC), composed of two group IV elements with four valence electrons. The other family is called *II–VI compounds* because they contain elements with two and six valence electrons. They come mostly, but not entirely, from columns IIB and VI of the periodic table. The most important elements in these compounds are:

Column IIB	Column IVB	Column VI
Zinc	Tin	Oxygen
Cadmium	Lead	Sulfur
Mercury		Selenium
		Tellurium

The II–VI compounds fall into two broad groups. Compounds of zinc and cadmium with group VI elements have large band gaps and can emit visible light. Compounds of lead and tin with group VI elements have small band gaps and emit in the infrared.

9.2.6 Organic Semiconductors

Organic LEDs (called OLEDs) are made from organic compounds that have electronic properties similar to inorganic semiconductors. They offer a number of potential advantages over inorganic semiconductors, including easy fabrication with inexpensive technology, compatibility with more substrate materials, and more freedom to tailor material properties including emitting wavelength. Their main application is in small displays like those on cell phones.

Two different classes of organic materials are used in OLEDs. Small-molecule compounds can be evaporated and deposited on a cooler substrate in ways similar to the techniques used for inorganic semiconductors, but at much lower temperatures. The process produces good devices, but is relatively expensive. Alternatively, long-term polymers can be formulated as liquids and printed onto surfaces through jets, as in an ink-jet printer. The process is cheap and easy, but the quality of the devices does not match those of small-molecule LEDs.

Organic semiconductor lasers have been made in the laboratory, but they require optical pumping.

9.3 LIGHT EMISSION AT JUNCTIONS

Henry J. Round was puzzled in 1907 when he saw light emission from an impure form of silicon carbide called carborundum. He knew the light arose from the junction between a metal conductor and the material, which we now know is a semiconductor, but he did not understand what produced it. Figuring that out took other researchers decades.

Today, we understand that phenomenon, called *recombination radiation,* which occurs when electrons in the conduction band of a semiconductor drop into holes in the valence band. Round saw light emission from a junction between a semiconductor and a conductor like the point contact junction of the first transistor. Modern LEDs and diode lasers emit light at a *p–n* junction between regions of *p*- and *n*-type semiconductor, which makes them more versatile, like the junction transistors that followed the point-contact type.

Recombination releases an amount of energy equal to the gap between the conduction and valence bands. That energy is released at all *p–n* junctions, but what happens to it depends on the nature of the band gap. In indirect-bandgap semiconductors, the energy is dissipated within the crystal. Direct-bandgap semiconductors instead emit a photon with energy equal to the band gap. Because the band-gap energy depends on the semiconductor composition, the emission wavelength also depends on composition.

Recombination is the basis of light emission from both LEDs and diode lasers. Let us look carefully at the differences, starting with the LED.

9.3.1 Light-Emitting Diodes (LEDs)

A light-emitting diode is a semiconductor diode that is forward-biased so carriers recombining at a junction spontaneously emit light. That light is spontaneous emission, incoherent light like that emitted from a light bulb.

The nominal wavelength of the spontaneous emission depends on the semiconductor bandgap energy, but its range is not as strongly constrained as in a laser. Figure 9-5 shows the range of wavelengths from three typical visible LEDs.

Table 9-1 lists a sampling of LED materials and the wavelengths they produce. The bandgap of a compound semiconductor depends on the mixture of materials it contains, so the output wavelength depends on their composition. Specific wavelengths are the peaks of particular blends; a range shows wavelengths available from various compositions.

LEDs radiate spontaneous emission in all directions from their junction layer, as shown in Figure 9-6. To get the most efficient output, the junction should be close to the surface of the device. This reduces the chance that the emitted light will be absorbed by other parts of the device. A transparent lens can be fabricated on top of an LED to concentrate the emitted light, usually perpendicular to the surface. Many LEDs are packaged in ways that concentrate light in one direction.

Unlike semiconductor lasers, LEDs lack the resonant cavities needed to build up stimulated emission, so LED emission is less in-

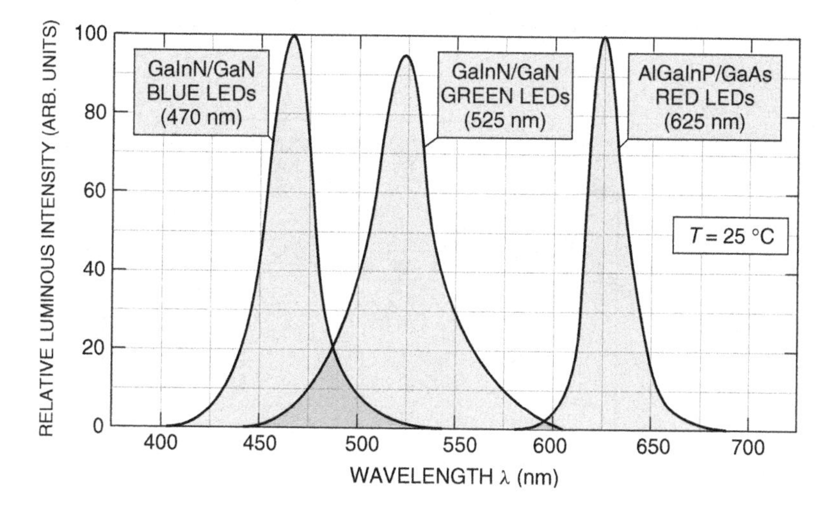

Figure 9-5. Bandwidths of typical commercial blue, green, and red LEDs, with curves drawn so all are about the same height. (Courtesy E. F. Schubert, www.lightemittingdiodes.org.)

Table 9-1. Emission ranges of some important inorganic LED materials

Material/substrate	Peak wavelength or range (nm)	Status
AlGaN/GaN	230–350	Developmental
InGaN/GaN	360–525	Commercial
ZnTe/ZnSe	459 (blue)	Developmental
SiC	470 (blue)	Commercial
GaP	550–590 (green-yellow)	Commercial
$GaAs_{0.15}P_{0.85}$	589 (yellow)	Commercial
AlGaInP/GaAs	625–700	Commercial
$GaAs_{0.35}P_{0.65}/GaAs$	632 (red)	Commercial
$GaAs_{0.6}P_{0.4}/GaAs$	650 (red)	Commercial
GaAsP/GaAs	700 (red)	Commercial
$Ga_{1-x}Al_xAs/GaAs$	650–900 (red and infrared)	Commercial
GaAs	910–1020 (infrared)	Commercial
InGaAsP/InP	600–1600 nm	Commercial

tense than a laser beam. LEDs also emit light across a wider range of angles than lasers. Both factors make them more desirable than lasers for some applications, especially illumination, displays, and other devices intended for viewing with the human eye. Because this is a book about lasers, we will not go into much detail on LEDs, but it is helpful to understand how they work and how they are used.

Early LEDs were intended for use as indicator lights or dot-matrix displays of single-color elements. The first pocket calculators displayed numbers on arrays of red LEDs.

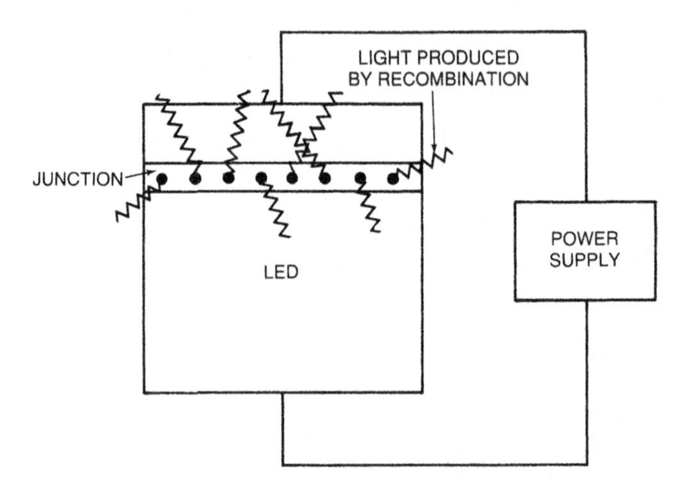

Figure 9-6. An LED emits light in all directions from its junction.

Now the range has broadened to include illumination and displays. Look carefully at red and green traffic signals, and you may see arrays of bright spots if they are made of LED arrays. So are red signal lights on new cars. Red, green, and blue LEDs can be combined to make white light sources for illumination; alternatively, a bright blue LED can illuminate phosphors that emit green and red light to create a white light source. Today's white LED lamps are used where weight and low power consumption are critical, such as headlamps for night-time campers and hikers. As they improve, their applications will expand into general illumination.

Organic LEDs can be made in arrays of red, green, and blue emitters to make bright color displays that consume little power. So far, they are limited to small areas, such as cell-phone displays, but larger displays have been demonstrated.

9.3.2 Semiconductor Diode Laser Concepts

It was a surprisingly small step from the first efficient LEDs to the first semiconductor lasers. Both were demonstrated in 1962 in the same material, gallium arsenide, and the LED was almost overlooked in the race to make the first semiconductor laser.

The simplest diode lasers are structurally similar to LEDs. Both generate light from recombination of electron–hole pairs at a forward-biased junction. Below the laser threshold, both generate spontaneous emission with an intensity that depends on the drive current. However, diode lasers have reflective surfaces that create optical feedback. The feedback has little impact when drive current is below the point needed to produce a population inversion, but is critical in crossing the threshold for laser operation.

At low drive current, electron–hole pairs (excitons) release their energy by spontaneous emission, as in an LED. As the drive current increases, it produces more electron–hole pairs that emit light spontaneously, increasing the likelihood that a spontaneously emitted photon will encounter and stimulate emission from an exciton that has yet to release its extra energy. Once the drive current reaches a high enough level, it produces a population inversion between the exciton state (the upper laser level) and the atoms with the extra electron bound in the valence band (the lower laser level). That leads to a cascade of stimulated emission as the laser crosses the threshold.

Because the excitons are in the thin layer of the junction plane, stimulated emission is most likely to build up along the junction. For this reason, the reflective cavity is aligned along the junction plane, with reflective surfaces perpendicular to the junction, as shown in Figure 9-7.

Semiconductors have a high refractive index, so an uncoated solid–air interface reflects much of the stimulated emission back into the semiconductor, as shown in Equation 5-4, providing feedback for the laser resonator. The large population inversion at high drive current makes gain high in semiconductor lasers, so cavities only a few hundred micrometers long can sustain oscillation. One facet often is coated to reflect all the incident light, so all the stimulated emission emerges from the other end, as shown in Figure 9-7.

Diode lasers have a well-defined threshold at which their output shifts from low-power spontaneous emission to higher-power laser operation, as shown in Figure 9-8. Below the threshold, the diode operates as a relatively inefficient LED. Above the threshold, the diode operates as a laser, converting a much higher portion of the input electrical power into light energy, as shown by the steeper slope.

Threshold current is an important factor in semiconductor laser performance. Electrical power needed to reach the threshold current winds up as heat that must be dissipated in the laser. So does the fraction of above-threshold current that is not converted into light. The extra heat is not just wasted power; it also degrades

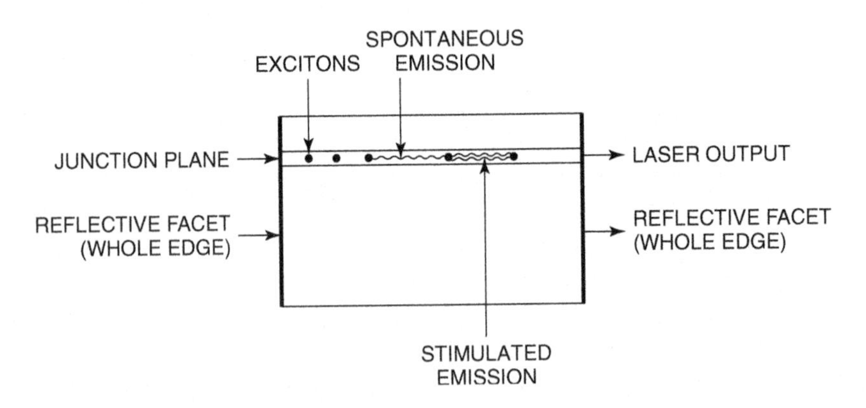

Figure 9-7. Reflection from facets forms a laser cavity that oscillates in the junction plane of a diode laser.

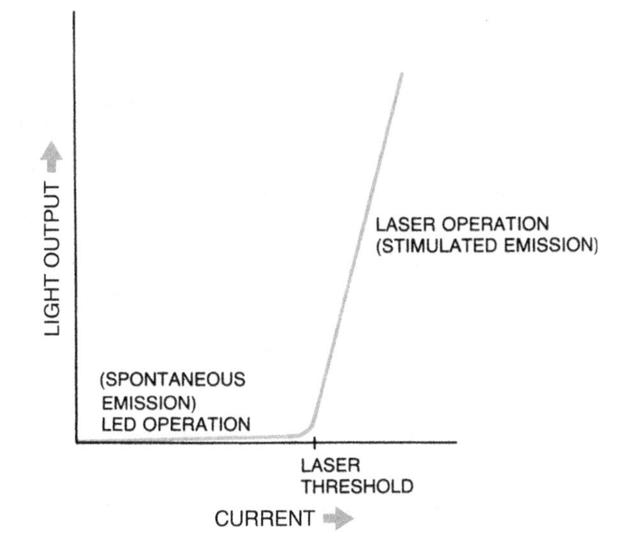

Figure 9-8. Light output of a diode laser rises sharply above the laser threshold current.

laser performance and tends to shorten its lifetime, so lower-threshold lasers tend to have longer lifetimes. High currents also stress the laser by putting highly concentrated power through the junction; this is measured as *threshold current density* (threshold current divided by the junction area operating as a laser) rather than the total threshold current.

Diodes built to operate as lasers are inherently less efficient as LEDs than those built to operate as LEDs. A main reason is packaging. As you saw earlier, spontaneous emission is emitted in all directions, and LEDs are designed to collect as much as possible of this light. However, edge-emitting diode lasers like the one shown in Figure 9-7 collect light only from the junction layer at the edge of the chip, a much smaller area. When a diode built to operate as a laser is below threshold, only a small fraction of the spontaneous emission it generates emerges through the edge of the junction plane; the rest is trapped within the packaged device.

9.4 LAYERS AND CONFINEMENT IN DIODE LASERS

The simple laser structure of Figure 9-7 had to be greatly refined to make diode lasers practical to use. The first diode lasers based on that structure had to be cooled to liquid-nitrogen temperature (77 degrees Kelvin), could only operate in pulsed mode, and had very high threshold currents. Refinements have increased efficiency and operating lifetime by confining both the drive current and the light energy to small regions of the semiconductor. The better the confinement, the lower the threshold current, the longer the lifetime, and the better laser operation. To understand the structures now used, we will examine one dimension of the laser at a time, starting with the layering structure of the semiconductor, then considering its width, and, finally, resonator structures.

9.4.1 Doping Layers and Substrates

A diode laser must contain at least three layers: a *p*-type layer, an *n*-type layer, and an intermediate active or junction layer where recombination causes light emission. The essential components of these layers are semiconducting compounds such as gallium–aluminum arsenide and dopants that create either extra carriers or holes.

Lasers are grown by depositing a series of thin layers on a substrate usually made of a semiconductor or sometimes an insulator such as sapphire. The most common diode-laser substrates are gallium arsenide and indium phosphide. Binary semiconductors are preferred as substrates because they can be produced in quantity at relatively low cost. Layers deposited on them must have an atomic spacing (often called *lattice constant*) closely matched to the spacing of atoms in the substrate. Typically, these thin layers are made of semiconductors with compositions slightly different from the substrate, such as aluminum–gallium arsenide on GaAs or InGaAsP on InP. The lasers are doped with elements with extra electrons or with fewer electrons to produce *n* or *p* carriers.

Typically, substrates are *n*-doped, with one or more *n*-doped layers deposited on them, then an undoped active layer, and, finally, multiple *p*-doped layers deposited on top of the active layer. Compositions and doping are typically varied to produce the desired properties.

9.4.2 Homojunction Lasers

The first semiconductor lasers consisted of two layers made from the same compound, generally gallium arsenide, one doped with a material that added extra electrons to the conduction band to make an *n*-type semiconductor, the other with a material that produced holes in the valence band to make a *p*-type material. A junction zone separated them. These devices were called *homostructure* or *homojunction* lasers because their two layers were made from the same compound semiconductor.

Homojunction diodes worked, but they did not work very well and required cooling even to produce pulses of light because they were not able to confine the locations of either current carriers or light very well.

9.4.3 Heterojunction Lasers

Improvements came from replacing the homojunction with a *heterojunction* between layers of semiconductors with different composition. The idea and its implementation in diode lasers earned Herbert Kroemer and Zhores Alferov the 2000 Nobel Prize in physics.

The appeal of a heterojunction comes from differences between the electronic and optical properties of the two materials. Those differences can control the flow of electronic carriers or light between the two sides of the junction. Figure 9-9 shows the

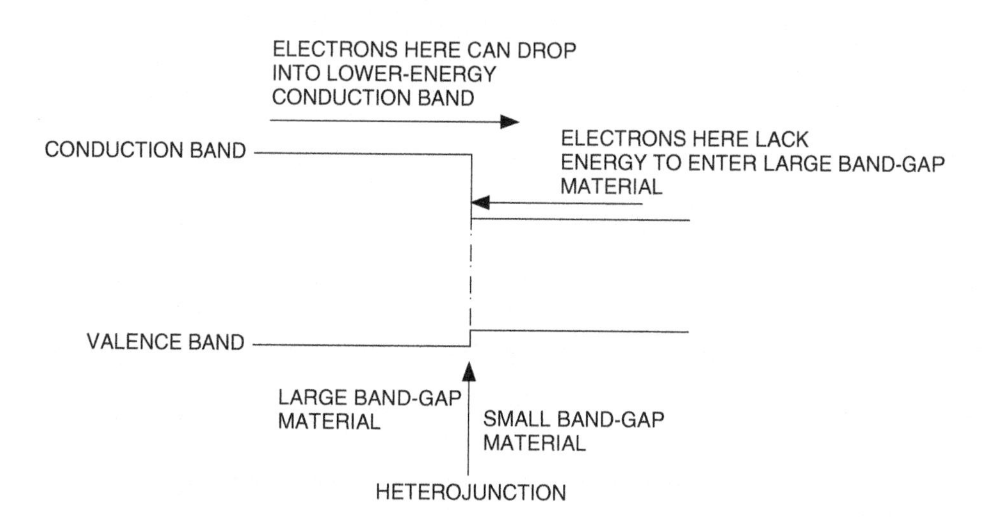

Figure 9-9. Energy levels at a heterojunction allow conduction electrons to drop to the low-band gap side, but not to return to the high-band gap side.

effects of differences in the band-gap energy on conduction electrons. The electrons in the high-band-gap material have enough energy to enter the lower-band-gap material, but the conduction electrons in the low-gap material do not have enough energy to reach the conduction band in the high-gap material.

Similarly, differences in the refractive index can block light in a high-index material from entering a low-index material, if the light strikes the boundary between the two materials at a steep enough angle. That produces total internal reflection, the same phenomenon behind optical fibers, which effectively guides light through the high-index material.

A *single heterojunction* between two materials improves the confinement of light and electrons enough for a diode laser to operate in pulsed mode at room temperature. This confinement is illustrated schematically by the curves at the right for three types of lasers in Figure 9-10. Note how the light and electrons occupy a smaller area in the single-heterojunction laser than in the homojunction laser.

If one heterojunction is good, two should better confine light and electrons, as you can see at the bottom of Figure 9-10. In this case, the active layer is GaAs, which has a lower bandgap energy than the GaAlAs layers above and below it. This design is called a *double-heterojunction* or *double-heterostructure* laser, and it was the first diode laser capable of continuous-wave operation at room temperature. The double-heterostructure design is the basis of today's semiconductor laser industry.

9.4.4 Lattice Matching and Strain

A crucial element in diode-laser fabrication is matching the atomic spacing of successive layers. Perfect crystals are arrays of regularly spaced atoms, but atomic spacing differs among compounds. Failure to match the atoms in successive layers can produce defects in the crystal, which degrade its optical, electronic, or mechanical properties.

Semiconductors can accommodate small differences in atomic spacing, which produce some strain within the crystal but not enough to cause damage. However, developers try to minimize strain by limiting differences in lattice spacing. This is particularly important in matching a deposited material to a substrate. As mentioned in Section 9.4.1, simple semiconductor compounds

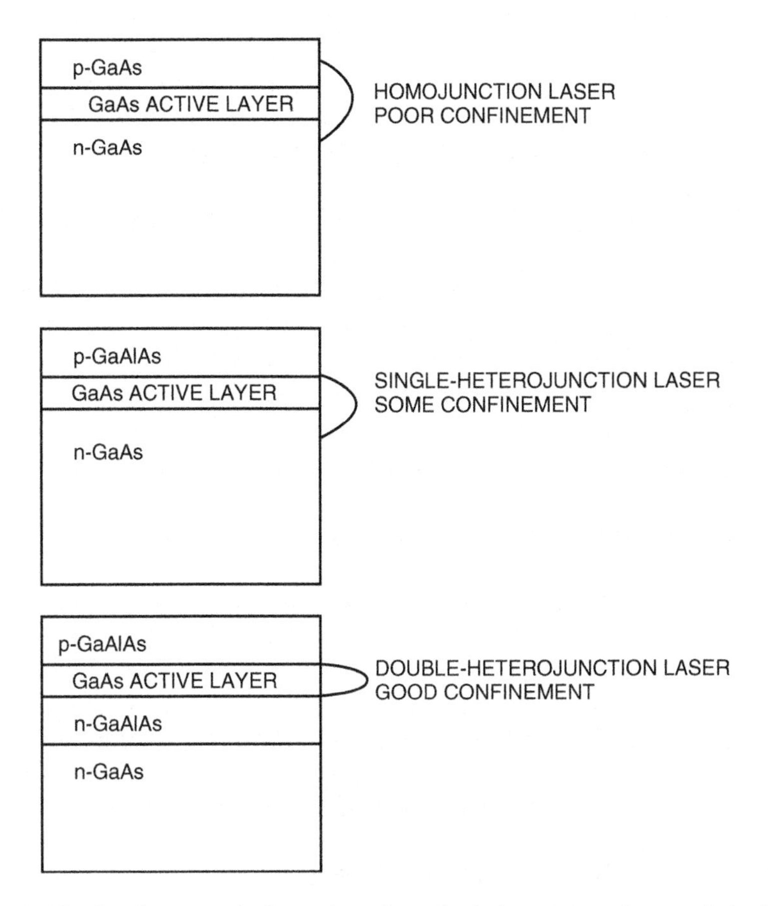

Figure 9-10. Confinement in homojunction, single-heterojunction, and double-heterojunction lasers. (Thickness of active layer is exaggerated to fit label.)

such as GaAs are preferred for substrates because they are much easier to make in quantity than compounds containing three or four elements, so layer compositions must be chosen to match an available substrate, typically GaAs or InP.

Somewhat larger differences in lattice spacing can be accommodated by depositing a series of layers with incremental steps in atomic spacing. The small increments produce small amounts of strain between adjacent layers, which the semiconductor can tolerate. Using such *strained-layer superlattices* broadens the range of materials usable in diode laser structures.

The lattice spacing depends on the nature and size of the atoms in the semiconductor, and varies with composition. Making heterostructures requires a way to change the band gap without introducing large changes in the lattice spacing. A major reason that GaAlAs compounds were widely used in the early stages of diode laser development is that the lattice spacing changes very little with aluminum content, so adding aluminum to increase the band gap did not cause lattice mismatches between layers of GaAlAs and GaAs.

The lattice spacing of other III–V compounds differs more with composition, so it is not possible to add just one element to change the band gap without also changing the lattice constant. Fabricating lattice-matched structures with different band gaps on an InP substrate requires a compound with four elements, InGaAsP. Adjusting both the In/Ga ratio and the As/P ratio makes it possible to produce InGaAsP layers with different band gaps but lattice spacing matched to each other and the substrate.

9.5 CONFINEMENT IN THE JUNCTION PLANE

The first diode lasers produced recombination energy and light across the entire junction plane. One of the refinements needed to allow continuous operation at room temperature was concentrating the drive current and stimulated emission to a narrow stripe in the junction plane, as shown in Figure 9-11. This confinement increases the power density, which improves efficiency and performance.

Narrow stripes limit the number of transverse modes the laser can oscillate in and, thus, improve beam quality. Typically, the stripes are only a few micrometers wide.

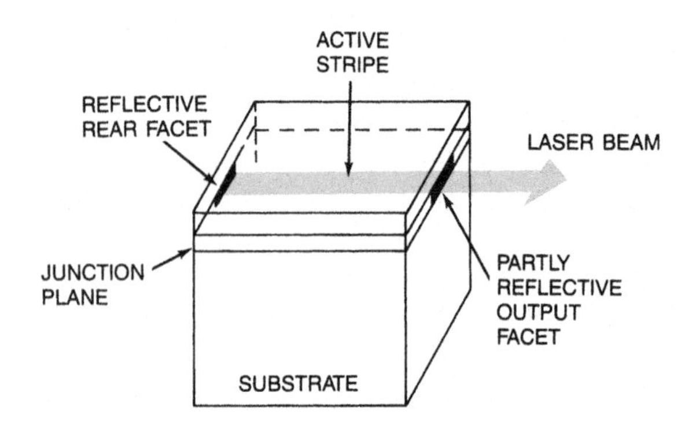

Figure 9-11. Stripe-geometry laser.

9.5.1 Gain- and Index-Guided Lasers

Stripe width can be limited in two ways: by restricting the flow of current to a narrow stripe, or by fabricating stripes of material with different refractive indexes in the junction plane. The two can be combined in a single laser.

In the *gain-guided laser* of Figure 9-12A, insulating regions at the top of the laser chip block current from flowing to either side in a complex double-heterojunction laser. The only path for the current is through a narrow stripe at the middle, which runs the length of the chip between the two cavity mirrors. Thus, recombination of current carriers and a population inversion occur only in the narrow stripe through which the current flows, so only that

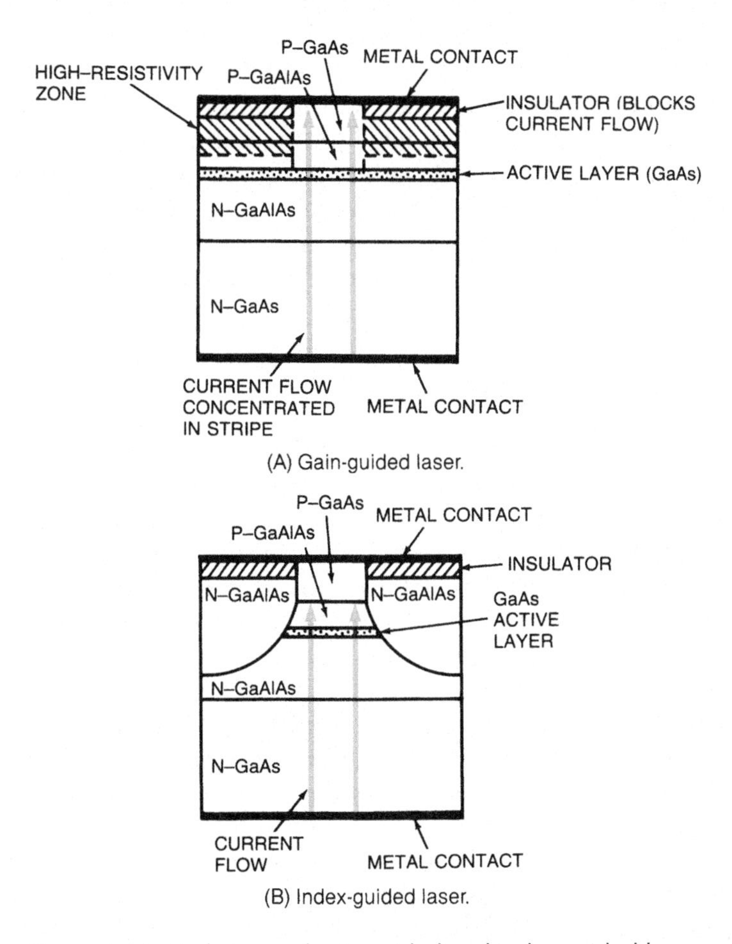

Figure 9-12. End views of gain-guided and index-guided lasers.

zone has gain. Because there is no gain at the sides, those regions do not emit light, even though no physical boundary separates the stripe from the rest of the active layer.

Index-guided lasers add another level of confinement by surrounding the stripe in the active layer with a material of a lower refractive index. As shown in Figure 9-12B, the resulting structure can be complex. In this case, the laser has been etched during fabrication to leave only a narrow stripe or mesa containing the GaAs junction layer that runs the length of the chip. Then *n*-type GaAlAs was deposited on either side of the stripe and covered with an insulator before depositing a metal contact. The insulator confines current flow through the mesa; the boundary with the GaAlAs confines light in the GaAs active layer the same way a double heterojunction confines light in the central active layer. The design shown in Figure 9-12B is an example of a *buried heterostructure* laser, in which the light-emitting strips are buried entirely by other materials, except at the light-emitting facets.

Index-guided lasers are used for most diode laser applications because they confine light better, producing better beam quality. Gain-guided lasers are simple to make, and their poorer confinement of light can be an advantage in generating high powers because spreading light over a larger area reduces the chance of optical damage to the emitting surface.

9.5.2 Broad-Area and High-Power Lasers

The power available from narrow-stripe diode lasers is limited by the generation of waste heat and by the damage that extremely high optical power can cause to optical surfaces. Diode lasers decrease in efficiency and age more rapidly as temperatures increase, so waste heat must be removed from lasers operating at high power. However, even active cooling cannot eliminate the danger of optical damage as the power per unit area becomes very high in the small emitting area of a diode laser.

One way to circumvent these limitations is to make the diode-laser stripe much wider—100 or 200 μm across. This produces gain in a larger volume and spreads the emission over a broader width of the edge of the active layer, reducing power density at the surface. Broad-area lasers can reach powers of a few watts.

Their main advantage is delivering high power from a single emitting aperture. The beam quality is not as good as that from a narrow-stripe laser, although the wide emitting aperture means that the beam does not diverge as rapidly. The high power from a single emitting area is an important advantage for diode pumping of solid-state lasers, and particularly of fiber lasers.

9.5.3 Multistripe Laser Arrays

Laser power from a single laser stripe is inherently limited, even from a broad-area laser. To obtain higher powers, many parallel laser stripes can be fabricated on the same monolithic semiconductor substrate, as shown in Figure 9-13. The combined output power can be much higher than from any single laser, although the beam quality is inevitably lower.

Multistripe diode lasers can produce the highest available power from any semiconductor lasers; they can be assembled into stacks of many wafers to multiply that high-power output. Repetitively pulsed arrays can generate the highest average powers, but continuous-wave versions also generate high powers. Although multistripe lasers efficiently convert input electric power into output light, they also produce comparable amounts of waste heat, which must be dissipated to avoid damage to the lasers.

Multistripe lasers are used to generate raw optical power. A primary application is pumping solid-state lasers, as described in Chapter 8. Although diode laser output can be used directly in some materials-working applications, their beam quality does not match that of diode-pumped solid-state lasers.

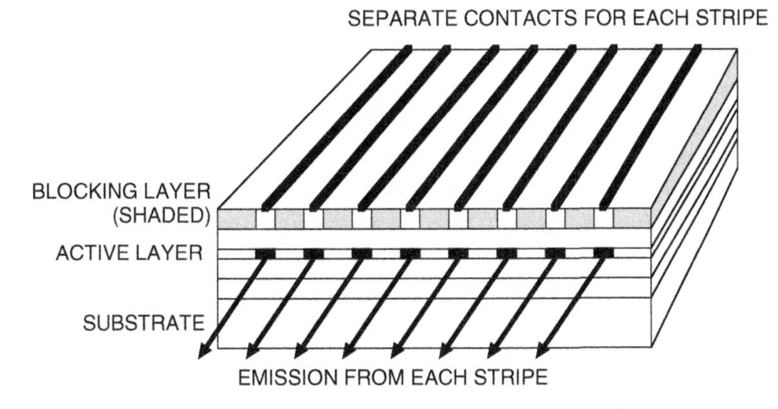

Figure 9-13. Monolithic diode-laser array has many parallel stripes.

9.6 EDGE-EMITTING DIODE LASERS

Diode laser resonators can be constructed in a number of ways. A logical way to group the various types of laser resonators is by their orientation relative to the active layer where the stimulated emission is generated.

Early diode lasers resonated in the plane of the junction layer, as shown in Figures 9-7 and 9-11. These lasers are called edge emitters because their output comes from the active layer at the edge of the laser chip. Their advantage is that they can extract laser gain from the length of the laser chip, typically a few hundred micrometers, to produce relatively high power. They take a few different forms.

An important alternative is the surface-emitting laser, in which the beam emerges from the surface of the chip. Surface emitters have other advantages in beam quality and come in two basic variations: vertical-cavity lasers that oscillate perpendicular to the junction layer, and hybrid lasers that oscillate in the junction plane but direct output through the surface. You will learn more about these types in Section 9.7.

9.6.1 Fabry–Perot Lasers

The simplest type of diode laser resonator is the Fabry–Perot cavity introduced in Chapter 3, a linear cavity with flat mirrors on the two ends. In a semiconductor laser, the linear cavity is the active stripe in the junction layer, and the flat mirrors are the facets on the edges of the chip. Typical diode laser resonators are only 300 to 500 micrometers long, but diode lasers have high gain, so milliwatt powers are easy and much higher powers are possible.

The cavity mirrors are formed by the boundary between the semiconductor crystal and the air. The refractive index of GaAs is 3.34 at 780 nm so according to Equation 5-4 roughly 30% of the light hitting the boundary is reflected back into the semiconductor, providing adequate feedback for laser oscillation.

A diode laser with such a simple cavity emits light from both facets. In practice, one facet is coated to reflect all or most of the incident light, and the beam emerges from the other facet. Light emerging from the rear facet can be monitored to control laser operation in devices such as fiber-optic transmitters.

Narrow-stripe lasers typically emit from an area less than a micrometer high and only a few micrometers wide, producing a single transverse mode, with a central intensity peak in the emerging beam. Wide-area lasers oscillate emit from a much wider stripe, up to about 100 μm, in multiple transverse modes. You will learn later in this chapter that these odd thin emitting areas strongly affect beam quality.

The resonators of Fabry–Perot diode lasers are short compared to those of gas lasers, so cavity modes are spaced more widely in diode lasers—about 0.2 nm for an 800-nm GaAlAs laser, and 0.7 nm for a 1550-nm InGaAsP laser. That means that Fabry–Perot lasers typically emit most of their light on one wavelength at any one time, but may have side bands. However, the gain curve of diode lasers is broad and the refractive index varies with temperature, which itself is a function of drive current. That means that a Fabry–Perot laser might *mode hop*—shift to another wavelength—when its intensity is modulated, as shown in Figure 9-14. Mode hopping is undesirable because it can generate noise.

Fabry–Perot lasers are widely used where low cost is important and precise wavelength is not critical, such as in CD players, laser pointers, and short-distance, low-speed communication links.

9.6.2 Distributed Feedback Lasers

Applications such as high-speed data transmission in fiber optics require limiting laser emission to a narrower range of wavelengths than possible with a Fabry–Perot cavity. This requires adding a

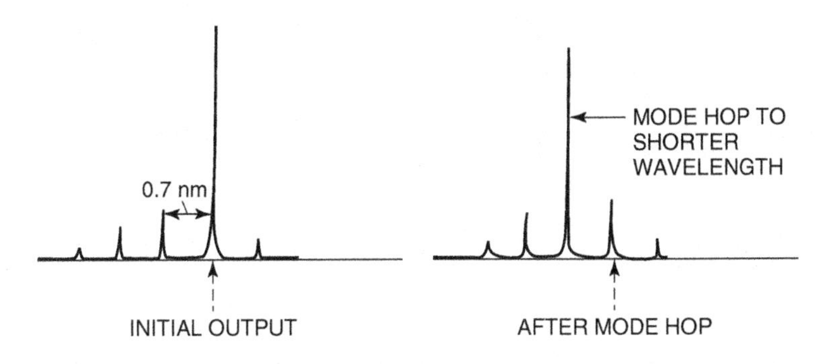

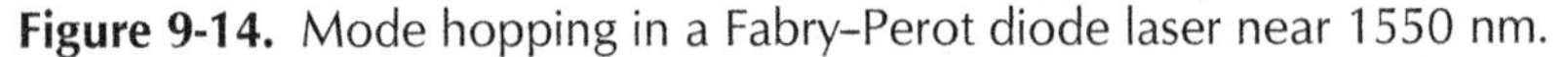

Figure 9-14. Mode hopping in a Fabry–Perot diode laser near 1550 nm.

mechanism that limits the range of wavelengths emitted by the laser. The leading approach is called a *distributed feedback* or *DFB* laser, shown in Figure 9-15A, in which a diffraction grating is fabricated in the base of the active layer. The regularly spaced grooves scatter light back into the active layer at a narrow range of wavelengths, so only those wavelengths receive the feedback needed for laser oscillation.

The spacing of the grating D selects the oscillating wavelength λ according to a formula that also depends on refractive index n and an integer m (in practice, 1 or 2) that denotes how light is being scattered by the grating:

$$D = \frac{m\lambda}{2n} \qquad (9\text{-}2)$$

Plugging in the numbers for a 1550-nm DFB laser made of InGaAsP ($n = 3.4$), we find that grating spacing D is 228 nm for $m = 1$

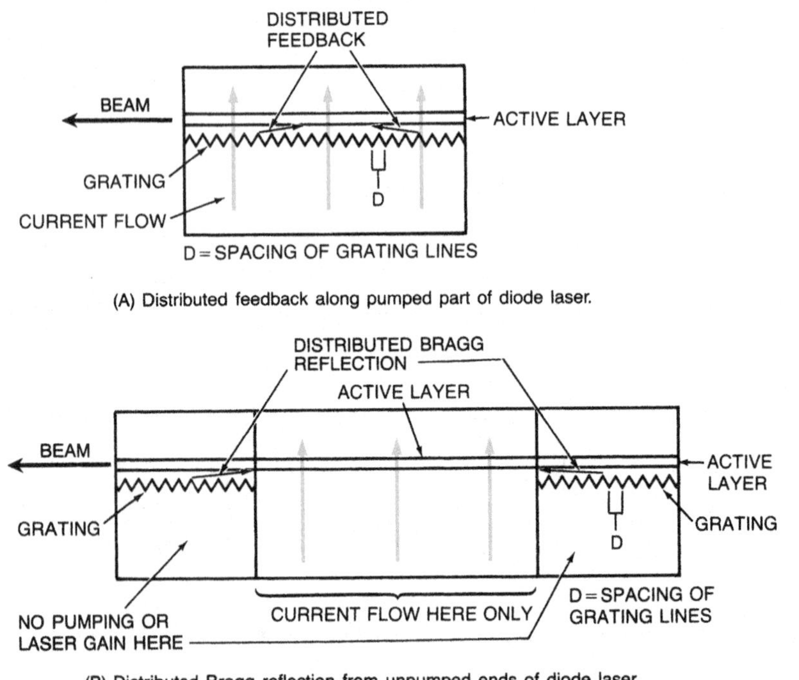

Figure 9-15. (A) Distributed feedback laser and (B) distributed Bragg-reflection laser.

and 456 nm for $m = 2$. Although oscillation wavelength shifts slightly because refractive index depends on temperature, DFB resonators maintain the laser in a stable single longitudinal mode, limiting the range of emitted wavelengths and preventing mode-hopping. Both are critical for high-speed fiber-optic transmission.

An important variation on the DFB laser is placing the grating in a part of the active layer where there is no laser gain, as shown in Figure 9-15B. This has the same effect, but the physics differ in detail. The result is called a *distributed Bragg reflection* or DBR laser.

9.6.3 External Cavity and Tunable Lasers

Diode-laser emission also can be limited to a narrow linewidth by placing it in an external cavity with suitable tuning optics. For example, one facet may be coated to transmit nearly all light emerging from the active layer into external optics that select the wavelength, as shown in Figure 9-16. The figure shows a very short external cavity but, in practice, longer cavities are used to limit output to a narrow range of wavelengths.

The selected wavelength can be tuned by adjusting the optics; in Figure 9-16, the grating is turned to feed light at different wavelengths back into the active layer of the laser. Such tunability is also an attractive feature of external-cavity semiconductor lasers.

Other types of tunable diode lasers have also been developed, largely for fiber-optic systems for which it is desirable to be able to switch emission wavelengths. One example is the *sampled grating, distributed Bragg reflector (SG-DBR)* laser. Like the example in Figure 9-15B, it has gratings on both sides of the active region of

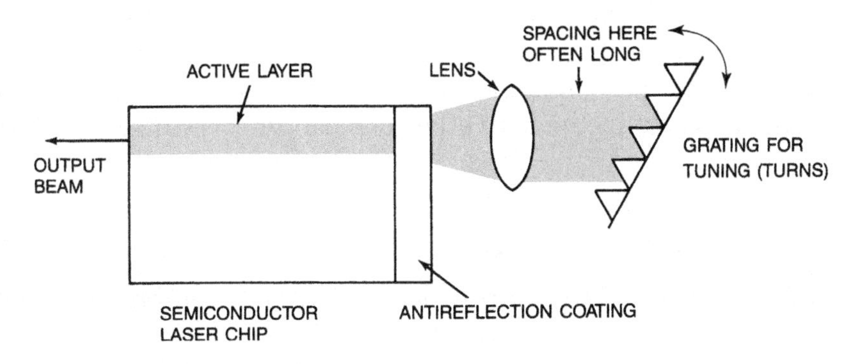

Figure 9-16. External-cavity semiconductor laser.

the laser, but in this case the spacings are different on the two sides. Slight adjustments of the two gratings tune them relative to each other, producing a disproportionately large change in the wavelength at which the laser oscillates.

9.6.4 Semiconductor Optical Amplifiers

A semiconductor diode can serve as an optical amplifier as well as a laser. Like a diode laser, a semiconductor optical amplifier is powered by a current flowing through the device, producing recombination at the junction layer. The difference is that the device amplifies light from an external source, rather than oscillating on its own. This requires coating the diode facets to suppress surface reflection so light makes a single pass through the diode rather than oscillating between reflective surfaces.

9.7 SURFACE-EMITTING DIODE LASERS

Diode lasers can also be designed to emit from their surfaces, which offer some important advantages over edge emitters.

Edge emitters must be diced to expose their edges, then packaged before they can be tested. Surface emitters can be tested before the wafer is diced, and are easier to package—a key advantage because packaging is the biggest cost in laser production. Surface emitters also can be integrated on a single substrate more easily.

The emitting areas on edge emitters are thin and wide, producing rapidly diverging beams that are hard to focus. Surface emitters can be designed with larger, round emitting areas, which produce better-quality beams with much lower divergence that are easier to focus. The larger emitting areas can also handle more power.

Two designs have emerged for surface emitters: one with a vertical cavity perpendicular to the active layer, and another that resonates partly in the active layer. Each deserves a separate description.

9.7.1 Vertical-Cavity Surface-Emitting Lasers (VCSELs)

Vertical-cavity surface emitting lasers (VCSELs) get their name because their resonant cavities are vertical, perpendicular to the active layer, as shown in Figure 9-17. Mirror layers are fabricated above and below the junction layer, with the beam emerging through the surface of the wafer; in practice, usually through the substrate, as shown in Figure 9-17.

VCSELs differ in profound ways from conventional edge emitters. Instead of oscillating along the long dimension of a long, narrow and thin slab of active layer, VCSELs oscillate perpendicular to the surface of a thin disk of active layer. VCSEL cavities also are shorter. These structural differences make VCSELs behave rather differently than edge emitters.

Overall gain within a VCSEL cavity is low because light oscillating between the top and bottom mirrors passes through only a thin slice of active layer. Although the gain per unit length is high in the active layer, the active layer itself is so thin from top to bottom that the total gain in a round-trip of the VCSEL cavity is small. To sustain oscillation, resonator mirrors on top and bottom of the VCSEL must reflect virtually all the stimulated emission back into the laser cavity.

The high-reflectivity mirrors needed for the laser cavity are fabricated in the semiconductor itself, by depositing many alternating thin layers of two different compositions of semiconductor

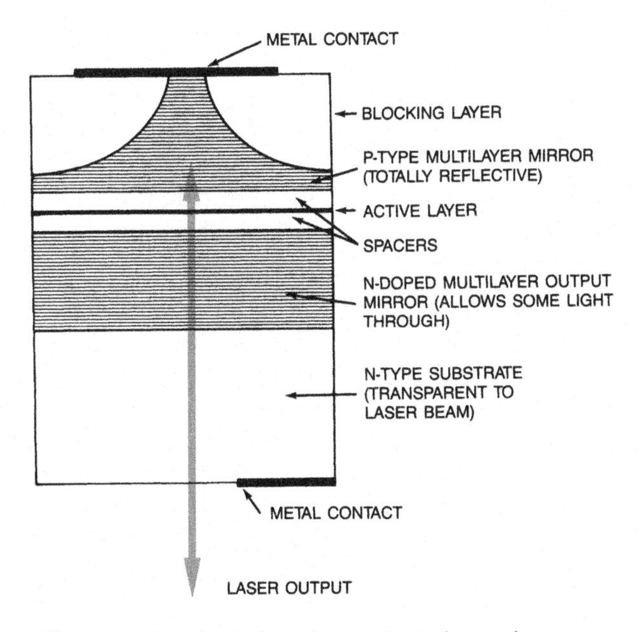

Figure 9-17. Vertical-cavity semiconductor laser.

with slightly different refractive index values. This multilayer structure forms a multilayer interference coating, described in Section 5.3.3, which can be designed to strongly reflect a particular wavelength. The reflector on the substrate side transmits a small fraction of the cavity light; the reflector above the active layer reflects all the light back into the cavity.

This structure is limited to generating powers in the milliwatt range, well below the maximum available from edge emitters. However, VCSELs have extremely low threshold current, making them significantly more efficient. Their high efficiency and low drive current also gives them a long lifetime.

The short length of the VCSEL cavity has another important consequence. Recall from Chapter 4 that an integral number N of wavelengths λ fit into a laser cavity with length L and refractive index n according to the formula

$$2nL = N\lambda \tag{9-3}$$

The shorter the cavity, the larger the difference between resonant wavelengths. That means that VCSELs are much less likely to hop to modes oscillating at different wavelengths than edge emitters. This improves their performance, and helps allow direct modulation by varying drive current for data rates to well above one gigabit per second.

The surface emission comes from a region that usually is circular and typically ranges from 5 to 30 micrometers in diameter. Unlike edge-emitting diodes, the beams are circular in cross section, an advantage for many optical applications. The output also can be coupled directly into optical fibers by putting the output face directly against the core of the fiber.

In principle, VCSELs can be made from any direct-bandgap III–V semiconductor using standard semiconductor manufacturing process to deposit the mirror layers as well as the p–n junction structure. Gallium arsenide VCSELs were easiest to develop because the refractive index of GaAlAs varies considerably with aluminum content, providing the refractive-index contrast needed for the multilayer mirrors. As a result, 850-nm VCSELs were the first to find wide applications in short-distance fiber-optic communication systems, where limited power was not a problem. More recently, VCSELs have been developed using InGaAsP compounds that emit at the 1300- and 1550-nm fiber-optic windows.

VCSELs are fabricated in arrays of many devices on a single wafer, but most of them are packaged individually rather than used in arrays. Because VCSELs emit from the top of the wafer, they can be tested before the wafer is diced into many individual devices and packaged. Edge emitters cannot be tested until the wafer is scribed and diced into individual components, raising the costs of testing and fabrication. This leads to lower testing and packaging costs for VCSELs, and lower prices.

The most important applications of VCSELs are in fiber-optic data links transmitting at gigabit speeds up to a few kilometers at wavelengths of 850 and 1300 nm. The milliwatt powers available from VCSELs are perfect for short distances, and direct modulation at 1 Gbit/s is easy and inexpensive, so they are the preferred laser type for these widely used fiber-optic links.

Tunable VCSELs have been made by suppressing reflection from the top or bottom and adding an external cavity. The principle is the same as external-cavity edge-emitters, but so far the power and applications have remained limited.

9.7.2 Horizontal Cavity Surface Emitters

Producing higher power from surface emitters requires a way to put more gain medium in the laser cavity. That can be done by deflecting light generated in the horizontal plane of the active layer so it emerges from the surface. A simple approach is to etch a 45 degree mirror to reflect light upward from the active layer. An alternative is to use a diffraction grating at the base of the active layer to scatter light out of the junction toward the surface.

Grating-coupled surface-emitting lasers have been demonstrated in a variety of configurations, including external-cavity and oscillator–amplifier versions. They also can be fabricated in arrays. At this point, they are largely developmental.

9.8 QUANTUM WELLS AND DOTS

Conventional semiconductor physics does not consider the quantum-mechanical properties of atoms and electrons, but quantum properties become significant on very small scales. A critical threshold is 50 nanometers, roughly the "wavelength" of an electron in a semiconductor. (Just as light has a dual existence as a wave and a particle, electrons can behave as waves, and thus have a wavelength of their own.)

Earlier you learned that double heterostructures can confine light and electrons so they can move in only two dimensions. When the double heterostructure is made thinner than 50 nanometers, its internal energy levels change, altering its confinement of electrons and making it a *quantum well.* Quantum wells confine electrons more tightly than bulk semiconductor layers. Stacking two or more quantum wells inside the active layer in a multiple quantum-well laser increases the quantum-well volume in which stimulated emission can occur. Photons have wavelengths of hundreds of nanometers in a semiconductor, so they are confined on a larger scale than electrons.

It's possible to make *quantum wires* that confine electrons within a linear structure smaller than an electron wavelength, so they can only move in one dimension, along the quantum wire. However, there is more interest in *quantum dots,* which confine electrons in all three dimensions on a scale smaller than the electron wavelength.

Quantum dots can be made in sizes from 2 to 50 nm, and have become a hot area in nanotechnology. They come in many forms, some of which can be embedded in the junction layers of diode lasers or LEDs. Recombination then occurs in the quantum dot, on sharply defined transitions that depend on the size and composition of the quantum dot. Quantum dot lasers have low thresholds and offer a new way to control the emission wavelength, but they have yet to achieve the high efficiency sought by developers.

9.9 QUANTUM CASCADE LASERS

A quantum cascade laser is a fundamentally different type of semiconductor laser in which energy is extracted step by step from electrons passing through a series of quantum wells in a semiconductor structure. Each of the quantum wells has energy levels designed to trap electrons passing through the structure, and an electric field is applied along the length of the structure. Electrons enter the quantum well and are trapped in an upper energy state, where they can be stimulated to emit light and drop to a lower energy level in the quantum well. From that state they can drop to a lower level and tunnel out of that quantum well and into another, as shown in Figure 9-18. The electrons drop successively lower in energy as they pass along the structure, like marbles rolling down stairs.

The slanted steps in Figure 9-18 show the energy level of the conduction band, which is lower in the quantum wells than in the intermediate zones. The energy levels within the quantum well are sublevels within the conduction band. The slant comes from the variation of the electric field along the length of the laser. The electric potential creates a gradient in energy levels that essentially pulls the electron through the structure, like a marble rolling down stairs. Ideally, the electron could be stimulated to emit energy each time it drops to a new quantum well, so one electron

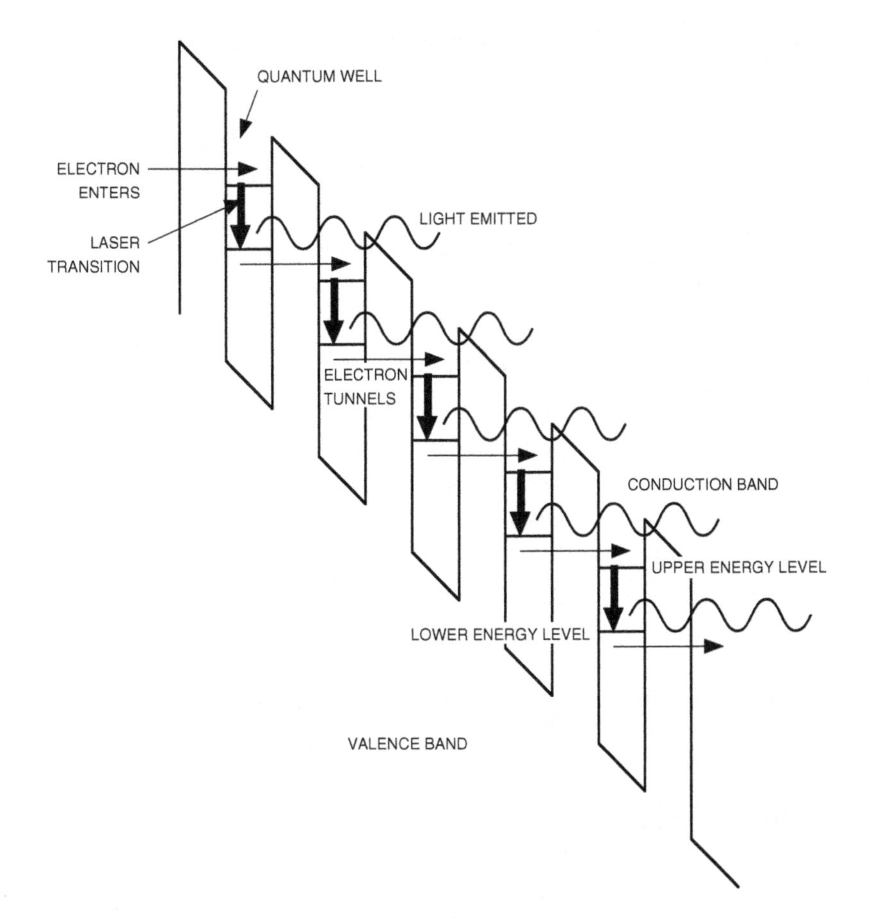

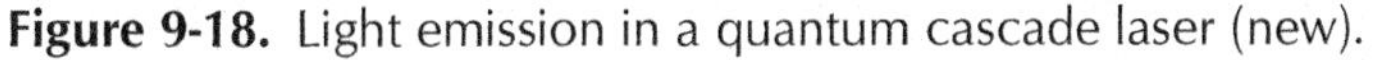
Figure 9-18. Light emission in a quantum cascade laser (new).

could emit many photons, but it is not that easy. Figure 9-18 shows a simplified view. In practice, each step shown represents a series of thin layers that combine to function as one level.

The energy levels in the quantum wells are tailored by selecting their composition and thickness. In practice, the separation of the levels produces laser emission at wavelengths from a few micrometers to more than 100 μm. Like other semiconductor lasers, quantum cascade lasers are made of III–V materials, but they are not junction lasers or diodes.

Quantum cascade lasers are important because they emit infrared wavelengths longer than are readily available from other semiconductors. Those wavelengths are important for sensing.

9.10 OPTICAL PROPERTIES OF DIODE LASERS

The optical properties of diode lasers depend on their structures, and both differ from those of other lasers. All diode lasers are physically small and emit from a small aperture. That affects beam properties including divergence, coherence, and mode structure. The details differ among types, particularly between edge emitters and VCSELs.

9.10.1 Beam Shape and Divergence

As you learned in Chapter 4, a beam emitted through a small aperture D has a relatively large divergence angle θ, because divergence is proportional to the wavelength divided by the aperture (i.e., to the number of wavelengths in the aperture). For circular apertures many wavelengths across, the divergence angle is

$$\theta(\text{radians}) = \frac{K\theta}{D} \qquad (9\text{-}4)$$

where K is a constant near 1 and depends on the beam profile.

Stripe-geometry, edge-emitting diode lasers emit from the edge of the active layer, a region a fraction of a micrometer high and several micrometers wide. This shape is quite different from the round emitting areas of most gas and solid-state lasers, so the formula for beam divergence is only an approximation. Nonetheless, measurements confirm that the beam spreads more rapidly in the vertical direction, where the emitting aperture is narrow, than in the horizontal direction, where the emitting aperture is wider, as shown in Figure 9-19. Typical values for beam divergence are 10 degrees (0.17 radian) in the direction parallel to the active layer and 40 degrees (0.70 radian) in the direction perpendicular to the active layer. This spreading angle is larger than the beam from a good flashlight, making the beam quite different from the tightly focused beam of a helium–neon laser or a solid-state laser.

Fortunately, external optics can correct for this broad beam divergence. A cylindrical lens, which focuses light in one direction but not in the perpendicular direction, can make the beam circular in shape. Collimating lenses can focus the rapidly diverging beam from an edge emitter so it looks as narrow as the beam from a helium–neon laser. Those lenses are hidden inside standard red semiconductor-laser pointers, so they produce tightly focused spots on the screen or wall.

VCSELs are a different matter because they emit from circular apertures generally at least 5 μm across, so their beams diverge symmetrically, with typical divergence of 10 degrees (0.17 radian). As for edge emitters, the addition of collimating optics can reduce beam divergence.

9.10.2 Bandwidth and Modes

Earlier sections have mentioned the oscillating modes of various types of diode lasers, but it is worth looking at them again and

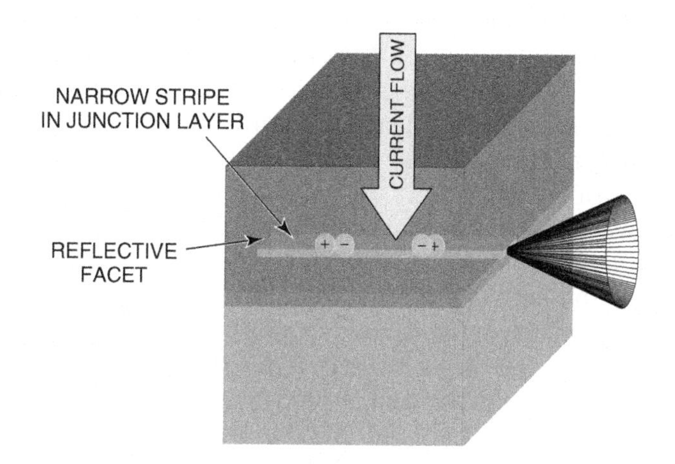

Figure 9-19. Beam divergence from an edge-emitting laser.

how they affect the spectral bandwidth of diode lasers. Spectral bandwidth is important in fiber-optic systems, because the slight variation in refractive index with wavelength can cause pulses to stretch out when traveling through long lengths of fiber, an effect called dispersion.

High-power diode lasers and arrays of many diode laser stripes oscillate in many modes, and the range of wavelengths they emit is usually limited to a nanometer or less, which is fine for most applications.

Narrow-stripe diode lasers typically oscillate in only one transverse mode, so they may be called "single-mode lasers," but Fabry–Perot lasers can oscillate in more than one longitudinal mode, and hop between modes when operating conditions change. The modes are 0.2 to 0.7 nm apart, and the range of wavelengths they emit is too broad for high-speed fiber-optic networks.

DFB, DBR, and external-cavity lasers limit laser oscillation to a single stable longitudinal mode, giving them the extremely narrow bandwidth required for high-speed fiber-optic networks.

The short cavity of VCSELs limits them to oscillating in a single longitudinal mode, but their bandwidth is not as narrow as DFB or DBR lasers, which internally stabilize the output wavelength. Large-aperture VCSELs oscillate in multiple transverse modes.

9.11 DIODE LASER MATERIALS AND WAVELENGTHS

Earlier in this chapter you learned that the wavelengths emitted by diode lasers depended on the composition of the semiconductor, and learned which families of semiconductors could emit light. Now that you have learned about the general workings of semiconductor lasers, let us look in more detail at the materials and their implications.

The two parameters that most directly affect the structure of diode lasers are the band gap of the material and the lattice constant or spacing between atoms.

When an electron and hole recombine, the energy they release—and thus the laser wavelength—depends on the gap between the conduction band occupied by the free electron and the valence band where the hole is. The band gap depends on the atomic composition of the semiconductor; in general, change the composition and you change the band gap, although there are some exceptions. Table 9-2 lists important types of semiconductor lasers and their usual wavelengths.

The band gap is also important in controlling electron behavior in a diode laser. If a material with a low band gap energy is sandwiched between layers with higher band gap energy (forming a double heterojunction), conduction electrons can be trapped in the low band gap material. Band gap energy is often given in electron volts; to calculate the equivalent wavelength (in nanometers) that a diode made of that material would emit, divide 1240 by the band gap energy E in electron volts:

$$\lambda = \frac{1240}{E} \tag{9-5}$$

As you learned in Section 9.4.4, the lattice constant is important in semiconductor fabrication. Adjacent layers should have nearly identical lattice constants so the atoms line up properly. If the atomic spacings do not match, the differences causes strain in the layers. Semiconductors can withstand a little strain, particularly in a series of thin layers, called a *strained layer superlattice,*

Table 9-2. Major diode laser materials and wavelengths

Material	Wavelength range	Comments
AlGaN	350–400 nm	Developmental at shorter wavelengths
GaInN	375–440 nm	Commercial; mostly in UV, violet, and blue
ZnSSe	447–480 nm	Laboratory
ZnCdSe	490–525 nm	Laboratory
AlGaInP/GaAs	620–680 nm	Commercial
$Ga_{0.5}In_{0.5}P$/GaAs	670–680 nm	Commercial
GaAlAs/GaAs	750–900 nm	Commercial
GaAs/GaAs	904 nm	Commercial
InGaAs/GaAs	915–1050 nm	Strained layer; commercial
InGaAsP/InP	1100–1650 nm	Commercial
InGaAsSb	2–5 μm	Some commercial
PbCdS	2.7–4.2 μm	Requires cooling
Quantum cascade	3–50 μm	Not a diode
PbSSe	4.2–8 μm	Requires cooling
PbSnTe	6.5–30 μm	Requires cooling
PbSnSe	8–30 μm	Requires cooling

but too much strain causes defects in the material, leading to device failure.

9.11.1 Semiconductor Properties and Composition

A pure semiconductor such as silicon has uniform properties, with the same band gap and lattice properties. A simple binary semiconductor such as GaAs or InP also has uniform properties.

You can get other properties by adding one or two other elements from the same groups of the periodic table to a compound semiconductor. This is equivalent to blending different materials to get a new compound with intermediate characteristics. For example, a mixture of 20% AlAs and 80% GaAs has band gap energy and lattice spacing partway between pure AlAs and pure GaAs. The formula for such a compound can be written $Ga_{0.8}Al_{0.2}As$ or simply GaAlAs. (The order of the two elements that replace each other—aluminum and gallium—is arbitrary; some people write AlGaAs.)

Things get more complex if the compound semiconductor contains four elements. Depending on the relative concentrations of the elements, such "quaternary" semiconductors can have properties somewhere in a broad range. For example, as shown in Figure 9-20, InGaAsP (more precisely $In_xGa_{1-x}As_{1-y}P_y$) can have lattice spacing and bandgap within an area defined by four possible binary compounds: InAs, InP, GaAs, and GaP. The material's properties can vary within the entire space because the In–Ga and As–P ratios can be adjusted independently. (The odd shape of the area in the figure arises because pure GaP is an indirect band gap material, unsuitable for lasers and somewhat different from the other binary compounds. The dashed line indicates the indirect band gap region.) Plots like Figure 9-20 are very helpful in understanding semiconductor properties; they are commonly drawn with lattice spacing on the bottom, but some have lattice spacing on the sides and band gap on the bottom.

The band gaps of different layers in a semiconductor device do not have to match, but the lattice spacing must be close to avoid flaws that degrade device performance. This means that all compositions in a bulk InGaAsP structure must fall roughly on a vertical line in Figure 9-20. The use of thin strained layers relaxes this constraint somewhat, as mentioned earlier.

Semiconductor laser structures are grown on substrates of simple-to-produce binary materials such as InP or GaAs, which

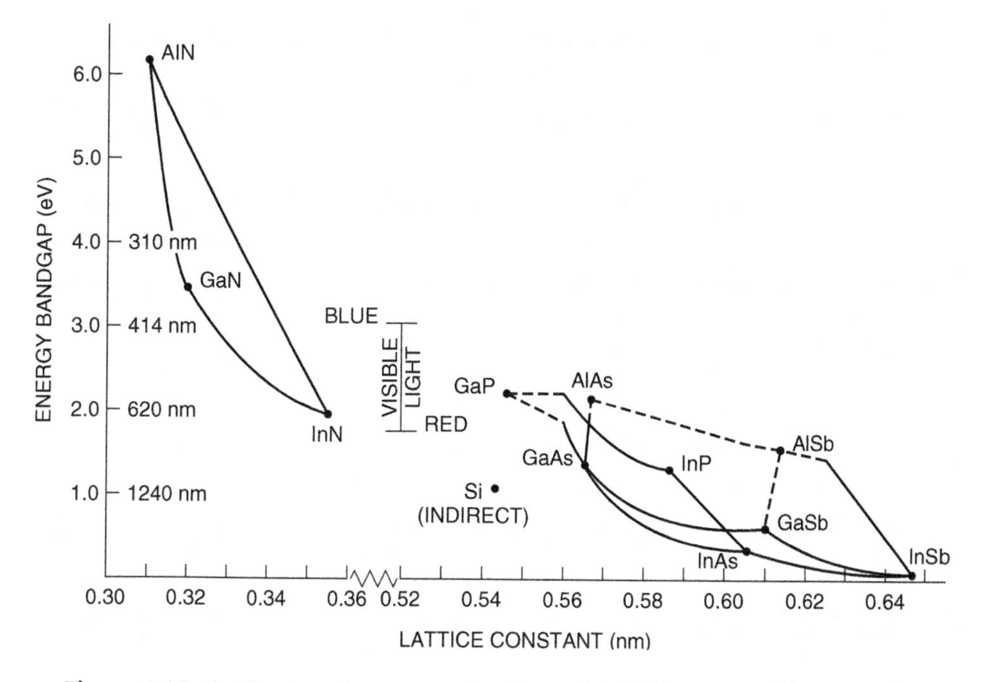

Figure 9-20. Lattice constant versus band gap for III–V semiconductors in the ultraviolet, visible, and near-infrared. Note discontinuity in lattice-spacing scale. Dashed lines are indirect band gap compounds.

account for most of the device volume. The choice of substrate restricts composition of other layers to materials with similar lattice spacing.

9.11.2 Blue, Violet, and Ultraviolet Diode Lasers (Nitrides)

One of the top laser breakthroughs of the 1990s was the development of blue diode lasers based on gallium indium nitride (GaInN). As shown on the left side of Figure 9-20, this compound can be used for blue and violet lasers; adding aluminum to the compound extends the band gap well into the ultraviolet, although diode lasers have been demonstrated only to about 340 nm. (LEDs have operated at wavelengths shorter than 280 nm.) The materials are often classed as gallium nitride (GaN) compounds. Some are fabricated on GaN substrates; others are deposited on sapphire (Al_2O_3) or silicon carbide (SiC) substrates.

So far, the most important applications of GaN materials have been in violet diodes emitting near 405 nm, which are used for high-density optical data storage in Blu-Ray and HD-DVD disks. Longer wavelength visible diode lasers can be made by adding indium and reducing gallium concentration. Blue diodes emitting at 440 nm are standard products and ultraviolet diodes emitting at 375 nm are also available.

Shorter-wavelength ultraviolet lasers become increasingly difficult because it's hard to make good electrical contacts with material containing more aluminum, and efficiency decreases as aluminum content increases. Available powers and device lifetimes drop as wavelength decreases below about 370 nm.

9.11.3 Red Diode Lasers

Although we take red diode lasers for granted today, the first diode lasers with beams easily visible to the human eye, at 670 nm, did not reach the market until the late 1980s. These lasers have active layers of $Ga_{0.5}In_{0.5}P$ surrounded by layers in which aluminum replaced some of the gallium to raise the band gap and improve confinement.

Red diodes also are available at 635 nm, where the human eye is more than 10 times more sensitive than at 670 nm. The 635-nm lasers have active layers of AlGaInP, in which the number of aluminum, gallium, and indium atoms together equal the number of phosphorous atoms, a blend that can be written as $Al_yGa_xIn_{1-x-y}P$. Like 670 nm diodes, they are grown on GaAs substrates.

Red lasers can generate moderate powers, to a few watts from an array, with the available power generally increasing at longer wavelengths.

9.11.4 Gallium Arsenide Near-Infrared Lasers

The oldest and best-developed family of diode lasers have active layers of GaAlAs or GaAs and are fabricated on GaAs. They are usually called "gallium arsenide" lasers. Pure GaAs active layers nominally emit at 904 nm, but replacing some of the gallium with aluminum increases the band gap to generate shorter wavelengths. Fortunately, aluminum and gallium atoms are nearly the same size, so GaAlAs can be easily lattice-matched to a GaAs substrate. However, flaws are more likely to develop in lasers with higher aluminum concentrations, so the shortest wavelengths from GaAlAs lasers typically are around 750 nm.

Hundreds of millions of inexpensive low-power 780-nm GaAlAs lasers are used each year in CD and CD-ROM players. Similar lasers are used in laser printers, short-distance fiber-optic communications, and in instruments.

High-power GaAlAs lasers are used extensively for pumping of solid-state lasers. The most widely used type are 808-nm lasers, for pumping neodymium-doped lasers. High-power pump diodes also are produced at 830 and 850 nm.

9.11.5 InGaAs Pump Lasers

Obtaining longer wavelengths from lasers in the GaAs family requires adding indium to replace some of the gallium and reduce the band gap. Fabricated on GaAs substrates, InGaAs lasers can generate high powers for pumping solid-state or fiber lasers. InGaAs emitting at 915 nm is widely used as a pump for ytterbium-doped fiber lasers. InGaAs emitting at 980 nm is widely used for pumping erbium-doped fiber amplifiers and erbium-fiber lasers. InGaAs emitting at 940 nm is used for pumping erbium-doped and erbium–ytterbium codoped fiber lasers and amplifiers.

Adding indium to increase the wavelength of the laser also increases the lattice constant, requiring fabrication of thin strained layers between the GaAs substrate and the InGaAs layers to prevent dislocations in the laser structure.

9.11.6 InGaAsP Lasers for Fiber-Optic Systems

The quaternary (four-element) semiconductor InGaAsP is a versatile laser material grown on substrates of InP. The sum of the number of indium and gallium atoms equals the total number of arsenic and phosphorous atoms. The chemical formula can be written as $In_xGa_{1-x}As_yP_{1-y}$, with the values of x and y varying independently. Adding either element changes the lattice constant, so successfully growing the compound on an InP substrate requires balancing the composition to match the lattice constant to the substrate.

InGaAsP active layers can emit at 1100 to 1650 nm when lattice-matched structures are grown on InP substrates. The major applications of InGaAsP lasers are in fiber-optic communications, so most lasers are manufactured for the fiber-optic bands around 1310 nm and from about 1480 to 1600 nm.

The type of InGaAsP lasers used depends on the number of wavelengths being transmitted through the fiber-optic system. Wavelength tolerances are ± 20 nm or so for systems transmitting only a single wavelength, which can use uncooled Fabry–Perot lasers. Systems transmitting several wavelengths about 20 nm apart (coarse wavelength-division multiplexing) also use Fabry–Perot lasers, but more care must be taken to see that their output falls into the proper wavelength channel.

High-speed networks pack wavelengths much closer together to increase capacity; this is called dense wavelength-division multiplexing. These wavelength slots are only a fraction of a nanometer wide, typically 0.8 nm, so they require narrow-line DFB or DBG lasers, which must be cooled to keep them at a stable temperature because laser wavelength varies slightly with temperature.

9.11.7 Other Diode Lasers

As you can see in Table 9-2 and Figure 9-20, diode lasers can be made from a variety of other semiconductor compounds.

III–V compounds of indium, gallium, arsenic, and antimony (InGaAsSb) have smaller band gaps than InGaAsP, and can be deposited on substrates of GaSb. Diode lasers made of these materials emit at wavelengths between 1.9 and 5 μm. Lasers at some of these wavelengths are available commercially; their main applications are in instruments.

Diode lasers have been demonstrated in compounds containing elements from groups II and VI of the periodic table. These II–VI compounds include zinc sulfide, zinc selenide, zinc telluride, cadmium selenide, and cadmium sulfide. However, GaN lasers proved more practical for short-wavelength diode lasers.

A family called "lead salt" diode lasers emits from 2.7 μm to about 30 μm in the infrared, depending on composition, as listed in Table 9-2. Made of compounds containing lead, sulfur, selenium, and other elements, they require cooling to cryogenic temperatures. Their prime applications were in research and precision measurement of the infrared properties of materials.

9.12 SILICON LASERS

If you have been following press reports over the past few years, you may wonder why silicon lasers have not been described yet.

The reason is that these much-hyped developments are not really diode lasers. They actually are optically pumped Raman lasers, similar to the fiber Raman lasers described in Chapter 8, or hybrids in which the emission comes from a III–V compound. These developmental devices are covered in Section 10.5.

9.13 PACKAGING AND SPECIALIZATION OF DIODE LASERS

Single unpackaged diode lasers look like tiny squares of metallic confetti. The lasers must be packaged for any practical applications. Some are packaged for general-purpose use, often in metal housings similar to standard transistor cans but with windows for optical output, such as shown in Figure 9-21. However, most are assembled in packages designed for specific applications. In general, testing and packaging the lasers can cost many times the

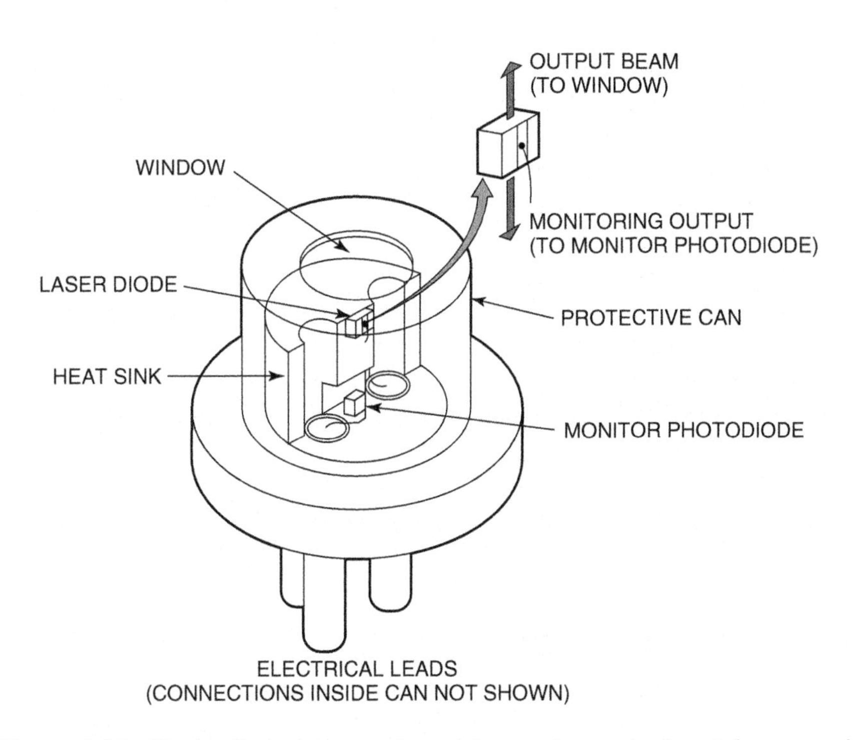

Figure 9-21. Single diode laser packaged in an electronic "can" for general-purpose use. (Courtesy Spectra-Physics, a division of Newport Corporation.)

price of bare diode lasers because it requires precision handling and alignment of the tiny chips.

Diode laser sales reached about $3 billion in 2006, accounting for more than half the global sales of all types of lasers, according to *Laser Focus World.* The total number of diode lasers sold was more than 800 million, of which some 720 million were used for optical storage—CD players, CD-ROM drives, and DVD drives and players. However, the average price for each optical storage laser was only $2, and total sales about $1.5 billion.

Sales of diode lasers for fiber-optic telecommunication systems came in a close second in dollars at about $1.15 billion, but was far behind in numbers at about 5 million. Third in dollar volume was pump diodes for solid-state lasers, at $163 million for 525,000 units. Let us look at what goes into lasers for each of these applications.

9.13.1 Optical Disk Lasers

The first CD players reached the market in 1980, and since then they have become ubiquitous and inexpensive. They incorporate a single GaAlAs laser emitting at 780 nm, packaged in a playback head that includes focusing optics and a light detector, which moves back and forth radially over the spinning disk. The original CD format was read-only, but writeable disks were added later. Writing requires more power than reading, and is not offered in the least expensive players.

DVD players and drives use 650-nm lasers, which can be focused to smaller spot sizes than 780-nm lasers. Smaller spot size and tightening other tolerances allow DVDs to squeeze in about six times more data than CDs. As in CD drives, the laser is packaged in an optical head for reading, and for writing if that option is included.

The emerging generation of HD optical disks uses 405-nm blue diode lasers, but so far the volume is small. The design approach is similar to that in other optical disk players.

9.13.2 Fiber-Optic Lasers

Diode lasers play a number of different roles in fiber-optic communications. Those used to transmit signals are packaged to fit into electronic equipment, and generally are mounted either with a fiber pigtail or in a connector housing that would bring a fiber close enough to the laser to collect light from it.

Inexpensive low-power GaAlAs VCSELs emitting at 850 nm are used in data communication systems spanning less than one kilometer. VCSELs emitting at 1310 nm are used in data communications systems spanning up to a few kilometers. These lasers often are not counted as telecommunications lasers.

InGaAsP diodes emitting near 1310 nm are used in some data communications, and in local communications systems, such as fiber-optic systems serving homes. InGaAsP diodes emitting from 1480 to about 1560 nm are used in regional or long-distance communication systems, with some installed in telephone company facilities that serve homes. Lasers used in dense wavelength-division multiplexing, where wavelength stability is required, are packaged so they operate at a stable temperature to maintain a constant wavelength.

InGaAs lasers emitting at 980 nm are used to pump erbium-doped fiber amplifiers used in long-distance telecommunications. Unlike the transmitter lasers, they are packaged to deliver higher-power pump light into one end of the fiber amplifier.

9.13.3 Pump Lasers

Packaging of pump lasers depends on power levels. Single-stripe or wide-area pump lasers may be coupled into optical fibers. Multistripe arrays or stacks of arrays are packaged in structures that focus light onto a laser rod (typically onto the side), and cool the diode stacks to maintain stable operation. One compact example is shown in Figure 9-22; it can emit a continuous 6-W beam. As you would expect, the higher the power level, the more cooling is required; the highest powers require active cooling systems.

9.13.4 Other Diode Laser Packaging

As visible diode lasers have come to replace helium–neon lasers in applications that require visible light, the diodes have been packaged with optics that focus the light to produce a tightly focused beam. Examples are laser pointers and supermarket scanners, which require highly directional beams. The user sees a small packaged laser that emits a red beam and functions much like a much larger and more expensive helium–neon laser.

Figure 9-22. Compact 6-W pump diode package with a mirror that directs output upward, packaged for surface mounting on a circuit board. (Courtesy Osram Opto Semiconductor.)

9.14 WHAT HAVE WE LEARNED?

- Semiconductor lasers are often called diode lasers or laser diodes.
- Diodes are two-terminal electronic devices that conduct current in one direction.
- Electrons in the valence band are bound to atoms. Electrons in the conduction band are free to move in a solid.
- Holes are vacancies in the valence band of a semiconductor. A hole moves when an electron from another atom fills the empty space in the valence band.
- *n*-type semiconductors are doped with elements that release electrons to carry current. *p*-type semiconductors are doped with elements that form holes.
- Current flow through a forward-biased diode causes electrons to recombine with holes at the junction layer between *n*- and *p*-type semiconductors.
- An electron–hole pair called an exciton exists briefly before the electron drops into the hole and releases its energy.

- LEDs emit recombination radiation when spontaneous emission occurs at the junction layer. The emission wavelength depends on the band-gap energy of the material.
- Semiconductors need a direct band gap to emit light efficiently. Silicon has an indirect band gap so it does not emit light efficiently.
- III–V compounds such as GaAs. InGaN, and InGaAsP are called compound semiconductors.
- Diode lasers are structurally similar to LEDs, but have resonant cavities. When drive current exceeds a threshold value, they generate stimulated emission.
- The current needed to reach laser threshold winds up as waste heat in a diode laser.
- A single-heterojunction laser improves confinement by depositing layers with different compositions on top of each other.
- Room-temperature continuous-wave emission requires a double heterojunction, with a junction layer of one composition sandwiched between layers of other composition.
- Confining light emission to a narrow stripe in the junction plane improves diode laser performance.
- Gain guiding confines current flow through a diode laser to guide light in a narrow stripe in the junction layer.
- Index guiding surrounds the active stripe in the junction laser with lower-index material to guide light.
- High-power diode lasers require either a broad stripe or arrays of many parallel stripes.
- A simple edge-emitting diode laser oscillates in the junction plane, with the edges of the chip serving as the resonator mirrors. This design is called a Fabry–Perot laser, because it uses two parallel surfaces as a resonator.
- The emitting area of a narrow-stripe laser is a few micrometers wide and a fraction of a micrometer high.
- Distributed-feedback and distributed Bragg-reflection lasers confine diode-laser oscillation to a narrow range of wavelengths.
- External-cavity lasers confine light to a narrow range of wavelengths that can be tuned by adjusting the optics.
- Semiconductor optical amplifiers resemble diode lasers, but reflection from edge facets is suppressed so they can amplify light but not oscillate.
- A VCSEL is a diode laser that oscillates in a vertical cavity, perpendicular to the junction plane, and emits light from its sur-

face. VCSELs are inexpensive to fabricate and package, and have low drive currents.
- Horizontal-cavity surface-emitting lasers oscillate in the junction plane, but include optics that direct their output through the surface of the chip.
- A quantum well is a double heterojunction thinner than 50 nm which confines electrons tightly in a layer with lower band-gap energy than the surrounding layers.
- A quantum cascade laser extracts energy in a series of steps from electrons passing through a series of quantum wells.
- Edge-emitting diode lasers have a large beam divergence, but external optics can focus the light into a narrow beam.
- Defects occur in diode lasers if the atomic spacing in successive layers is not matched or managed carefully.
- GaInN diodes emit blue, violet, and ultraviolet light.
- Diode lasers with active layers of AlGaInP emit at 635 nm in the red.
- GaAlAs lasers emit in the near-infrared; their main use is in CD and CD-ROM players. They also can emit high power for pumping other lasers at 808, 830, and 850 nm.
- InGaAs lasers emitting at 980 nm are widely used as pump lasers.
- InGaAsP diode lasers emit at wavelengths of 1100 to 1650 nm; their main applications are in fiber-optic communication systems.
- Packaging determines the function of a diode laser; it often costs more than making the laser itself.
- Diode lasers are simple to modulate by changing their drive current.

WHAT'S NEXT?

In Chapter 10, we will describe types of lasers that are in development or do not fit into the three major categories of Chapters 7, 8, and 9. These include tunable organic-dye lasers, free-electron lasers, extreme-ultraviolet lasers, and silicon lasers.

QUIZ FOR CHAPTER 9

1. In order to make a good diode laser, a material should be
   a. A semiconductor with an indirect bandgap
   b. A semiconductor with a direct bandgap

   c. An insulator with an indirect bandgap
   d. An insulator with a direct bandgap
   e. A conductor
2. Which of the following is a quaternary III–V semiconductor?
   a. InGaAsP
   b. PbSnSSe
   c. GaAlAs
   d. GaAs
   e. NSbAsP
3. The active layer of a diode laser is made of $Ga_{0.8}Al_{0.2}As$. What percent of the atoms in the active layer are of each element?
   a. 80% gallium, 20% aluminum, 100% arsenic
   b. 80% gallium, 20% aluminum, 0% arsenic
   c. 40% gallium, 10% aluminum, 50% arsenic
   d. 20% gallium, 80% aluminum, 100% arsenic
   e. 33% gallium, 33% aluminum, 33% arsenic
4. An exciton is
   a. A free electron in the conduction band
   b. A free electron in the valence band
   c. The energy released by the combination of an electron and a hole
   d. An electron–hole pair in an excited state because the electron has not combined with the hole
   e. A molecule that exists only in the excited state.
5. What gives a double-heterostructure laser better efficiency than a homostructure laser?
   a. Reverse biasing
   b. Better confinement of conduction electrons
   c. Restriction of current flow to the active layer
   d. Lower levels of spontaneous emission
   e. A homostructure laser is more efficient
6. Which type of laser is made from GaAlAs?
   a. Fabry–Perot edge emitter
   b. VCSEL
   c. External cavity
   d. Distributed feedback edge emitter
   e. All of the above
7. Which kind of laser has the shortest optical cavity?
   a. Helium–neon gas
   b. Edge-emitting diode
   c. External cavity diode

d. VCSEL
e. Impossible to tell

8. You are building a video player for use with high-definition disks. What kind of laser do you need?
   a. InGaN diode
   b. GaAlAs edge emitter
   c. GaAlAs VCSEL
   d. InGaAsP distributed feedback
   e. Quantum cascade
9. A semiconductor has a band gap of 1.5 electron volts. At what wavelength will it emit light if it can operate as a laser?
   a. 1500 nm
   b. 1000 nm
   c. 827 nm
   d. 667 nm
   e. 333 nm
10. Which type of semiconductor lasers emit a single longitudinal mode?
    a. Homojunction
    b. Buried-heterostructure
    c. Monolithic arrays
    d. Distributed-feedback
    e. Those built for compact disc players
11. What family of diode lasers emits the shortest wavelength?
    a. GaAlAs
    b. AlGaInP
    c. InGaAsP
    d. GaInN
    e. AlGaN
12. What type of diode laser is most likely to be used in a red laser pointer?
    a. GaAlAs
    b. AlGaInP
    c. InGaAsP
    d. GaInN
    e. AlGaN

CHAPTER 10

OTHER LASERS AND RELATED SOURCES

ABOUT THIS CHAPTER

Gas, solid-state, and semiconductor lasers are widely used in a variety of applications, and most types are available commercially. However, other types of lasers (and some important laser-like sources) either remain in the research stage, are largely limited to research, or somehow do not fit into the three main categories. This chapter describes both lasers that have been around for many years, such as the tunable dye laser, and emerging types such as extreme-ultraviolet sources, the free-electron laser, and silicon lasers.

10.1 TUNABLE DYE LASERS

Tunable wavelength is an invaluable feature for lasers used in research, and scientists have long been willing to put up with considerable inconvenience to be able to adjust laser wavelength. Today, the titanium–sapphire laser described in Chapter 8 is the most important tunable laser. However, for many years the most important tunable lasers were based on organic dyes dissolved in liquid solvents, which are still manufactured and still used for some applications.

Tunability is important to researchers who want to be able to probe the effect of small changes in wavelength, or want wavelengths not readily available from other lasers. The first materials that laser physicists found suitable for tunable lasers were organic dyes, a family of large and complex molecules with complex sets

Understanding Lasers: An Entry-Level Guide, Third Edition. By Jeff Hecht

of electronic and vibrational energy levels that makes them brightly colored. The vibrational energy levels create many sublevels of the electronic states, as in solid-state vibronic lasers, so transitions can occur across an unusually wide range of energies, and laser gain can occur across a range of wavelengths. Although any one dye has a limited range of gain, many different dye molecules are available, and their combined emission ranges cover the near-infrared, visible, and near-ultraviolet.

Laser dyes require optical pumping with light from an external laser or another bright optical source. Normally, laser dyes are dissolved in liquid solvents, so the active medium is a liquid. Laser action has been reported in vapor-phase dyes, but they do not work well. Recently, solid dye lasers have been developed, in which the dye is embedded in a transparent solid. That's more convenient, but heat dissipation and replenishment of the dye molecules, which tend to break down during laser operation, are issues.

10.1.1 Properties of Laser Dyes

The organic dyes used in lasers are large and complex molecules with complex spectra. The dye molecules are fluorescent, meaning that they can absorb a short-wavelength photon and almost immediately "fluoresce," spontaneously emitting light at a wavelength longer than the absorbed photon. Upper-state lifetimes are typically a few nanoseconds, and bright pump light is needed to produce a population inversion. The laser transition is between a level in an electronically excited band to a vibrationally excited sublevel of the ground electronic state, as shown in Figure 10-1. Dyes have many sublevels so closely spaced that they merge together to form a continuum, and the light can be tuned continuously across the bands.

Laser dyes have gain bandwidths of 10 to 70 nm, with typical ranges 20 to 40 nm. Gain peaks in the center of the range, and drops to the sides. Tuning across wider ranges requires switching dyes, and some dye lasers have multiple dye cells to make this easier. The actual tuning range of a particular dye may vary when the dye is pumped with different light sources.

Most dyes belong to one of several families of organic compounds, and seemingly countless variations are possible just by adding an atom or two at different places on the molecule. For ex-

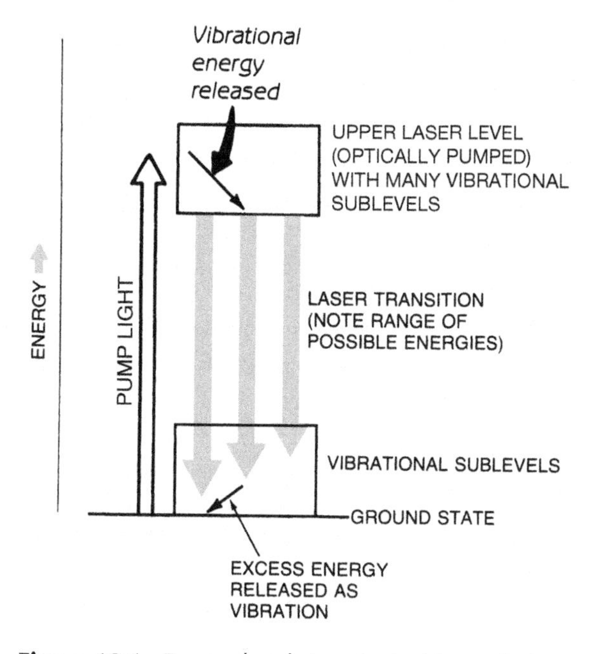

Figure 10-1. Energy levels in a typical laser dye.

ample, there are about 100 dyes in the coumarin family of dyes shown in Figure 10-2. The question marks show where atoms can be added to make new molecules.

Organic dyes have important limitations. They degrade after tens or hundreds of hours of use, even when the dye solution is flowed through the cell to reduce exposure to intense pump light. Most do not dissolve readily in water, and require organic solvents which are mostly flammable and toxic. Dye solutions are generally classified as hazardous waste both because of the dye and the solvents.

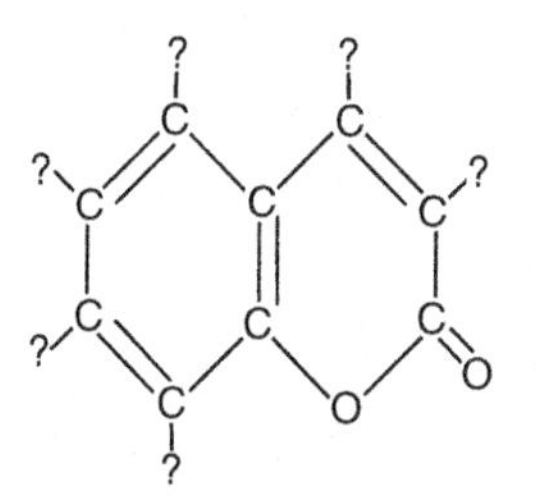

Figure 10-2. Chemical structure of coumarin laser dyes.

10.1.2 Dye Laser Structure

Dye lasers focus pump light from an external source onto the dye solution, which is located in a resonant cavity that includes optics to tune the laser wavelength as well as the usual cavity mirrors. In low-power dye lasers, the dye solution may be sealed in a closed cell that is illuminated by pump light. At higher powers, the dye solution typically flows through the pumping region to avoid rapid degradation. Continuous-wave dye lasers require steady illumination with a pump beam tightly focused to achieve high power densities, and the dye solution flows in an unconfined jet through the pump volume.

Some dye lasers have cavities designed to allow oscillation at a wide range of wavelengths, but most have cavities designed to tune the laser wavelength. Figure 10-3 shows three types of tunable cavities used with dye lasers, which differ in the choice of tuning element.

Prisms and diffraction gratings disperse light at different angles as a function of wavelength, so wavelength is tuned by turning either these elements or some other optical components to direct a particular wavelength through the resonant cavity. In Figure 10-3A, the grating is turned to change the wavelength, but in Figure 10-3B the prism stays fixed while the rear cavity mirror moves to select a particular wavelength refracted by the prism. In both cases, the selected wavelength resonates in the cavity, and other wavelengths are deflected out of the cavity.

In Figure 10-3C, the cavity is tuned with a wedge-shaped etalon, an optical element that functions as a miniature optical cavity to limit laser oscillation to a narrow range of wavelengths. A particular wavelength is selected by changing how the light passes through the etalon in the laser cavity, which in turn changes the distance light travels between the reflective surfaces of the etalon. The figure shows two possibilities. Flat etalons are tuned by turning them to change the distance the beam passes between the two surfaces. Wedge-shaped etalons are tuned by moving them up and down to change the distance the light passes between the two surfaces.

Simple prisms and gratings can limit dye laser linewidth to less than 0.01 nm. Adding an etalon can further restrict the linewidth, and adding further equipment and control systems can limit emission range to much less than one part in a billion (10^9). Tunable laser cavities can also take odd shapes such as the ring

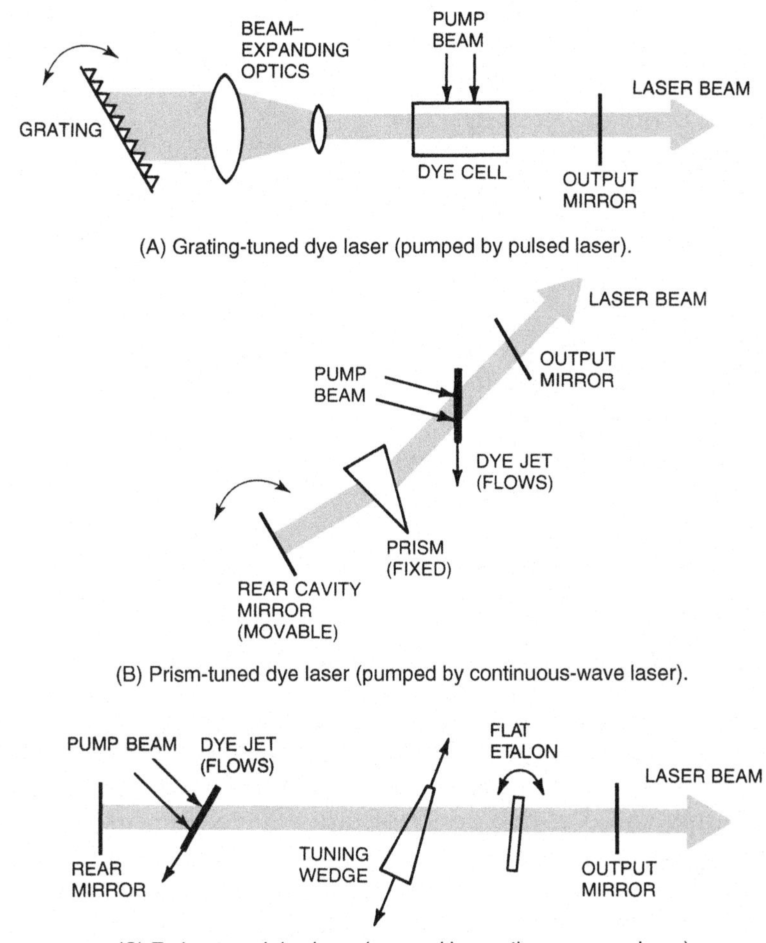

Figure 10-3. Three examples of wavelength-tunable laser cavities used with dye lasers.

cavity shown in Figure 10-4. Internal optics restrict laser oscillation to one direction, and other components limit oscillation to a narrow range of wavelengths. This design can be mode locked to generate a series of extremely short pulses.

Note that Figures 10-3 and 10-4 do not include all the details required for operation of tunable dye lasers; they are simplified to help you understand their operational principles. Similar cavities can be used with other tunable lasers, such as titanium–sapphire lasers.

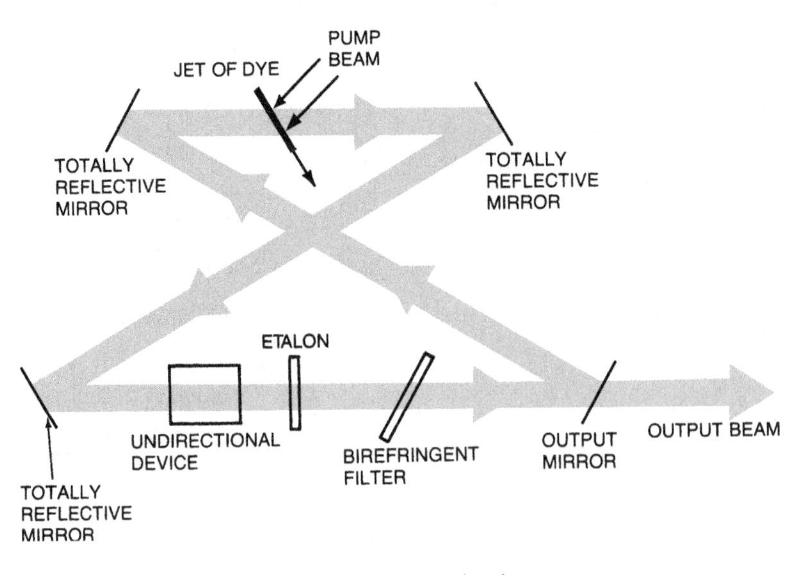

Figure 10-4. A ring dye laser.

10.1.3 Dye Laser Pumping

Dye laser pumping requires high intensities of light at wavelengths shorter than the desired emission wavelength to excite dye molecules to the upper laser level efficiently. Over the years, both flashlamps and a wide variety of lasers have been used to pump dye lasers.

Flashlamps generate intense pulses of visible light suitable for exciting many laser dyes. Linear flashlamps can be focused onto a tube containing dye solution in an elliptical cavity, as for lamp-pumped solid-state lasers. Alternatively, special coaxial flashlamps can be made, which surround an internal tube containing liquid dye. The big advantage of flashlamp pumping is that it does not require an external laser, making it simpler to operate and reducing capital costs. Flashlamp-pumped dye lasers are used in medical treatment requiring specific wavelengths not available from other laser sources.

Many types of pulsed lasers can pump dye lasers, offering high peak power at specific wavelengths that can be matched to dye absorption bands. The second, third, and fourth harmonics of neodymium lasers are widely used today. Other types that have been used include nitrogen and excimer lasers. They are mainly used in research.

Continuous dye pumping is possible with argon lasers or harmonics of neodymium lasers, but CW dye lasers have fallen out of favor because most researchers find tunable solid-state lasers easier to use. CW dye lasers have been used both to make measurements that require a continuous beam, and to generate trains of short pulses.

10.2 EXTREME-ULTRAVIOLET SOURCES

Although many lasers have been developed at near-ultraviolet wavelengths of 300 to 400 nm, conventional laser operation becomes increasingly difficult at shorter wavelengths. As transition energy increases, population inversions become harder to produce and excited-state lifetimes drop. Conventional mirrors and optical materials do not work, as even air becomes opaque at wavelengths shorter than about 200 nm.

The short-wavelength end of the ultraviolet spectrum was long neglected, but it now has become important both for research and for its potential use in increasing the component density in integrated circuits, where features are now fabricated with 193-nm ArF excimer lasers. Several laser-based approaches have produced wavelengths shorter than 50 nm, as shown in Figure 10-5, which

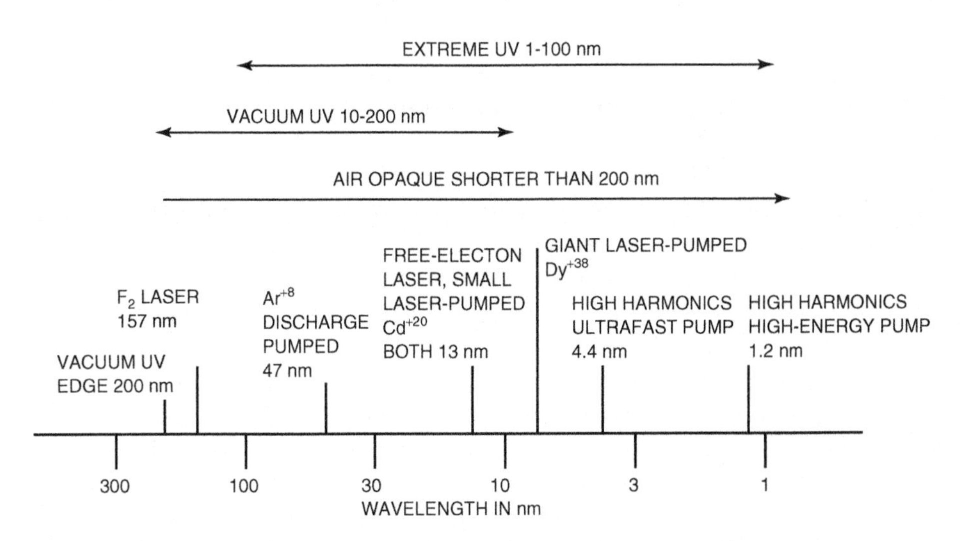

Figure 10-5. Shortest wavelengths from extreme-ultraviolet sources.

also lists the shortest wavelength from a conventional laser, the 157-nm molecular fluorine laser. This section describes the physics involved in generating these extremely short wavelengths, and the types that rely on conventional lasers or electronic transitions. The free-electron lasers described in Section 10.3 also can generate very short wavelengths, but they rely on different physical principles.

10.2.1 Extreme Ultraviolet Transitions

Like visible transitions, extreme-ultraviolet lasers operate on electronic transitions, but of a different type. Visible transitions involve electrons in the outer shell of an atom, where inner-shell electrons shield them from the strong attraction of the atomic nucleus. Extreme-ultraviolet transitions involve electrons dropping to inner-shell orbits, where they are more closely bound to the nucleus. As a result, transition energies are much larger—tens or even hundreds of electron volts, not the 1 to 5 eV typical for outer-shell transitions. Figure 10-6 shows the difference for selenium. Inner-shell transition energies E typically are given in electron volts; you can convert them to wavelength in nanometers using the formula

$$\lambda(nm) = \frac{1240}{E(eV)} \qquad (10\text{-}1)$$

The terminology in this part of the spectrum is not rigidly defined. This book follows the common convention that calls wavelengths of 1 to 100 nm *extreme ultraviolet.* Wavelengths shorter than 200 nm are often called the *vacuum ultraviolet.* The region from 1 to 10 nm (or sometimes from 1 to 100 nm) is also called *soft X-rays,* so extreme-ultraviolet lasers in this range have also been called X-ray lasers. Wavelengths shorter than 1 nm are called *hard X-rays,* but no laser techniques have yet reached that range.

Whether you call the radiation extreme ultraviolet or soft X-rays, it behaves quite differently from visible light. It has enough energy to strip electrons from atoms or molecules, so it called *ionizing radiation,* and can damage biomolecules and living cells. Soft X-rays can see through things that look opaque to our eyes, like soft tissue, but they are blocked by bone and other materials

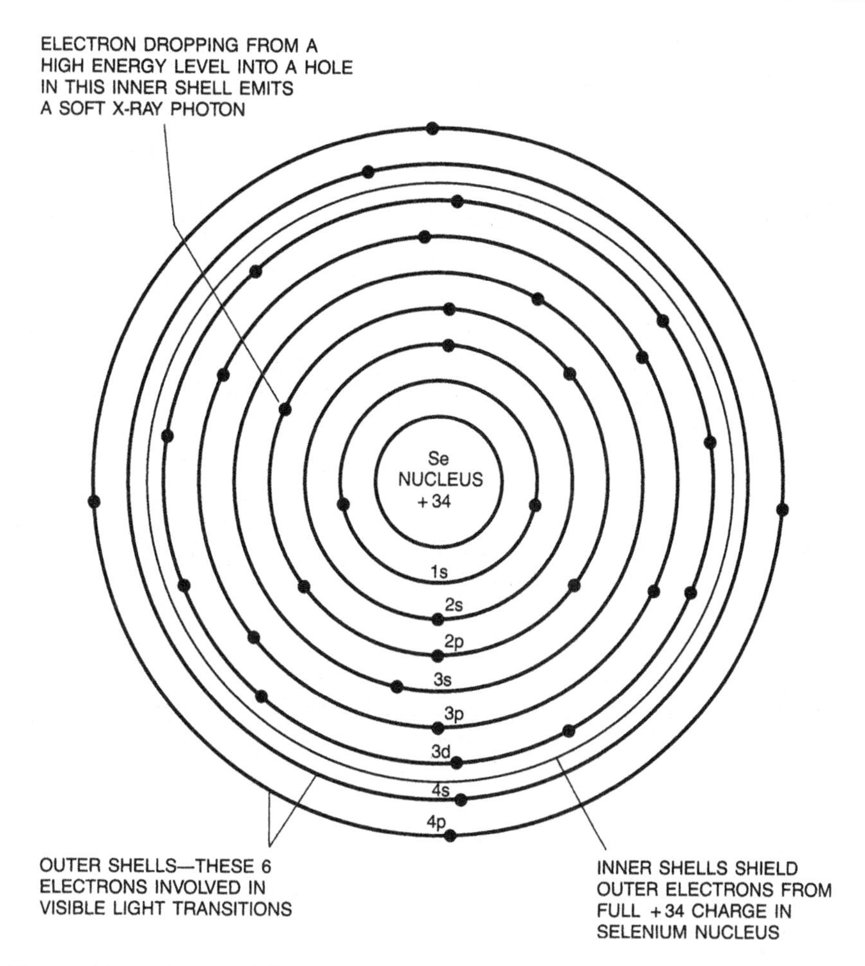

Figure 10-6. Electron falling into an inner shell of selenium emits an extreme-UV photon.

containing heavy elements. Soft X-rays also cannot penetrate through long lengths of air. Another important difference is that most materials have refractive index very close to one in the extreme ultraviolet, so they can neither reflect or refract light well, so different types of optics are needed.

10.2.2 Laser-Pumped Extreme-Ultraviolet or X-Ray Lasers

The first technique to produce laser-like emission in the extreme ultraviolet was firing tightly focused high-energy laser pulses onto solid targets. The pump pulses vaporized the spot they hit, ionizing atoms with such force that they stripped away many of the outer-shell electrons. The ions then recaptured the electrons, producing population inversions on short-wavelength laser transitions dropping to inner electron shells.

First demonstrated in 1984 with selenium by Dennis Matthews at the Lawrence Livermore National Laboratory in California, this technique produces single-pass amplification by stimulated emission, because resonator mirrors are not available for extreme-ultraviolet wavelengths. Researchers have progressed to shorter wavelengths since the first demonstrations.

Figure 10-7 shows how Livermore demonstrated the first definitive extreme-UV laser; current experiments are similar in concept. A short, powerful pulse from a powerful laser built for fusion experiments was shaped into a long, thin line that was focused onto a thin foil of selenium. The pulses, lasting about a nanosec-

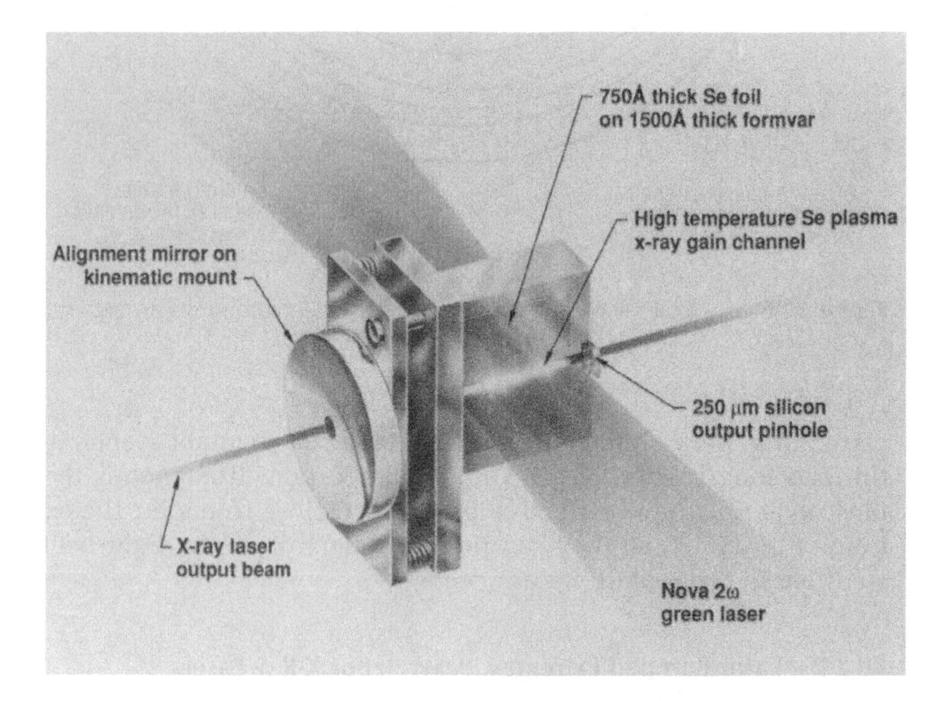

Figure 10-7. A small, hot plasma generated by a laser pulse emits an X-ray laser pulse. (Courtesy of Lawrence Livermore National Laboratory.)

ond and reaching peak powers of terawatts (10^{12} W), vaporized and ionized the selenium, producing a linear plasma. Electron recombination produced stimulated emission at 20.6 and 20.9 nm. The long, thin plasma concentrated the stimulated emission in a beam, but there was no resonant cavity because suitable mirrors were not available.

Over the years, a series of experiments with other large fusion research lasers has generated gain at wavelengths as short as 6 nm from dysprosium from which 38 of its 66 electrons were removed (Dy^{38+}), first reported in 2001 by the Rutherford Appleton Laboratory in Britain. However, fusion lasers are huge, so other researchers have developed techniques that work with smaller laboratory-size lasers. Instead of producing and exciting the laser plasma in a single step, these techniques reduce energy requirements by first ionizing the plasma, then exciting it by aiming a pump beam along its length. Using this approach, Jorge Rocca of Colorado State University has produced laser action, defined as saturated gain, in cadmium atoms from which 20 of the 48 electrons have been removed (Cd^{20+}), using only a laboratory-scale laser.

10.2.3 The Bomb-Driven X-ray Laser

During the 1980s, Livermore also conducted highly controversial experiments which sought to excite powerful X-ray laser emission using the blast of X-rays from a nuclear explosion. The goal was to develop a space-based laser battle station that could focus X-ray pulses onto targets hundreds or thousands of kilometers away to defend the United States against nuclear missile attack. Edward Teller, the father of the American hydrogen bomb, promoted the idea as a "third-generation" nuclear weapon, and it became part of Ronald Reagan's "Star Wars" program for space-based missile defense.

Livermore conducted several X-ray laser experiments during a series of underground nuclear tests, but the details remain highly classified. It now appears that the tests did produce amplified stimulated emission at X-ray wavelengths, reportedly at 1.4 nm. However, they apparently did not generate the tightly focused and extremely intense beams needed for missile defense. The project was eventually cancelled, amid charges that its results had been misrepresented.

10.2.4 Discharge-Pumped Extreme-Ultraviolet Ion Lasers

A newer and much less forceful approach to the extreme ultraviolet is to extend the discharge excitation techniques used for ultraviolet rare-gas ion lasers. This idea is to fire high-energy electrical pulses into a gas, stripping off several outer electrons to produce highly excited states, producing a short-lived population inversion. Experiments at Colorado State University have discharged peak currents of 25 kiloamperes through thin tubes up to 30 centimeters long filled with argon, producing Ar^{+8} ions and compressing them into a column only 200 to 300 cm wide. The excited ions emit 47-nm ultraviolet light, which is amplified in a single pass through the plasma.

10.2.5 High Harmonic Generation

The shortest wavelengths of laser-like output don't directly come from lasers. They are high-order harmonics of pulses from visible or near-infrared lasers. The laser generates pulses with extremely high peak power, which are focused into a gas. At laser intensities of 10^{13} to 10^{16} watts per square centimeter, the electromagnetic fields in the near-infrared laser pulse become strong enough to ionize atoms in the gas. The resulting free electrons oscillate back and forth in the strong electromagnetic fields, generating high-order odd harmonics of the input laser frequency.

Steady refinements have pushed this approach well below 10 nm, corresponding to extremely high harmonics. Pumping with ultrafast pulses from a comparatively small laser, Margaret Murnane and Henry Kapteyn at the University of Colorado have generated wavelengths as short as 4.4 nm, corresponding to the 115th harmonic of the input laser frequency. The shortest wavelength on record, 1.2 nm, required higher-energy but longer-duration pulses from the massive Vulcan laser at the Rutherford Appleton Laboratory in Britain, corresponding to harmonics higher than the 850th order.

10.3 FREE-ELECTRON LASERS

The most unusual type of lasers to have produced extreme-ultraviolet beams are free-electron lasers, in which the light emission comes from a beam of "free" electrons, unattached to any atom, passing through a magnetic field that varies periodically in space in particular ways that coax the electrons to release some of their energy as light. Electron accelerators called synchrotrons also can emit light deep into the ultraviolet, but that light is not coherent or monochromatic. The beam from a free-electron laser behaves much like an ordinary laser, although it is generated in an entirely different way.

Free-electron lasers can operate across a wide range of the electromagnetic spectrum, covering wavelengths from about a millimeter to the extreme ultraviolet. We will introduce them first, then go back to their use as extreme-ultraviolet sources.

10.3.1 Structure of a Free-Electron Laser

A free-electron laser has three essential components: a powerful electron accelerator that accelerates a beam of electrons to high speeds, an array of magnets called a *wiggler* or *undulator,* and a set of cavity optics that form a laser resonator. The magnet array is built so the magnetic field varies periodically along its length, and that magnetic field bends the electron beam back and forth.

Electron accelerators are massive but fairly standard pieces of scientific equipment; what matters in a free-electron laser is what happens once the electrons have been accelerated, shown in Figure 10-8. It is easiest to visualize the magnetic field as an array of permanent magnets aligned in alternating directions, but electromagnets also can be used. If the north pole of the first magnet is

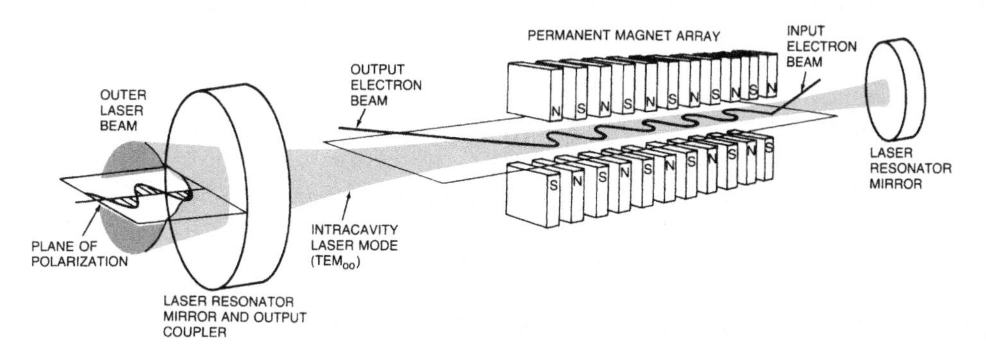

Figure 10-8. Electron beam passing through a wiggler generates a laser beam in a free electron laser. (Courtesy of the University of California at Santa Barbara Quantum Institute.)

above the electron beam, as shown in the figure, the south pole of the second magnet must be above the electron beam, then the north pole of the third, and so on.

The electron beam enters the magnet array at right at a slight angle above the axis of the laser resonator. The magnetic field from the first magnet bends the electron beam in one direction, then the opposite-polarity field from the second magnet bends it back in the other direction. This process repeats until the electron beam passes out the other end of the magnet array, so the beam wiggles or undulates its way through the magnets.

Every time the magnets bend the path of the electrons, the electrons release some energy as light. When the electrons are bent in the other direction, they then reabsorb most of the light. However, the electrons can amplify light if its wavelength λ is close to a resonance that depends on the speed v of the electrons along the laser axis, the speed of light c, and the period of the wiggler magnet p, defined as the distance from one north pole to the next along the array:

$$\lambda = \frac{p}{2[1 - (v^2/c^2)]} \tag{10-2}$$

The $[1 - (v^2/c^2)]$ term also appears in Einstein's theory of special relativity, and is needed because the electrons are moving at relativistic speeds.

Equation 10-2 highlights the basic physics of a free-electron laser, but if you want to calculate the laser wavelength, it is more practical to use a different version of the formula based on the wiggler period, the electron's accelerated energy E (measured in million electron volts), and its rest mass, 0.511 million electron volts:

$$\lambda = \frac{0.131p}{(0.511 + E)^2} \tag{10-3}$$

This tells us that the wavelength gets shorter as the magnet period decreases and electron energy increases.

10.3.2 Tunability of Free-Electron Lasers

The formula for free-electron laser wavelength depends on two quantities that might be adjusted: the period of the wiggler magnet and the energy of the electrons passing through it. In principle, the wavelength could be tuned from microwaves to soft X-rays, but no single device could do that.

So far, we have considered the wiggler magnet as a stack of permanent magnets with alternating polarity, but the key element of the wiggler is not the physical magnet but the magnetic field, which varies periodically along its length. That means you do not need a physical magnet, just a periodic magnetic field that could come from any source, including a light wave. Thus, a light wave could create a wiggler field for a free-electron laser with periodic variations down to the wavelength of light, much shorter than the smallest magnets you could build. With such a short period, you could produce very short wavelengths from a free-electron laser.

Varying the energy of injected electrons also changes the free-electron laser, with high-energy electrons producing shorter-wavelength emission. Electron energy typically is easier to vary than the spacing of magnets.

10.3.3 Types of Free-Electron Lasers

Although the physical principles of free-electron lasers are the same across their wide operating range, the actual devices differ considerably, and individual devices are built to operate over limited parts of the spectrum.

Long-wavelength free-electron lasers operate in the Raman or "collective" regime, where the interactions are best described as among many particles. In this case, the electron energy is relatively low (under 5 million electron volts), the current densities are relatively high, and the output wavelengths longer than around 100 μm. Gain is high and output power can be high.

Interactions involving individual particles are more important at shorter wavelengths, which are generated using electrons with higher individual energies. This requires bigger accelerators, although they produce weaker currents. Relatively few devices are in operation; the list that follows is a sampling:

- Vanderbilt University's free-electron laser in Tennessee, emitting pulses spanning the range from 2.1 to 9.8 μm.
- The Thomas Jefferson National Accelerator Facility in Newport News, Virginia, which has produced a continuous beam of 14.2 kilowatts at 1.6 μm. Powers of hundreds of watts are available at the long end of the infrared.

- The Duke University Free Electron Laser Laboratory in North Carolina runs on a storage ring capable of handling electrons to 1.2 billion electron volts, and emits from 400 to 190 nm in the ultraviolet.
- The free-electron Laser at Hamburg (FLASH) operates at DESY, the German Electron-Synchrotron Laboratory in Hamburg. It uses billion-electron-volt electrons to produce femtosecond pulses at wavelengths from 6.5 to 47 nm. The shortest wavelengths so far are third and fifth harmonics at 4.6 and 2.7 nm, respectively.

Free-electron lasers now under construction or in the planning stages are supposed to generate even shorter wavelengths and higher powers. In 2009, the Linac Coherent Light Source is to begin operation at the Stanford Linear Accelerator Center, generating hard X-ray pulses at 0.15 to 1.5 nm, the shortest-ever laser wavelengths. Yet despite promising results, the technology remains challenging, especially at short wavelengths, where large, powerful, and very expensive electron accelerators are needed as well as special magnets and other complex equipment. For the foreseeable future, short-wavelength free-electron lasers will remain research tools available only in a few specialized facilities.

10.4 SILICON LASERS

Silicon lasers have become a holy grail for the developers of integrated electronic circuits. As the speed and complexity of integrated circuits increases, electronics engineers are having a hard time keeping up with the need to communicate signals on and between chips. They would like to integrate optics into chips so signals can be processed electronically but communicated optically. Silicon lasers would be ideal because they could be fabricated in the same process as the electronic components.

As you learned in Chapter 9, the fundamental problem is that silicon has an indirect band gap, so recombination of holes and electrons in silicon releases most of its energy as heat rather than as photons. Developers have explored a number of approaches, and so far the most encouraging results come from hybrid approaches involving optical pumping of silicon or combining III–V semiconductors with silicon.

10.4.1 Indirect Bandgap Problem and Recombination Radiation

As you saw in Chapter 9, the fundamental limitation on light emission from silicon diodes is that it takes milliseconds for the electrons to change their momentum enough to drop from the conduction band into the valence band. During that period, electrons are likely to encounter defects in the silicon that make them release their energy in forms other than light. Neither efforts to change electron momentum faster or purification of the silicon to prevent electrons from encountering defects have proved practical.

Another alternative is physical or quantum confinement of electrons to reduce the chance that they will encounter defects before releasing their excess energy as light. Internal nanostructures can provide physical confinement and change energy-level structures, producing some light emission from cryogenically cooled samples pumped by light. Placing quantum dots in the active layer where recombination is occurring also can trap electrons, and a number of experiments have produced light emission. Electrical excitation has worked, but the results are much better when the material is excited by light from an external laser. So far, success has been limited.

10.4.2 Silicon Raman Lasers

The first pure silicon laser, demonstrated in 2004, was based on stimulated Raman scattering in a silicon waveguide. Raman oscillation was at 1675 nm with pumping at 1540 nm. Both pulsed and continuous-wave versions have been demonstrated.

A silicon waveguide can be integrated with electronic circuits, but an obvious drawback is that a silicon Raman laser requires optical pumping with an external source.

10.4.3 Hybrid III–V/Silicon Laser

The most promising approach to silicon lasers so far is a hybrid laser that bonds a stack of compound semiconductor layers containing a *p–n* junction to a silicon waveguide. Current carriers recombine with the active layer and produce photons, but only a thin layer of compound semiconductor separates the active layer from the silicon waveguide, as shown in Figure 10-9. Photons are large compared to the active layer, so light leaks into the silicon waveguide, a process called *evanescent coupling,* which is en-

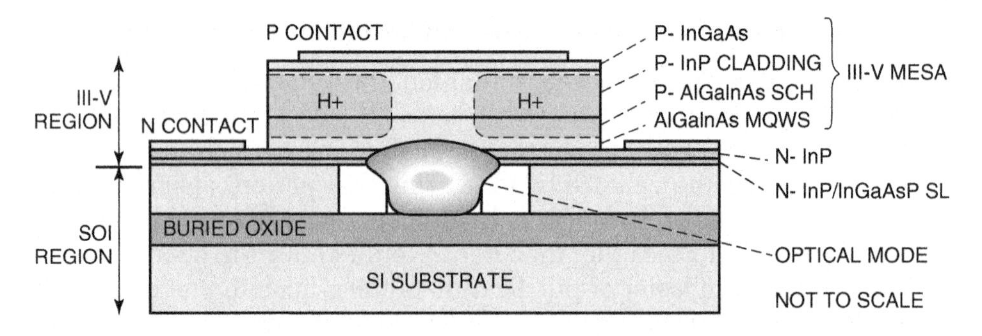

Figure 10-9. A hybrid InGaAsP/Si waveguide laser generates light in the compound semiconductor and transfers it to the silicon waveguide. (Courtesy of Intel Corp. and University of California Santa Barbara.)

couraged by the design. The silicon waveguide has lower loss than the active layer, so most of the light ends up in the silicon.

Experiments have demonstrated laser operation, first when the III–V structure was pumped optically, and later when a current was passed through it. More than 90% of the light from the laser ended up in the silicon waveguide. Fabricating many silicon waveguides side by side produces an array of hybrid silicon waveguide lasers.

Strictly speaking this device is not a silicon laser. The silicon generates none of the light; it collects and transmits light in the 1550-nm band generated in the AlGaInAs active layer. The compound semiconductor is complex, as shown in Figure 10-9, and must be bonded to the silicon surface, so it can't be fabricated monolithically. Nonetheless, the hybrid silicon laser is a promising approach to generating light from silicon integrated circuit chips.

10.5 WHAT HAVE WE LEARNED?

- Complex organic dyes in liquid solutions fluoresce across a range of wavelengths, making them useful in tunable lasers.
- Laser cavities developed for tunable dye lasers can be used with other tunable lasers.
- Organic dye lasers require optical pumping with light from an external laser or a flashlamp.
- Extreme-ultraviolet sources emit wavelengths shorter than about 100 nm, where the atmosphere is nearly opaque.

- Atoms stripped of several electrons in a high-energy electric discharge can emit extreme-ultraviolet laser beams at wavelengths as short as 47 nm.
- High-order harmonics of visible or near-infrared pulses, generated in gases exposed to extreme laser intensities, can reach wavelengths less than 10 nm.
- Laser ionization of metal films can strip more than a dozen electrons from metals, producing gain deep into the extreme ultraviolet.
- Free-electron lasers extract energy from a beam of electrons passing through a magnetic field that varies sinusoidally in strength. They can be tuned across a broad range of wavelengths by changing the electron energy or the period of the magnetic field.
- Free-electron lasers draw on well-developed technology for particle accelerators.
- Bigger accelerators are needed for free-electron lasers generating short wavelengths.
- Silicon lasers are being sought to integrate with silicon electronics.
- The indirect band gap of silicon makes conventional diode lasers extremely difficult in silicon.
- Stimulated Raman scattering in silicon can be used to make a laser, but it requires an external source of pump light.
- A hybrid laser can be made by bonding a diode laser made from a compound semiconductor to a silicon waveguide in a way that couples light generated at the junction of the compound semiconductor into the silicon waveguide.

WHAT'S NEXT?

The next three chapters explore the applications of lasers first in low-power applications, then in high-power applications, and, finally, in research.

QUIZ FOR CHAPTER 10

1. What property gives organic dye lasers their wavelength tunability?
   a. The many vibrational sublevels of electronic transitions
   b. The use of organic liquids as solvents

   c. The high gain of the optically pumped dyes
   d. Photodissociation of the dyes under intense pump light
   e. The liquid nature of the laser medium
2. Typical tuning bandwidth of an individual laser dye is
   a. 1 nm
   b. 5 nm
   c. 5–10 nm
   d. 10–20 nm
   e. 20–40 nm
3. Typical linewidth of the output of a dye laser is
   a. 1 nm
   b. 0.01 nm
   c. 1 MHz
   d. 0.1 MHz
   e. determined by the nature of the laser cavity and tuning optics
4. What type of laser is now used to write the features on state-of-the-art integrated circuits?
   a. Argon-ion laser
   b. Free-electron laser
   c. Argon fluoride laser
   d. X-ray laser
   e. Neodymium–YAG laser
5. The shortest wavelength produced from a discharge-pumped laser is 47 nm. What is the energy of a 47-nm photon in electron volts?
   a. 2.6 eV
   b. 12 eV
   c. 26 eV
   d. 124 eV
   e. 1240 eV
6. Why might you argue that the extreme-ultraviolet emission from laser-produced plasmas in the Livermore experiments was *not* laser action?
   a. Because the experiment did not produce stimulated emission.
   b. Because no mirrors were used to produce feedback and laser oscillation.
   c. Because the wavelength was too short to be a laser.
   d. Because the excitation laser vaporized the target and left nothing behind.
   e. Because the laser vaporized the power meter.

7. The atmosphere absorbs X-rays. Why would an X-ray laser be considered for use as a an antimissile weapon?
   a. The laws of physics do not apply to weapon systems.
   b. The nuclear bomb that powered the laser would blow a hole in the atmosphere.
   c. The X-ray laser would be fired in space against targets above the atmosphere.
   d. X-rays can deactivate nuclear weapons through the air.
   e. The opacity of the atmosphere at X-ray wavelengths was discovered later.
8. How is energy extracted from free electrons in a free-electron laser?
   a. Their paths are bent back and forth when they pass through a periodic magnetic field.
   b. The free electrons induce currents in the magnets they pass, and the magnets emit light.
   c. The electrons release energy after they evaporate from the magnets.
   d. The energy is extracted when electrons bounce off a resonator mirror and lose energy.
   e. The electrons transfer energy to helium atoms in the air, which then transfer the energy to another species.
9. What is the most critical limitation of a silicon waveguide Raman laser?
   a. The gain is low.
   b. The Raman shift changes the wavelength to the infrared.
   c. It has to be pumped with light from an external source.
   d. It requires a high-power current source.
   e. The silicon has to be deposited on the waveguide in a separate step.
10. What happens to the light in a hybrid III–V/silicon laser?
    a. Light emitted by a III–V junction is coupled into an adjacent silicon waveguide.
    b. Light emitted by a silicon junction is coupled into an adjacent III–V waveguide.
    c. Light emitted by a III–V junction is transmitted through the bottom surface of a silicon waveguide.
    d. Light emitted by a III–V junction optically pumps a silicon diode.
    e. Light emitted by a III–V junction is converted to an electrical current by a silicon photodiode.

CHAPTER 11

LOW-POWER LASER APPLICATIONS

ABOUT THIS CHAPTER

The final three chapters of this book cover laser applications, which are many and diverse. To keep the topic manageable, we will break them into three broad groups: those requiring little power (such as playing CDs), those requiring high power (such as cutting and welding), and those in scientific research, such as studying the properties of atoms and molecules. The division between low and high power is somewhat arbitrary, based on how the lasers affect the materials they illuminate. The intent is not to list everything, but to list the most important, interesting, and illustrative applications to help you understand laser applications in general.

This chapter covers applications that require low laser power, typically well under a watt. A laser with such little power does not have a dramatic effect on the objects it strikes, but it may make minor changes, such as exposing photographic film. As we will see, there are many types of low-power laser applications. In some, the laser is little more than a high-performance light bulb, but for others special features of laser light, such as coherence or tight beam collimation, are essential. In most cases, the laser light is communicating or processing information, whether it is playing music from an audio CD, sending signals through a fiber-optic communication network, or printing information on a page.

11.1 ADVANTAGES OF LASER LIGHT

Theodore Maiman's first laser made headlines in the summer of 1960. As word of the laser spread, it seemed that every scientist and engineer wanted their own laser. Many had no clear purpose in mind, but the laser seemed like a neat toy, and curiosity lurks deep in the souls of scientists and engineers. They built their own lasers, or bought lasers from the handful of little companies that sprang up to make them. Then they zapped just about everything they could find. They informally measured laser power in "gillettes"—how many razor blades a laser pulse could pierce. It was a fertile time for experiments, new ideas, and accidental discoveries but, inevitably, many of the experiments failed, and for a while the laser became "a solution looking for a problem," a description that Maiman's assistant Irnee D'Haenens coined at the dawn of the laser age.

Much has changed since the first wave of laser enthusiasm. Many jobs now demand lasers. In Chapter 12, we will describe how laser power can drill holes, perform surgery, and change materials in other ways. Chapter 13 will describe laser applications in scientific research. In this chapter, we will concentrate on the many uses of low-power lasers.

What can low-power lasers do that other light sources can't? Often, lasers merely serve as well-behaved, long-lived light bulbs. However, other jobs may require special properties of laser light, such as coherence, or the tight alignment of laser beams.

Low-power lasers have a number of important advantages over other light sources:

- Lasers produce well-controlled light that can be focused precisely onto a small spot.
- Low-power laser beams, focused tightly in a small spot, can reach a high power density.
- Laser beams mark a straight line.
- Laser beams don't press on objects they contact, so they don't deform them or affect their motion.
- Lasers emit a very narrow range of wavelengths, and for most practical purposes are monochromatic, or single-color.
- Most laser light is coherent.
- Lasers can generate extremely short pulses.
- Many lasers produce steady powers for tens of thousands of hours or more.

- Diode lasers are very compact and inexpensive.
- Diode lasers can be modulated directly at high speeds by changing drive current.
- Diode lasers efficiently turn electrical energy into light.

On the other hand, lasers have some significant drawbacks:

- Lasers do not emit white light.
- Light is concentrated in the beam, not spread out.
- Only certain wavelengths are available, particularly at low cost; some wavelengths are only available from expensive lasers.
- The concentration and collimation of laser light makes it hazardous to the eye.

You can get a feeling for these advantages and disadvantages by comparing a laser to a light bulb. A light bulb illuminates a room with uniform white light; a laser beam illuminates a single spot with a single color. If you want to illuminate a room, you want a light bulb. If you want to illuminate a single spot in a way that makes it easy to see, you want a laser pointer.

In the rest of this chapter, we will show how the advantages and disadvantages of low-power lasers lead to a range of laser applications.

11.2 READING WITH LASERS

Low-power lasers are widely used to read printed symbols. The laser spot scans across a surface, and a detector measures changes in the reflected light as the beam moves. Computers or special-purpose electronics decode the pattern of reflected light to interpret the symbols.

In principle, lasers can read any pattern printed so that the symbols reflect different amounts of laser light. If the laser beam is red, the pattern should be printed so that some parts (the ink) absorb red light and other parts reflect it. Black and white inks are a simple example, but a red laser beam cannot read a pattern printed in red ink on white paper because both the ink and the paper reflect white light. One Boston-area milk company learned this the hard way when they printed milk cartons with red ink, making the striped product codes invisible to the red laser beam used to read them.

In principle, this scanning system can be used to read any suitable printed symbols, including text, as long as suitable software is available to decode the changes in reflected light. In practice, the main applications are reading symbols designed especially for use with laser scanners. The main examples are bar codes consisting of stripes of different widths, and two-dimensional codes that are patterns of light and dark zones.

11.2.1 Bar Code Scanners

Most bar codes used with laser readers are variations on the Universal Product Code originally developed for automated checkout in supermarkets. That system was developed to read codes printed on food packages moved quickly past a scanning window at any angle, although some care is needed to make sure the code is visible to the scanner.

Figure 11-1 illustrates the workings of a supermarket scanner. A beam deflector—here, a rotating mirror—scans the laser beam

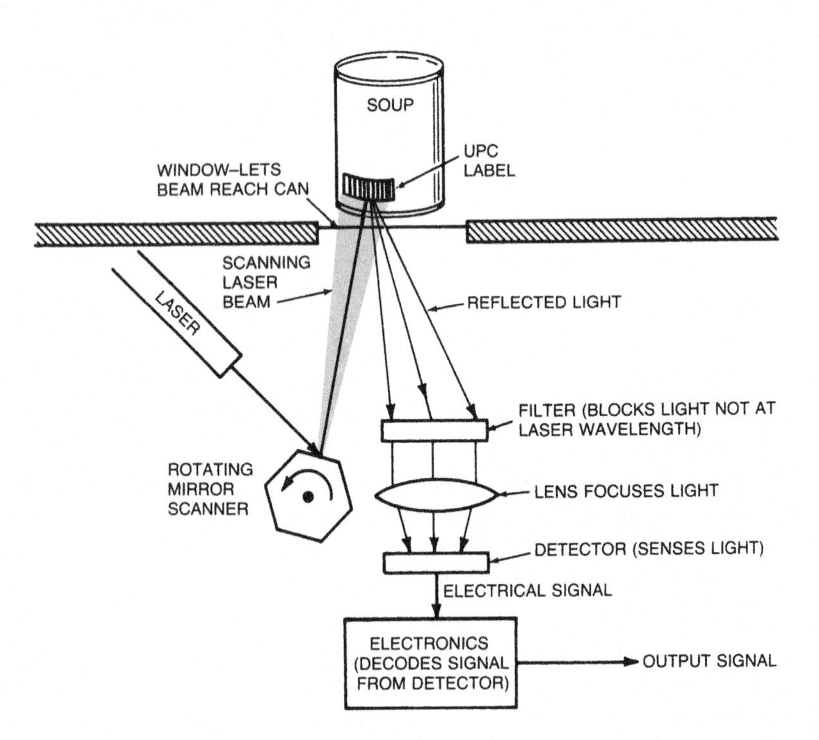

Figure 11-1. A laser bar-code reader.

repeatedly in a carefully controlled pattern across the zone where the bar code appears. The optical system collects reflected light, filters out light other than the laser wavelength, and focuses the remaining light onto a detector, which generates an electrical signal proportional to the amount of light reaching it. That raw signal then is processed to interpret the information encoded by the bar code.

Our diagram does not show other lighting, but we can't ignore it in the real world. In a supermarket, for example, bright overhead lights shine down onto the checkout counter, illuminating packages at the same time as the laser. However, the laser light is concentrated at a single red wavelength, and the room light is spread across the visible spectrum. Using a filter that transmits the laser wavelength but blocks other ambient light makes the bar code readable without interference.

Supermarkets were the first to adopt laser scanners for automated checkout after the Universal Product Code was developed in the 1970s. Now, most large retail stores use similar systems. In some stores, the laser system is installed under the counter, with the beam emerging through a window to illuminate packages as they pass along the counter. In other stores, clerks bring hand-held readers equipped with red lasers or LEDs close to the bar code on the package. The bar code identifies the product and the size of the package; the store's computer then looks up that item and adds its price to your bill. Supermarkets print special bar codes that indicate the price and weight of meat and produce packaged at the store.

The UPC symbol originally was designed to be read by a 632.8-nm helium–neon laser, but now most readers use red diode lasers close to that wavelength. Variations on the pattern are now standard on many products including books and magazines, and on many airline luggage tags. Other kinds of bar codes are used for other applications, such as tracking and checking out library books.

11.2.2 Two-Dimensional Bar Codes

Two-dimensional codes are patterns created by varying between light and dark zones in both length and width. They come in many varieties; Figure 11-2 compares one example used by a package delivery company to a UPC-style bar code printed on a

Figure 11-2. Two-dimensional bar code compared with a standard one-dimensional UPC symbol from a magazine.

newsstand magazine. Two-dimensional bar codes can encode more information than a linear bar code or UPC symbol.

11.2.3 Other Laser Readers

Lasers can read many other types of printed symbols. For example, some voting machines use laser readers to interpret black marks put on ballots by voters. Lasers have also been used to scan text for optical character recognition.

11.3 OPTICAL DISKS AND DATA STORAGE

The biggest single application of lasers in terms of numbers is in playing and/or recording information on optical disks. This covers audio CDs, video DVDs, recordable CDs and DVDs, and other optical disk formats such as the HD-DVD and Blu-Ray for high-definition video. In 2006, some 720 million diode lasers were sold for these systems, nearly 90% of all diode lasers sold, according to *Laser Focus World*. Those lasers are relatively inexpensive, with average selling price only about $2, but so many are sold that they accounted for just under half of the $3.1 billion in diode lasers sold in 2006.

11.3.1 Optical Disk Technology

The common concept at the root of all optical disk technology is shown in Figure 11-3. An optical head focuses light from a semiconductor laser onto the surface of a rapidly spinning disk. Data is recorded as tiny spots on circular tracks on the surface, each recording a single bit. For playback, the laser illuminates the surface, and reflected light is focused onto a detector in the optical head, which generates an electrical output that is decoded to play back the data on the disk.

Optical storage can cram tremendous amounts of data onto a small surface area because the laser light can be focused to a spot roughly a wavelength across—about 1 μm for the 780-nm diode lasers used in CD players. The format originally was developed to reproduce music in digital form; digital data storage was added later. This allows up to 72 minutes of music to be stored on a single 12-centimeter CD, or 700 megabytes on a single CD-ROM (the data counterpart of a musical CD). The amount of music can be increased by an order of magnitude using digital data compression

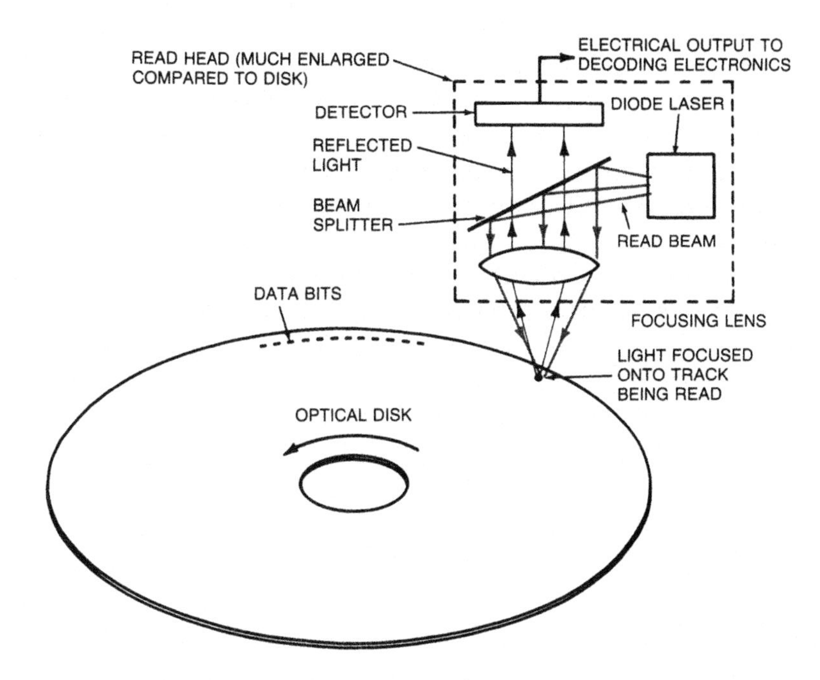

Figure 11-3. Optical disk system.

to produce MP3 files, but the resulting files cannot be played by standard CD audio players.

The original CD and CD-ROM specifications covered only prerecorded disks, but recordable versions were developed and are now in wide use. We will look at them after describing the higher-capacity cousins of CDs.

11.3.2 DVDs and Other Video or High-Capacity Disks

Optical disk storage capacity depends on how densely data spots can be written and read. The density is inversely proportional to the square of the wavelength, reducing wavelength by a factor of two would multiply storage capacity a factor of four. The development of shorter-wavelength diode lasers and of optics that can focus the beam to tighter spots has led to two generations of optical disks with greater storage density. Red diodes emitting at 650 nm are used in DVDs, which originally were developed to distribute digital videos and later expanded to include computer data. DVDs retain the 12-cm format of the CD, allowing development of players that can handle both formats. DVDs hold about two hours of video, or 4.38 gigabytes of data, although the nominal capacity is often listed as 4.7 gigabytes.

Although DVDs record data digitally, the video they store is the standard-definition video of analog television signals. High-definition digital television requires squeezing several times more data on a single disk. Two competing standards were introduced: Blu-Ray and HD-DVD, both based on 405-nm violet lasers and 12-centimeter disks, and Blu-Ray won.

The optical improvements are not enough to account for the much greater storage capacity of DVD, HD-DVD, and Blu-Ray disks. The higher-capacity disks use digital data compression to squeeze more information onto the disks. The compression algorithms examine successive video frames and decide what information is essential and what can be discarded because it has not changed significantly. Although compression is nominally "lossless," audiophiles complain that compressed MP3 files cannot match the quality of uncompressed CDs.

11.3.3 Writing on Optical Disks

Many CDs and DVDs are prerecorded and replicated, using materials that can be played repeatedly, but are not sensitive to light so no data can be written on them. Blank CDs and DVDs are also available, manufactured with an internal layer of light-sensitive organic dye that changes its reflectivity after being heated by a laser beam. Read-only drives lack the laser power to heat the material. The higher-power lasers in read-write drives are operated at high power to write and at lower power to read.

A number of writable optical disk formats have been developed over the years. The current types of disks that allow writing only once, called CD-R (CD-Recordable) and DVD-R disks, are the descendants of disks called WORMs, for write-once, read many. Early WORM disks came in a variety of sizes and formats, but computer users and manufacturers settled on the 12-cm CD size, and retained that size for DVDs.

Rewritable optical disks have also been developed, using other materials that change their phase when heated by a laser beam, changing their reflectivity in ways that can be erased and overwritten many times. The current generation are called CD-RW (CD-rewritable) and DVD-RW, and they are not used as widely as write-once disks. A major concern is their archival lifetime, which is shorter than CD-R. Prerecorded disks have the longest lifetimes.

Earlier rewritable optical disks were based on a different technology, called magneto-optical because they used both magnetic and optical effects. The disk was coated with a partly transparent material that could be magnetized if it was heated beyond a certain temperature, and retained that new magnetization when cooled to lower temperatures. Data recording was accomplished by heating the coating to the critical temperature with a laser beam, then magnetizing it by exposing it to a magnetic field from a magnet on the other side of the disk. Later heating could erase the disk, but a lower-power laser beam would not disturb the data. Magneto-optical disks never captured a large share of the data-storage market, but Sony still uses magneto-optical recording in its MiniDisc system, which records on a 6.4-cm disk and is still popular in Japan.

11.4 LASER PRINTING AND MARKING

The first laser printers introduced in the 1970s were massive machines that printed at very high speeds. Much less expensive laser printers based on diode lasers were introduced in the 1980s for use with personal computers, and laser printers have been coming down in price ever since. Once limited to black and white, laser printers now can also print in color.

Laser printing is based on a process called raster scanning, illustrated in Figure 11-4. Input information controls the power emitted by the laser at left, turning the light off and on as the beam is scanned back and forth across a screen or page, producing a series of dots on the page. Then the page is moved or the laser is tipped a bit to scan another line. Line by line, a series of scans across the page builds up a picture. In the case of Figure 11-4, the dark spots are written when the light is on, and nothing is written so the paper remains white when the light is off.

This basic concept appears in many variations. The most common is the laser printer of computer output, now used with many personal computers. Others include laser systems for producing printing plates and for marking products with identification codes.

11.4.1 Laser Computer Printers

Laser printers used with personal computers are based on the same printing technology as an office photocopier. In an ordinary copier, a bright light illuminates the page to be copied, and the reflected light is focused onto a rotating drum coated with light-sensitive material that carries a static charge. The light reflected from the page discharges the static charge as the drum turns past it. Next the drum passes another point where the regions that still carry an electric charge attract a dark fine-grained material called

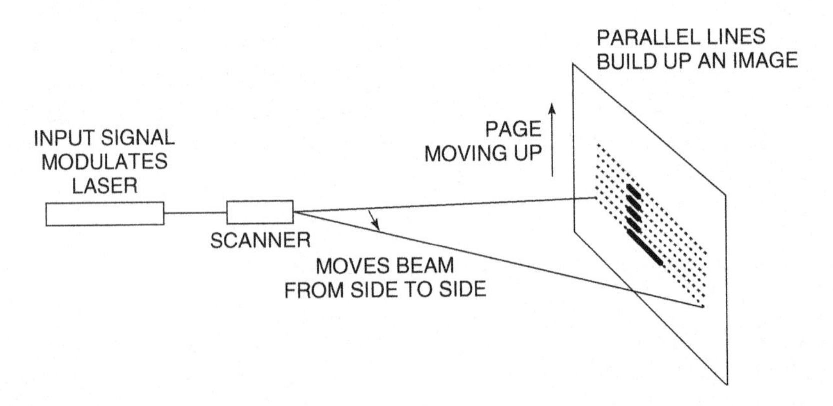

Figure 11-4. Raster scanning a laser beam across a page.

a "toner," forming an image of the original page. (The dark toner adheres to the parts of the drum where the dark areas of the original page were imaged.) The drum then transfers the toner image to a blank sheet of paper, copying the original page. A final stage bonds the fine dark toner to the paper.

A laser printer contains a similar drum arrangement, as shown in Figure 11-5, but in this case there is no original. The computer generates a pattern of light and dark regions, which modulates a laser beam that raster-scans the drum's coated surface, writing a pattern of charged and uncharged areas that corresponds to the printed page. The rotating drum then picks up the toner, and transfers the image to plain paper as in a conventional copier. The toner cartridge essentially transfers to the printed page the image that the laser writes on the toner cartridge.

With typical resolution of 1200 dots per inch, the output of a black-and-white laser printer looks almost as good as a typed page. In fact, the copy for whole books, including all text and art, can be produced using a laser printer.

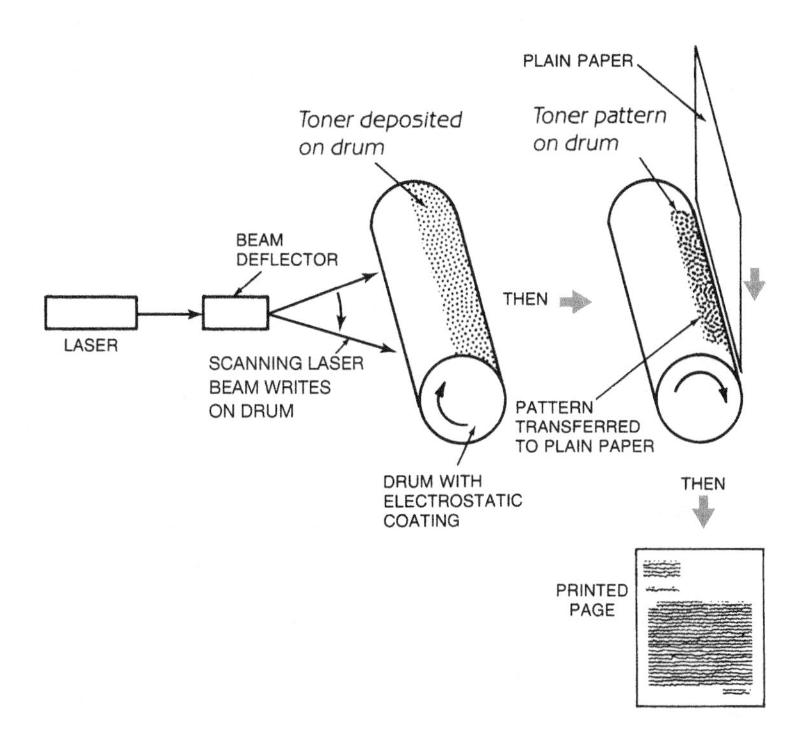

Figure 11-5. Elements of a laser printer.

Laser printers once printed only black toner on white paper. Color versions are now readily available, which work by making three passes using different colored toners, one for each of the three colors needed to generate the full range of colors for reflected images. Because they are more complex, color printers are more expensive than black and white ones, and comparable models have somewhat lower resolution—600 dots per inch is typical.

11.4.2 Laser Prepress Equipment

Laser systems also play important roles behind the scenes in the publishing industry. Computer-to-plate or computer-to-press systems use lasers to write final versions of pages to be printed. The laser may write on photographic film that is used to transfer an image to the plates used in actual printing, or may write directly on the printing plates.

Another laser application in printing is to prepare color photographs for printing in magazines or books. Color printing processes create full-color images by running pages through a press three times, each time adding one of three colors needed to build up a full-color image. Printing that way requires a process called color separation, which analyzes the colors in the original photograph and uses that information to prepare separate plates for the three colored inks that together make a color image. Lasers can scan the original picture to produce those color separations.

11.5 FIBER-OPTIC COMMUNICATIONS

Nearly five million diode lasers were sold in 2006 for use in fiber-optic telecommunications, not including short data links. These lasers are produced in smaller quantities than optical disk lasers and must meet much more stringent requirements, so their average price is about $240, bringing total sales to $1.15 billion. That reflects the critical importance of lasers in the fiber-optic systems that provide the backbone of the global telecommunications network. Thanks to fiber optics, the costs of long-distance and international phone calls have dropped, their quality has improved, and the Internet connects people around the globe.

Fiber optics came into widespread use for long-distance communications in the 1980s as companies built national networks and the first transatlantic fiber-optic cable was installed. Rapid advances in the technology multiplied transmission capacity during the 1990s, leading to a boom in construction of long-distance fiber-optic systems that slowed abruptly in the early 2000s when the industry realized it had installed more fiber transmission capacity than it needed. Some new systems are being installed today as steady increases in demand are catching up to capacity.

11.5.1 Lasers In Fiber-Optic Communications

Lasers provide the optical signals transmitted through fiber-optic cables. Almost all of the lasers used in fiber-optic systems are diode lasers, which can be coupled directly to the light-guiding cores of optical fibers. Fiber lasers and fiber amplifiers are used in some special cases, as described below.

Lasers are packaged in transmitters that deliver a modulated optical signal to the transmission fiber. Lasers can be directly modulated by changing the drive current at rates to 1 or 2.5 gigabits (billion bits) per second. At higher data rates, or when the signals must travel very long distances, the lasers generate steady beams that are modulated in intensity by a separate modulator before the beam enters the fiber. The light then travels through the fiber to a receiver at the other end, where it is detected and converted into electronic form, as shown in Figure 11-6.

That figure shows a point-to-point fiber link that carries a signal from one point to another. Such a fiber link is one connection in the global telecommunications network, an array of connections that transmit signals around the globe. In practice, separate fibers carry signals in opposite directions, so you need at least two fibers to connect two points.

11.5.2 Wavelength-Division Multiplexing

The simple fiber-optic link of Figure 11-6 carries light from a single laser transmitter. However, the glass in the fiber transmits a much broader range of wavelengths than emitted by any single laser, so it can simultaneously transmit signals on many different wavelengths, a practice called *wavelength-division multiplexing*

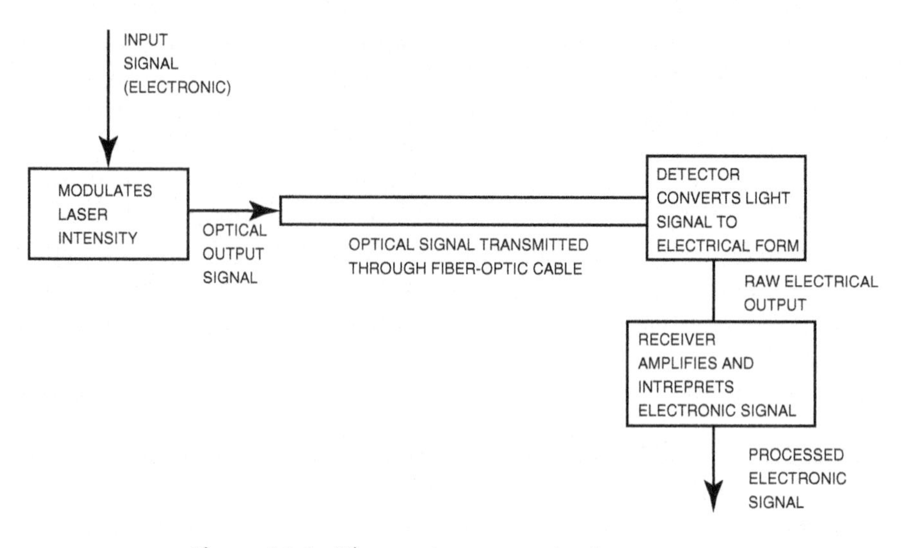

Figure 11-6. Fiber-optic communication system.

(WDM). From a communications standpoint, wavelength-division multiplexing is like assigning separate frequencies to radio stations broadcasting in your area. Each radio station transmits on its own frequency channel, in a way designed to avoid interfering with stations transmitting on other frequencies in the same area. In the same way, wavelength-division multiplexing creates many optical channels through the same fiber.

The advantage of wavelength-division multiplexing is its ability to multiply the transmission capacity of a single fiber. The maximum practical data rate for a single fiber-optic transmitter on most fibers in the long-distance network is about 10 Gbit/s, limited by properties of the transmitters and fibers. However, the capacity can be multiplied by adding additional transmitters at different wavelengths, and optics to combine and separate the signals at the two ends of the fiber. Thus, a single fiber can carry 10 signals at 10 Gbit/s, for a total capacity of 100 Gbit/s.

How far can it go? Wavelength-division multiplexing allows today's long-haul fiber-optic cables to carry dozens of separate signals in each fiber. Upgrading transmitters, receivers, and fibers allows transmission at data rates of 40 Gbit/s per wavelength. Laboratory demonstrations have squeezed more than 10 trillion bits per second (10 terabits) through a single fiber, enough to carry more than 100 million phone calls simultaneously.

11.5.3 Optical Amplifiers in Telecommunications

So far, we have considered light transmission through a fiber without regard to distance. However, distance matters because the fiber attenuates signals, so laser signals fade with distance. Even if you can stretch transmission to 100 kilometers, at some point you run out of signal. To go further, you need to amplify the signal, which is where optical amplifiers come in.

The standard type of amplifier in fiber-optic systems is the erbium-doped fiber amplifier described in Section 8.9. Erbium amplifies light in a band from about 1530 to 1570 nm, which includes the wavelengths at which optical fibers are the most transparent and can carry signals the furthest. (That match was sheer good luck because erbium works exceptionally well as an optical amplifier.) Pass an attenuated signal in that band through an erbium fiber amplifier and it amplifies it by a factor of 100 or more so the signal can travel through another length of fiber, as shown in Figure 11-7.

Better yet, an erbium fiber amplifier amplifies all the signals in its operating range, so if you start with 10 wavelengths between 1530 and 1570 nm, all 10 of them will be amplified. It takes some optical tricks to make the amplification uniform across that band, but it's possible. That means fiber amplifiers can stretch transmission ranges for wavelength-division multiplexing as well as for single-channel systems.

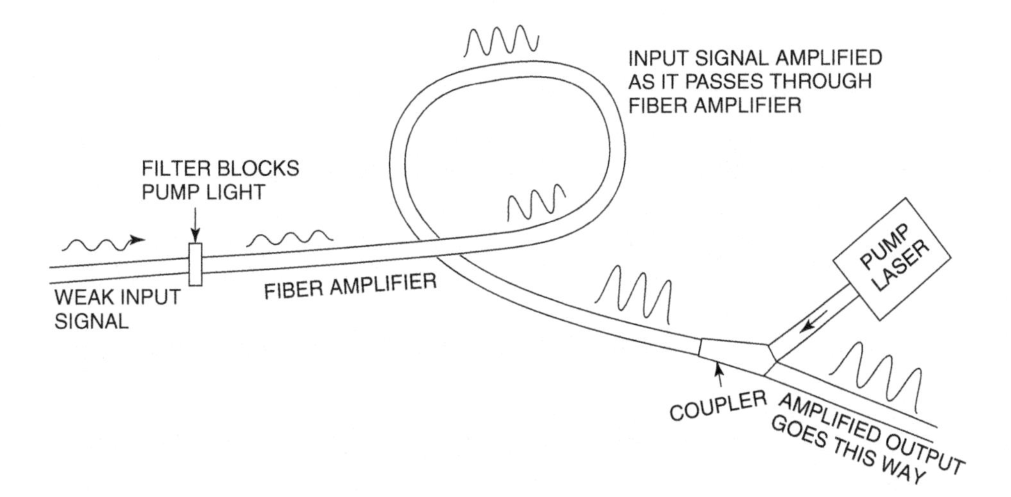

Figure 11-7. Fiber amplifier boosts strength of an optical signal.

11.5.4 Types of Fiber Networks and Lasers

Fiber-optic communication networks are built on many scales. High-speed telecommunications networks span the planet, but fiber networks are also used to carry signals for entertainment systems inside some high-end cars. Different types of lasers are used for different types of fiber networks:

- *Long-distance telecommunications networks* use InGaAsP lasers emitting precise narrow wavelength ranges in the 1550-nm range and erbium-fiber amplifiers.
- *Regional and metropolitan telecommunications networks* use InGaAsP lasers emitting in the 1300 or 1550 nm bands, but do not require such narrow linewidths. Some may require lasers emitting within 20-nm bands for coarse wavelength-division multiplexing.
- *Fiber-to-home networks* like Verizon's FiOS system use InGaAsP lasers emitting at 1490 and 1550 nm, sometimes with a fiber amplifier at 1550 nm, to transmit signals to homes. InGaAsP lasers emitting at 1310 nm transmit signals upstream from homes to the switching office.
- *High-speed data links* use inexpensive 850-nm GaAs VCSELs to span short distances within buildings, and 1310-nm InGaAsP VCSELs or edge emitters to span longer distances.
- *Automotive networks* use LEDs rather than lasers in their transmitters because the signals do not have to go far.

11.5.5 Laser Communications through Air and Space

Laser beams can also send signals through the air and space, although most optical communications are through fiber.

In some cases, it is simpler to modulate a laser beam and send the signal through the air than try to make the connection through fiber. For example, if a company had offices in two buildings across the street from each other, it might be simpler to send signals through the air than to lease capacity on a telephone company cable. Radio waves are another alternative, but optical systems can carry very high data rates.

Laser communications through space offers higher capacity than radio links, but took a long time to develop because of the need to direct laser beams accurately enough to send signals between pairs of satellites or spacecraft that are constantly moving.

Recently, operational laser links have been demonstrated between satellites.

11.6 LASER MEASUREMENT

Lasers have proved invaluable for a diverse range of measurements. Laser instruments have long been used in surveying and construction alignment. Scanning laser instruments can measure the profiles of three-dimensional objects. Pulsed lasers can measure distances. This section briefly describes important measurement techniques. It does not try to cover the many ways in which lasers can be used to perform measurements inside specialized instruments.

11.6.1 Laser Surveying and Construction Alignment

Using a laser beam to draw a straight line may sound trivial, but it can simplify many tasks in building construction, surveying, and even agriculture. Early laser instruments for these applications used the red beams of helium–neon lasers, but modern instruments have shifted to red diode lasers.

An important construction application is defining a plane surface to aid in lining up mounts for suspended ceilings or partitions. A laser is mounted in a tripod, with the beam directed up into a prism that bends the beam so it emerges out the side. The prism rotates in a full circle, sweeping the beam around the walls. Adjust this laser plane generator properly, and it sweeps the laser spot around the walls at the same height. Construction workers mount the hangers for suspended ceilings at this level, so they are all even around a room. Turn the laser plane generator 90 degrees, and it can mark places for partitions.

Surveyors use red laser beams to define straight lines for land measurements.

In agriculture, laser beams define the gradients of irrigated fields. The slope should be large enough that the water does not form puddles, but slow enough that it does not run off too fast. A tripod-mounted laser can draw a straight line at the desired angle (for example, at 1 degree from the horizontal). Then the farmer can mount a sensor on his grading equipment to automatically keep the blade at the right height as it moves around the field.

11.6.2 Laser Radar and Distance Measurement

Radar (radio detection and ranging) has long measured distance with microwaves. The radar transmitter emits a short pulse, and an antenna watches for reflections of the pulse. The time it takes the reflected pulse to return measures the distance to an object. For example, a return time of one microsecond means that a pulse traveling at the speed of light made a round trip of 300 meters, so the object is 150 meters away.

Laser pulses can be used in the same way as microwave pulses to measure distance. Some police forces use compact laser radars to time how fast cars are moving (by measuring the change in distance between successive pulses) to foil speeding motorists with microwave radar detectors. This application is called both laser radar and LIDAR (for light detection and ranging).

Pulses should be short, or have very sharp rise times, for accurate distance measurements because you need an accurate measurement of the round-trip time. If your laser fires a one-microsecond pulse, the pulse is spread along a distance of 300 meters, leaving the object's position uncertain to 300 meters. That means it is impossible to measure the range to an object 150 meters away with a 1-μs laser pulse, but you could get a more accurate position using a one-nanosecond pulse, which is only 0.3 m (30 cm) long. In practice, accuracy also depends on pulse timing and measurement electronics.

Laser ranging is the measurement of distance to objects using laser radar. During the Apollo era, laser ranging was used to measure the distance to the moon by firing red pulses from a ruby laser though a telescope and timing how long the pulse took to return from a retroreflector that astronauts had placed on the moon. Laser ranging from retroreflectors on satellites can precisely locate points on the earth for geophysical research. A laser instrument called the Mars Orbital Laser Altimeter in the Mars Global Surveyor mapped elevations on the Martian surface by firing 10 pulses per second from a neodymium laser and measuring return times.

Armies use laser rangefinders to measure the distance to potential targets. In modern systems, the rangefinder may provide data directly to a gunnery computer to pinpoint the target. Simple laser rangefinders are used in proximity sensors installed on certain anti-aircraft missiles to detect when they are close enough for their warheads to destroy the target. When the rangefinder measures reflection from a target within that range, it triggers detonation of the warhead.

11.6.3 Laser Surface Profiling

Laser ranging can be combined with other techniques to precisely measure the surface contours of objects in three dimensions. A variety of approaches are used. For example, picosecond laser pulses could be fired at each point on the surface and return time measured to calculate the distance from the laser. A 1-ps pulse is only 0.3 millimeter long, so distances can be measured reasonably precisely. Laser beams can also be scanned across the object and their angles of reflection measured to determine the shape of the surface.

A big advantage of laser surface profiling is its ability to collect data on delicate objects without touching them. For example, paleontologists can scan a delicate fossil to build up a three-dimensional computer model of the object, then manipulate that model on a computer rather than handle the delicate object. Once the object has been scanned, the digital file can be copied so others can study the digital model. If the original object was distorted, for example, by the pressure of sediments on a fossil, the digital model can be manipulated to remove the effects of the distortion. Three-dimensional digital models also can be used to produce replicas of the original object using lasers, as described in Section 12-4, or other techniques.

11.6.4 Interferometric Measurements—Counting Waves

The most precise optical techniques for distance measurement rely on the interference of light waves, described in Section 2.1.3. The trick behind *interferometry* is to measure distances in units of the wavelength of coherent light.

To visualize the idea of interferometry, consider the arrangement shown in Figure 11-8A, where a beamsplitter divides a laser beam into two equal halves, directing each half at a different mirror. The mirrors reflect the laser light back to the beamsplitter, and part of the light from each mirror reaches the detector. In this case, the two waves arrive precisely in phase, their amplitudes line up precisely, and they constructively interfere to produce a bright spot.

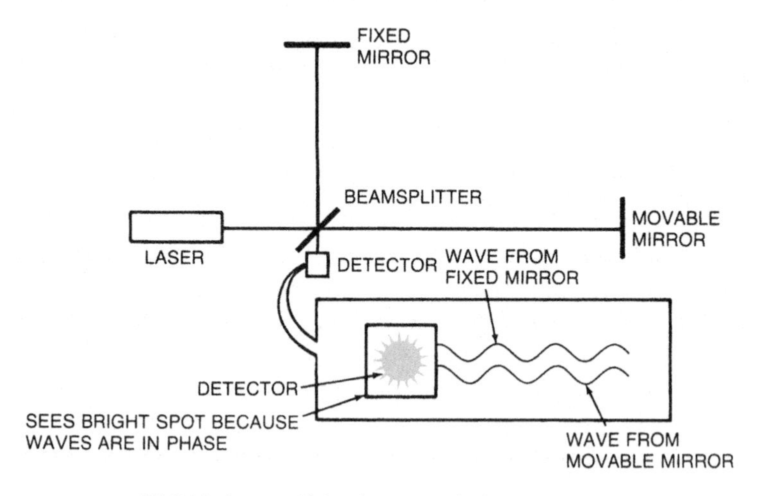

(A) Original setup with interference producing a bright spot.

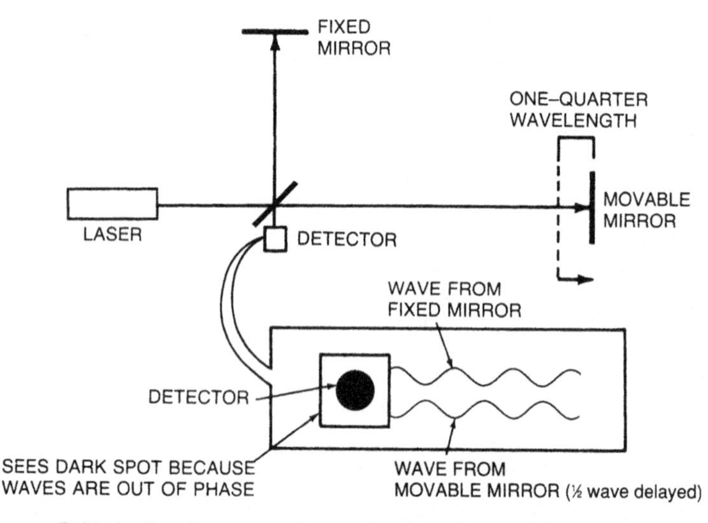

(B) Moving the mirror one-quarter-wavelength produces a dark spot.

Figure 11-8. An interferometer measures distance in units of wavelength. (A) Original setup with interference producing a bright spot. (B) Moving the mirror one-quarter wavelength produces a dark spot.

Now, suppose we move the mirror on the right just one-quarter wavelength farther from the laser, so the wave reflected from that mirror has a path a half-wavelength longer than it previously was, as shown in Figure 11-8B. That shifts the light from the mirror at right half a wavelength out of phase with the light from the mirror at the top, so their amplitudes are now 180 degrees out of phase and add to zero, producing a dark spot at the detector. Move the mirror at right another quarter-wavelength away from the laser, and it will produce another half-wave shift, so the two beams are in phase at the detector, and the dark spot turns light. Counting the number of times the spot turns from light to dark and then back tells us how much the mirror moves in units of the wavelength. Remember that with this arrangement the distance the mirror moves is half the number of wavelengths the round-trip distance changes.

The measurements can be quite precise because the wavelength is small. The detected spot shifts from light to dark when the mirror moves only one-quarter of a wavelength. When using a helium–neon laser emitting at 632.8 nm, that means we can detect movement of only 158.2 nm. If the mirror was moved 1 μm it would go through just over three full light–dark–light cycles, each corresponding to a half wavelength.

Interferometry also can be used to measure distance changes across a broad area. In that case, the interference pattern becomes visible as a set of light and dark fringes, tracing paths over the area being studied. Each fringe indicates a change in distance between the two surfaces of one-quarter wavelength.

11.7 LASER POINTERS, ART, AND ENTERTAINMENT

Lasers can entertain as well as measure. One of the simplest laser applications is the laser pointer, which is part entertainment and part a tool for presenting information. Laser beams also can be pointed at objects or into the sky for entertainment, or used to project displays.

11.7.1 Laser Light Shows and Displays

Laser light shows have been around for decades but, like fireworks, they remain entertaining. Millions of people around the world have seen brightly colored laser beams light up the night sky, a theatre, or a planetarium.

The concept is simple. Just bounce visible laser beams off moving mirrors, which scan patterns in the air, on the clouds, on nearby buildings, or on the screen. The scanning patterns can be controlled by music, by a computer program, or by someone operating the lasers (who becomes the optical analog of a musician). It is fascinating to watch the patterns unfold, and the display is safe as long as the beam is kept out of people's eyes. In practice, the key to a good laser display is imaginative control of the scanning system.

Many types of visible lasers have been used in displays. For many years, light shows used argon to generate green and blue beams, and krypton lasers to produce red beams, either from separate lasers or from mixed-gas lasers containing both. Today, solid-state lasers are more common because they are easier to use and more durable.

11.7.2 Laser Pointers

The simplest thing a laser beam can do is point, and as red diode lasers have become cheap, red laser pointers have become so inexpensive that the cheapest models have become giveaway items. Their nominal application is to project bright red spots onto screens during presentations, so speakers can illuminate the points they are making. In practice, red laser pointers are also used as toys, for things like playing with cats.

A new generation of green laser pointers have become available, and their prices have dropped as low as around $50. That is a remarkable achievement for a diode-pumped, frequency-doubled neodymium laser, emitting at 532 nm. The eye is about 10 times more sensitive in the green than in the red, so green laser spots look much brighter at equivalent powers. They are also helpful for people who suffer red–green color-blindness and cannot see the spots projected by red lasers. (The first person I saw with a green laser pointer was a prominent solid-state laser physicist who is color-blind and was delighted to have a pointer he could see.)

The greater visibility of green beams enables other applications. Amateur astronomers have embraced them as a way to point out bright objects in the sky to other observers, because the green beam is more visible in a dark night sky than red light.

Unfortunately, green laser pointers also have a downside. Inexpensive models have become readily available on the Internet with power much higher than the upper limits established by laser safety codes. These are potentially dangerous to the eye, and staring into the beam could cause serious eye damage. (See Appendix A for information on laser safety.)

Even worse, these green lasers can shine enough light into the cockpit of an airplane to dazzle the pilot's eyes at night. Most pilots are not looking down, but those coming in for a landing are particularly vulnerable. This has led to a few dumb stunts and public warnings about green lasers. It also landed one man in jail for a particularly stupid stunt—shining his green laser onto a police helicopter that was looking for the person who had been shining a green laser at planes landing at a nearby airport.

11.7.3 Laser Displays

It is possible to use lasers to project bright, full-color video-like displays onto a screen. Many such systems have been demonstrated, using separate red, green and blue laser beams that are scanned repeatedly across the screen. The big advantage is that the laser spots can be very bright, making impressively bright displays.

However, the laser displays are also very expensive, and in a sense they are too bright, because they must obey regulations developed to prevent inadvertent exposure to laser intensities that might damage vision. So far, such displays have not reached the commercial market.

11.8 LOW-POWER DEFENSE APPLICATIONS

Low-power lasers have found a number of important applications. Laser gun sights and laser target designators aim weapons. Laser countermeasure systems are designed to divert attacking missiles. And lasers are used in battle simulation systems for training.

11.8.1 Laser Gun Sights

Lasers can be aligned along gun barrels to aid in aiming. These systems act like laser pointers, with the bright spot pointing out where the gun is aimed. Police may use them as warnings, to show criminals that they are literally in police sights. Hunters also use them. Versions are available based on red, green, and blue lasers, but the red and green are preferred; the eye is less sensitive to blue than to green.

11.8.2 Laser Target Designators

A laser target designator is a different type of aiming system intended to guide smart bombs and missiles to their targets. A soldier on the ground or in a plane fires the designator, marking the target with a series of coded infrared pulses recognized by a smart bomb or missile. Older designators used neodymium lasers emitting at 1064 nm, but modern ones use lasers emitting longer wavelengths that pose less eye hazard.

The smart bomb or missile contains a sensor that looks for the characteristic series of pulses emitted by a target designator. (The coding prevents the bomb from being misguided by bright lights, fires, reflections of the sun, or other sources of light.) The simplest such sensors focus light from the target onto detectors divided into four quadrants. As long as equal amounts of light fall onto each quadrant, the bomb is on course. If one quadrant starts getting more light than the others, the bomb corrects its course to balance the light reaching all the quadrants.

First used in the Vietnam War, laser-guided bombs have proved much more accurate than conventional unguided bombs. However, the soldier holding the designator must remain in a line of sight to the target in order to keep it "marked" with the laser signal until the bomb hits it. Unfortunately, if the soldier can see the target, the target can also see the soldier, and fire bullets back at the source of the laser beam, if they recognize they are being marked.

Another hazard is eye damage during training exercises, in which modern soldiers spend more time than in battle. This has led to deployment of new designators operating at wavelengths longer than 1500 nm, which pose little eye hazard because they do not reach the retina.

11.8.3 Laser Battle Simulation

Lasers also play an important role in training in the Multiple Integrated Laser Engagement System, called "Miles," which equips soldiers with diode lasers and sensors. The lasers are attached to various weapons, and each fires a characteristic sequence of pulses. One code indicates that the pulses come from a rifle, another a bazooka, and a third denotes heavy artillery. Sensors are strapped on trucks and tanks as well as soldiers.

When the war games start, the soldiers fire laser pulses at each other, and the sensors keep score. A laser-simulated rifle shot can "kill" a soldier. Tanks, however, can only be knocked out by certain types of weapons. (To keep things honest, when the sensors on a tank detect a "kill," they turn off the controls and fire a plume of purple smoke to tell everyone on the battlefield that the tank is dead.)

Similar laser battle simulation systems are used today by armies around the world. Perhaps someone who saw one of them invented "laser tag" games, which don't actually use lasers.

11.8.4 Laser Countermeasures and Aircraft Defense

Lasers can also be used to foil attacking weapons, particularly missiles aimed at aircraft. Heat-seeking missiles that home on aircraft engine exhaust are readily available around the world and are relatively inexpensive. They are guided by infrared sensors, which look for particular infrared wavelengths produced by hot gases emerging from engines.

Laser countermeasures have been developed for military aircraft. Mounted in the plane, they scan for enemy missiles, and if they see one, they fire an infrared laser beam at the missile. The laser beam is brighter in the infrared than the engine exhaust, so it confuses the sensors in the missile, diverting them away from the aircraft. The Department of Homeland Security is investigating civilian versions of this system to protect airliners from terrorists with heat-seeking missiles, but the high cost of deploying the lasers on all large civilian airliners is raising concerns.

11.9 SENSING AND SPECTROSCOPY

A broad range of low-power laser applications have emerged from research on the interaction between light and matter. The study of those interactions and how they depend on wavelength is called *spectroscopy,* and it has been a particularly fruitful field for lasers. Once we know how materials react to light at various wavelengths, we can devise ways to sense the presence of various materials and measure the quantities present.

From an applications standpoint, both detection and measurement are important. Specificity is critical. Whether you are looking for a specific biomolecule to diagnose the presence of disease or warn of chemical attack, you want to make sure you are not getting false alarms from misidentifying other molecules in a complex environment.

Lasers are only one of many tools used in spectroscopy and sensing, so this section will concentrate on laser-based techniques.

11.9.1 Fluorescence Spectroscopy

Intense light can excite atoms or molecules to short-lived, high-energy states that quickly release much of the energy they absorbed at a longer wavelength. This process is called fluorescence and occurs in many materials. Fluorescence spectroscopy is attractive for measurements because the wavelengths at which fluorescence appears is useful for identifying materials.

Figure 11-9 gives a simple example of how fluorescence spectroscopy can be used to check newly made clothes to see if they are contaminated with oil that would make them unsellable. Pulses from a xenon–chloride excimer laser at 308 nm in the ultraviolet cause the oil to fluoresce at 450 nm in the blue part of the spectrum, so checking for fluorescence at that wavelength would reveal contamination. We can do that by moving the clothes on a conveyor belt past the excimer laser, which fires every time an

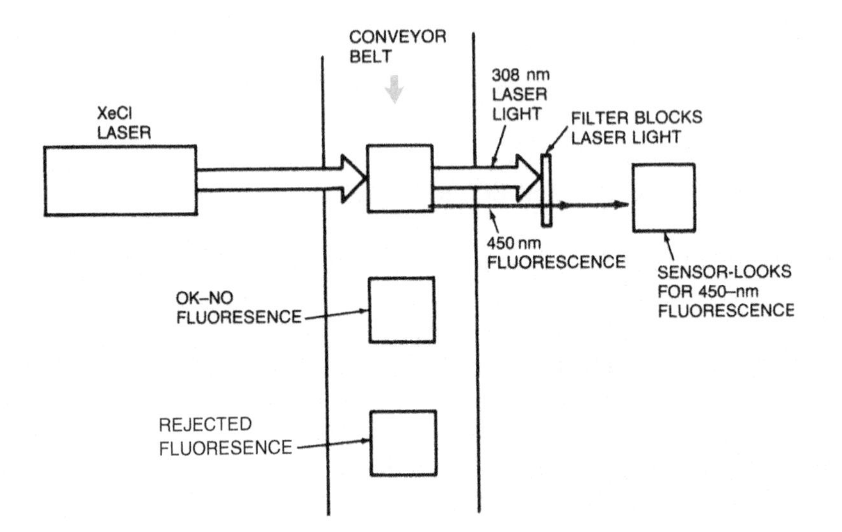

Figure 11-9. A simple fluorescence spectrometer senses 450-nm fluorescence to detect contamination.

item passes the laser. To check for fluorescence, we mount a detector that looks at the clothes through a filter that transmits only light at 450 nm. If more than a certain amount of light passes through the filter, the detector triggers a warning that marks the clothing for rejection because it is contaminated with oil.

The same idea can be used to detect other contaminants, but we may have to use other lasers and look for other fluorescence wavelengths. We also can do more with fluorescence spectroscopy than our simple example indicates. We can, for example, measure the amount of fluorescence to measure how much material is present. That could test quickly for certain water pollutants.

In practice, it is important to be sure that contaminants are being identified accurately. For example, white cloth used for pockets might fluoresce at 450 nm, potentially fooling the sensor into thinking a pair of pants was contaminated. To avoid that, sensors could monitor several different wavelengths, and reject an item only if all the responses matched. For example, fluorescence at 450, 462, 501, and 512 nm might indicate that the oil was present, but the cloth used for the pockets might fluorescence at 450, 480, and 630 nm.

11.9.2 Absorption Spectroscopy

Absorption spectroscopy identifies materials by the wavelengths they absorb rather than those they emit. It essentially checks what wavelengths have been subtracted from the illuminating light.

You do not have to have a laser to perform absorption spectroscopy. You can start with an ordinary light source that emits the full spectrum of visible light, then spread out the transmitted spectrum to see what wavelengths have been absorbed. However, the laser gives much finer resolution, so you can identify the precise wavelengths that have been absorbed. With an ordinary light source, you might know only that light was absorbed near 750 nm. A laser could tell you that light was absorbed at two wavelengths in that range: 749.87 and 750.13 nm. Because the laser can concentrate light at a single wavelength, it also can spot absorption lines too weak to see using other techniques.

Tunable lasers are the most common types used in absorption spectroscopy. Often, they are "swept" in wavelength, by tuning throughout their emission range. For example, a titanium–sapphire laser could be swept from 740 to 770 nm once every 30 seconds, as shown in Figure 11-10. If the wavelength varied uniformly with time, you could measure the wavelength at which light was absorbed by timing when the sensor measured absorption. For a 30-second scan at a uniform rate, detecting absorption 10.13 seconds after the start of the sweep would mean the wavelength was 750.13 nm. Tunable semiconductor lasers can be swept across that wavelength range even faster.

Infrared absorption spectroscopy at wavelengths of a few micrometers is particularly useful for organic compounds because absorption bands arising from molecular vibration and rotation—the most useful for identification—lie in the infrared. Laser sources are available in that region, but nonlaser sources are easier to use when the extremely high resolution of laser sources is not essential.

11.9.3 Laser Remote Sensing

Many important applications of laser spectroscopy are in remote sensing, which monitors distant objects or senses conditions remotely. For example, laser remote sensors might probe for ozone

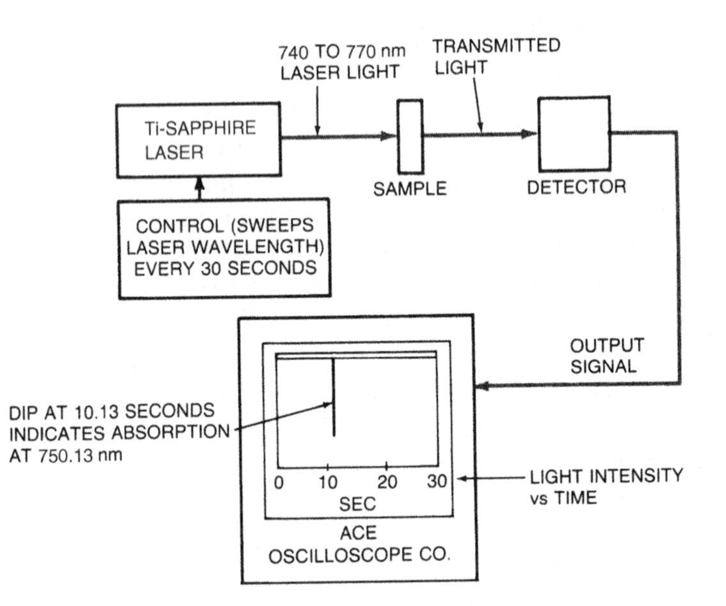

Figure 11-10. Absorption is measured by sweeping Ti-sapphire laser wavelength from 740 to 770 nm every 30 seconds.

in the upper atmosphere by transmitting a wavelength known to be affected by ozone. The laser would fire a pulse through the atmosphere, then look for returns reflected by ozone molecules. The return time would reveal the altitude of the ozone; the intensity of the return would indicate the amount of ozone. Likewise, a laser could be aimed at the plume of gas emerging from a distant smokestack to look for the emission of particular pollutants, or a laser tuned to particular wavelengths could monitor for chemical weapons or biological agents on a battlefield.

Remote sensing is a broad field that does not require lasers. Spy satellites and earth-resources satellites both perform remote sensing by photographing the Earth's surface looking for particular features. Such sensing is called passive because the sensors merely watch. Laser remote sensing is called *active sensing* because it probes objects with the laser beam and monitors their response.

Laser remote sensing can work in a variety of ways. Typically, developers find a wavelength that is either absorbed by the material being sought or causes fluorescence and look for a suitable signal showing the material's presence. To verify their measurements, they may probe with a second wavelength that does not cause absorption or cause fluorescence from the material.

The design of sensing systems depends on both the material being sought and on the application. To measure the emission of a known pollutant from a smokestack, all you need is a laser tuned to a particular absorption line unique to that pollutant and suitable detectors that can measure the absorption. Developing a military or security system to warn of the presence of dangerous chemical agents is more difficult because you cannot be sure what agents might be used, but you need an immediate warning if any might be present. Then the system should identify the agent to indicate what protective action should be taken. Other considerations include such issues as materials that might falsely trigger the sensor, the consequences of false positives, and balancing the problems of false positives against the dangers of missing dangerous materials.

11.9.4 Close-Up Laser Sensing

Lasers also can be used for a variety of close-up sensing, detection, and measurement applications, and often are packaged into measurement instruments. For example, fluorescence induced by blue or ultraviolet light can reveal important details in living cells viewed under microscopes, so some high-performance microscopes have optional laser accessories to help visualize internal cell structures.

Lasers can be used for some simple spectroscopic measurements of important medical parameters. For example, oxygenated blood absorbs light at different wavelengths than blood lacking oxygen, so it is possible to measure oxygen levels in the blood by shining red light through tissue, typically a fingertip.

Some biological tests may use lasers to read out results. For example, the best way to identify specific microbes is by testing with an antibody known to react with that microbe. A reaction indicates the microbe is present, but the reaction may not be easy to detect. Laser illumination might detect fluorescence if the antibody has reacted with the microbe, but not if there is no reaction. Similarly, medical patients could be given fluorescent dyes that are collected by cancer cells, then illuminated with lasers that could excite the fluorescence.

Measuring the scattering of light by particles in a gas or transparent liquid can indicate both particle velocity and size, which are important for applications from medicine to aeronautics. The scattering of femtosecond laser pulses that illuminate the skin can reveal structures in the outer layer of the skin, and potentially identify pathologies; the technique is called *optical coherence tomography,* and, like X-ray tomography, the data it generates can be used for three-dimensional imaging.

Simple tasks such as counting blood cells can be automated with laser-based sensors. The blood sample can be passed through a transparent tube so thin that only a single blood cell can fit through it at one time. If the tube were illuminated by a laser, the passage of a cell would block some of the light, triggering a sensor to count the cell. Similar laser measurements could be used on assembly lines to verify the presence of parts and to show that they are the correct size and orientated properly.

11.9.5 Spectroscopy in Scientific Research

The cutting edge of spectroscopy is in scientific research, where lasers offer extremely narrow linewidth and the ability to measure small differences in wavelength or frequency, such as the difference between the same transition of two different isotopes of the same element. You will learn more about lasers in research in Chapter 13.

11.10 HOLOGRAPHY

Holography is best known as a method of producing truly three-dimensional images with a laser, but it is not that simple. Although virtually all holograms are made today with lasers, Hungarian-born engineer Dennis Gabor invented holography in 1948, a dozen years before Maiman made the first laser. Working in Britain, Gabor originally conceived of holography as a way to improve resolution of the electron microscope.

Conventional photography records just the amplitude of light waves. Gabor's idea was to also record the phase of the waves, information lost in normal photography. With a record of both phase and amplitude, he, in principle, could reconstruct the entire original light wave. He was not thinking of three-dimensional images; his goal was higher-quality, two-dimensional images, and his early holograms were of flat two-dimensional transparencies.

Three-dimensional holograms require coherent light, and were not invented until 1963 by Emmett Leith and Juris Upatnieks. Holography had made little progress in the years after Gabor's work, and Leith and Upatnieks had not heard of Gabor when they began their research. They moved beyond Gabor by proposing a new way to record holograms, which overcame problems that had seriously limited Gabor's holograms, but required light that remained coherent over a distance at least as large as the object being recorded. They picked a red helium–neon laser as their source, and helium–neon lasers are still valued for their coherence length, which is much longer than that of diode lasers.

Their process of recording a hologram is shown in the top part of Figure 11-11. Light from a laser is split into two beams by a beamsplitter. One beam, called the *object beam,* illuminates the object and is reflected. The other beam, the *reference beam,* follows a separate path. The two are both directed onto a photographic plate, where the coherent beams form an interference pattern that is recorded as a hologram.

What the photographic plate records is actually an intensity pattern, light and dark areas formed by constructive and destruc-

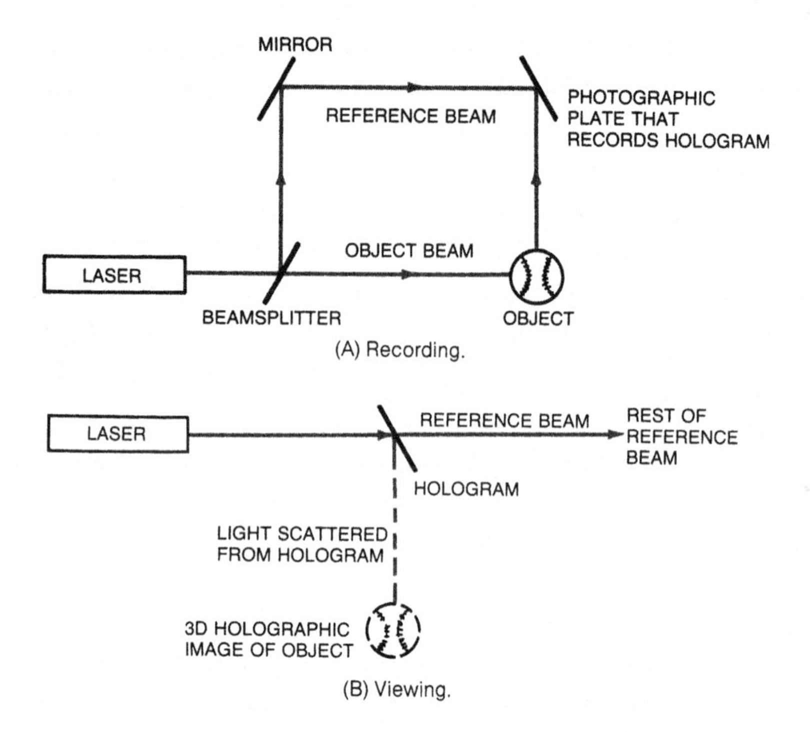

Figure 11-11. Recording and viewing a hologram in coherent laser light.

tive interference of light when the object and reference beams overlap. If you illuminate the hologram with a beam of light following the same path as the reference beam, scattering of the new reference beam from the interference pattern recorded in the hologram reconstructs the light waves originally reflected from the object. If you look at that light, you see a three-dimensional image of the original object, floating in space at the object's position, as shown at the bottom of Figure 11-11.

The first laser holograms had to be viewed in laser light to look three-dimensional. However, later advances made the images viewable in three dimensions with ordinary white light. The most widely used white-light holograms today are called *rainbow* holograms because their color ranges across the spectrum depending on the viewing angle. Rainbow holograms can be mass-produced by pressing them into metallized plastic, and have been widely used on bank cards to demonstrate authenticity, on stickers, and on the covers of books and magazines. It is easy to miss their three-dimensional nature unless you look carefully, and you will

d. Violet lasers are needed only for recording data; Blu-Ray disks can be played in ordinary CD or DVD drives.
e. The violet beam is scattered inside the player and makes the front glow.

4. A CD is played with a 780 nm laser and stores 700 megabytes. If the only difference between a DVD player and a CD player was that the DVD was played with a 650 nm laser, how much data could a DVD hold?
   a. 840 Megabytes
   b. 1.008 Gigabytes
   c. 2.1 Gigabytes
   d. 4.7 Gigabytes
   e. 10 Gigabytes
5. What other factors allow more data to be stored on DVDs?
   a. None
   b. Magnetic recording material
   c. More sensitive recording material
   d. Better optics and electronic data compression
   e. Better optics only
6. If the only difference between a CD and a Blu-Ray player was the use of a blue laser in the Blu-Ray player, how much data could be recorded on the Blu-Ray disk?
   a. 1.3 Gigabytes
   b. 2.5 Gigabytes
   c. 4.7 Gigabytes
   d. 8.5 Gigabytes
   e. 15 Gigabytes
7. How does a color laser printer print full color with only a laser to write on only one electrostatic drum?
   a. The laser writes on different spots which respond to different colors.
   b. The laser writes on different spots, which convert the toner to different colors before it is transferred to the paper.
   c. The printer contains three colors of toner which the laser transfers to the drum as the paper passes through three times.
   d. An inkjet stage sprays ink onto the paper, then the laser vaporizes the unneeded color.
   e. The printer uses a special paper that responds directly to laser light by changing color.
8. A standard digitized version of a telephone voice signal has a

data rate of 64 kilobits per second. How many of these voice signals can be combined to make a single data stream at 10 Gbit/s?

a. 178
b. 1780
c. 17,800
d. 178,000
e. 1,780,000

9. The average distance of the moon is 384,000 km from the Earth. If you are building a laser radar to precisely measure the distance from the Earth, how much time do you have to wait for the reflected light to return?
   a. 1 millisecond
   b. 1 second
   c. 1.28 seconds
   d. 1.82 seconds
   e. 2.56 seconds
10. You have a laser gun sight aligned perfectly with the barrel of your rifle and have the spot lined up to a target 300 meters away. Why do you miss and how?
    a. The rifle bullet cannot reach that distance.
    b. You forgot to account for the force of gravity on the bullet so it hit below the laser spot.
    c. You forgot to account for the force of gravity on the light, so the bullet hit above the laser spot.
    d. You did not account for the curvature of the earth.
    e. Murphy's Law says something always will go wrong.

CHAPTER 12

HIGH-POWER LASER APPLICATIONS

ABOUT THIS CHAPTER

Chapter 11 described major uses of low-power lasers, loosely defined as those that do not change objects not intended to be light-sensitive. This chapter covers major applications of high-power lasers, which cause significant changes to the materials they illuminate. The applications practical today include cutting, welding, drilling, and marking in industry, some specialized tasks in the manufacture of electronic components, and a variety of medical treatments. Several other applications for high-power lasers remain in various stages of development, including causing and controlling chemical reactions, laser weapons, and thermonuclear fusion.

12.1 HIGH- VERSUS LOW-POWER LASER APPLICATIONS

The line between high- and low-power laser applications is based on how the laser beam affects materials. If a laser cuts or welds a material, it clearly is making a major change, whether the material is sheet metal in a factory or living tissue during surgery. Marking serial numbers or product codes on equipment is not quite as clear, but is considered materials-working if the laser beam etches away surface material to mark the object.

On the other hand, it is reasonable to consider laser applications low-power if they change properties of light-sensitive materials, such as laser printing and optical data storage. Part of our rationale is that ordinary light can make similar changes, like the

Understanding Lasers: An Entry-Level Guide, Third Edition. By Jeff Hecht

exposure of photographic film. Part is that some of these applications are closely related to other low-power applications; for example, optical data storage by writing on disks is one of a family of optical disk applications that logically belong together. A final element of the rationale is the need to make some separation lest these two chapters merge into a single monstrous and unmanageable whole.

The last chapter showed the diversity of low-power laser applications. Less diversity is evident in the use of high-power lasers. Many applications are variations on the basic theme of laser materials working—heating an object to temperatures sufficient to vaporize and remove its exposed surface. From this standpoint, many medical applications are merely specialized materials working (a technically accurate view, if a bit callous to the patient), and laser weapons perform materials working on unfriendly objects.

Many interesting high-power lasers remain in development. Although some ideas are relatively new, some high-profile ideas have been in development for decades, notably high-energy laser weapons and laser-induced nuclear fusion. Research and development on those two long-running efforts have paid important dividends for other high-power laser applications by increasing power levels and steadily improving technology.

12.2 ATTRACTIONS OF HIGH-POWER LASERS

In Chapter 11, we listed attractions of low-power lasers. Let us take another look at the attractions of lasers, this time concentrating on those important for high-power applications:

- Lasers produce well-controlled light that can be focused precisely onto a small spot.
- Lasers can generate light in short pulses, concentrating energy to produce very high peak power levels for short intervals, affecting materials more strongly than steady power.
- Power densities can be tremendous if high-power pulses are focused onto small areas.
- Laser beams do not push the objects they strike, allowing noncontact processing.
- Lasers can be controlled directly by robotic systems.

- High powers are available at various wavelengths, so laser types can be picked to match the strongest absorption by various materials.
- Laser light travels a well-defined straight line.

This list is not identical with the one in Chapter 11, but some points are quite similar. On the other hand, the list of disadvantages of high-power lasers is rather different:

- High-power lasers can be very expensive.
- Metals strongly reflect light at many wavelengths.
- Powerful lasers are not available at most wavelengths, so they cannot be matched to the strongest absorption bands of all materials.
- Laser light cannot cut deeply into materials because ablated material gets in the way.
- Interaction with the air limit transmission of high-power beams over long distances.
- Laser light lacks the momentum of a physical projectile like a bullet.

These advantages and limitations shape the high-power laser applications covered in the rest of this chapter.

12.3 MATERIALS WORKING

Materials working involves cutting, welding, drilling, and otherwise modifying industrial materials, including both metals and nonmetals. The 1960s engineers who played at measuring laser power in gillettes were, at least in some primitive sense, exploring one aspect of materials working (drilling holes in razor blades). Materials working has become the largest single market for lasers (in total dollars), exceeding $1.68 billion in 2006 according to *Laser Focus World*. To understand how lasers work on materials, we will start with two of the earliest applications, which used lasers to drill very different materials, then move on to other applications.

12.3.1 Drilling Diamond Dies

One way to make thin metal wires is by pulling the metal through tiny holes in diamond "dies." Diamond is an excellent die material because it is the hardest substance known. It also conducts heat well, so it can dissipate the frictional heat generated by pulling metal through the hole.

The problem is that diamond is so hard that conventional drills will not penetrate it. Only diamond bits can drill into diamond, and those diamond bits get dull as they drill. Laser light, however, does not get dull. If a laser beam is focused onto a tiny spot in the diamond, it heats carbon atoms so they evaporate or combine with oxygen in the air. Deposit enough laser energy, and you can make a hole through the diamond.

Engineers discovered that lasers could drill holes in diamond dies during the 1960s, and ever since this has been a standard example of laser applications. It is not a high-volume application, but it is one where there are no attractive alternatives.

12.3.2 Drilling Baby-Bottle Nipples

The rubber used to make baby-bottle nipples is on the other end of the hardness scale from diamond, but that doesn't make it easy to drill holes in it. The holes in a nipple must be small for milk to flow properly, so tiny wire pins used to be pushed through the molded nipples. However, that's like trying to punch holes in Jell-O with a thin soda straw. The flexible rubber can catch, bend, and break the fine pins.

A laser beam is an excellent alternative because nothing solid contacts the nipple to push it out of shape. The laser pulse simply burns a tiny hole through the rubber, without touching anything to the nipple. Like drilling holes in diamond dies, this laser application has been around for many years because it is the best solution to a tricky problem.

12.3.3 Noncontact Processing

As we have indicated, an important factor in drilling both diamond dies and baby-bottle nipples is that laser drilling is a noncontact process. There is no physical laser "bit" that can get dull while drilling diamond. Nor can the laser beam distort the rubber nipple or be caught while drilling the hole.

The noncontact nature of laser processing is important for many types of laser drilling, cutting, and welding. Interestingly, many materials most often machined by lasers tend to fall into categories similar to diamond and rubber—either very soft or very hard. Titanium, a light metal used in military aircraft, is too hard to cut easily with conventional saws, but can be cut readily with a carbon dioxide laser. Other advanced alloys and hard materials also are often cut with lasers. On the other extreme are some rubbers, plastics, and other soft materials that usually are cut with lasers.

12.3.4 Wavelength Effects and Energy Deposition

The fundamental interaction in laser materials working is the transfer of laser energy to the object, causing a change such as evaporation of surface material. The details are complex, but we can describe in general two key factors: the effect of laser wavelength and how energy deposition affects the material.

Energy transfer to the object depends on absorption at the laser wavelength, which varies considerably among materials. Table 12-1 lists absorption of some common metals and a few other materials at some laser wavelengths. In general, metals tend to reflect more light at longer wavelengths, so they absorb a larger fraction of laser energy at shorter wavelengths, and most metals are easier to cut with shorter wavelengths. Titanium is an exception to the pattern because is absorbs relatively high 8% of light from a carbon dioxide laser, and is almost impossible to cut with conventional saws. In contrast, skin and white paint absorb more

Table 12-1. Surface absorption (percent) at important laser wavelengths

	Laser and wavelength			
Material	Argon-ion, doubled Nd or Yb (near 500 nm)	Ruby (694 nm)	Nd or Yb (1064 nm)	CO_2 (10.6 μm)
Aluminum	9%	11%	8%	1.9%
Copper	56%	17%	10%	1.5%
Human skin (dark)	88%	65%	60%	95%
Human skin (light)	57%	35%	50%	95%
Iron	68%	64%	~35%	3.5%
Nickel	40%	32%	26%	3%
Seawater	low	low	low	90%
Titanium	48%	45%	42%	8%
White paint	30%	20%	10%	90%

light at longer wavelengths, so carbon dioxide lasers cut such materials better, along with plastics. A number of other factors also affect the use of lasers. For example, aluminum is rarely cut with lasers because it reflects strongly across the spectrum and is soft enough to be cut easily with conventional saws or blades.

Absorption at the solid surface does not tell the whole story, however. Once it absorbs some energy, the material melts, then vaporizes. The absorption of most materials increases with temperature, so the first bit of heating is the hardest, and liquid metals often absorb more energy than the solid.

Vaporization can create problems by forming a layer of vapor or plasma between the laser and the object, as shown in Figure 12-1. This plasma layer absorbs some laser energy, blocking it from reaching the object. Sometimes, turbulence produced by heating the object can remove some or all of this plasma, but in other cases the plasma must be removed or blown away to improve energy coupling.

Another complication comes as the laser beam starts penetrating deep into the material. As laser vaporization excavates a deeper hole, the newly vaporized material gets in the way. It must go further to escape from a deep hole or cut, and some of the material may be deposited in or around the hole. At best this can reduce the efficiency of laser cutting or drilling. At worst, it can stop the process altogether.

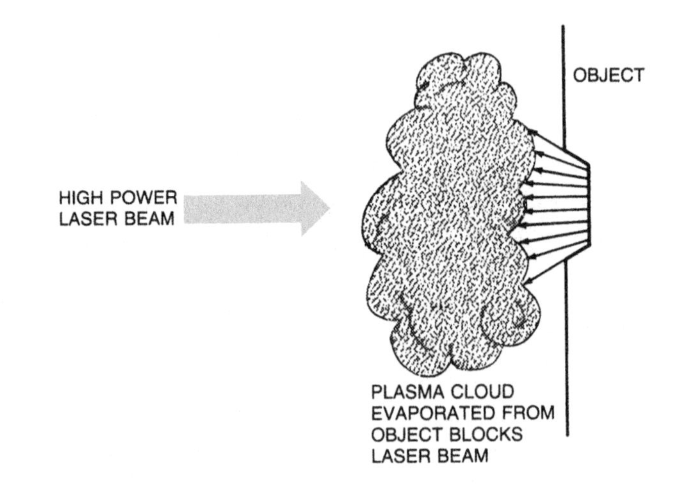

Figure 12-1. Formation of a plasma blocks the laser beam from reaching the object.

The optics that focus the laser beam also influence how deeply the laser can penetrate. Figure 12-2 compares how beams are focused by lenses with short and long focal lengths. The short-focus lens focuses light into a tighter spot, but the beam spreads out faster beyond the point where the focal spot diameter is smallest, reducing the concentration of light deeper in the hole. The longer-focus lens does not produces as tight a focal spot, but the beam does not spread out as fast, so it can drill deeper holes and cut thicker sheets of material.

12.3.5 How Laser Drilling Works

Drilling means making a single hole through an object. Laser drilling is done with one or a series of short laser pulses, each of which removes some material, making the hole deeper. The process requires high peak power, and works well for short repetitive pulses. (If the energy is delivered over too long an interval in a long pulse, the peak power may not reach high enough levels to vaporize and remove material from the hole.) The number of pulses required depends on factors including wavelength, peak power, the nature of the material, and repetition rate. For most efficient

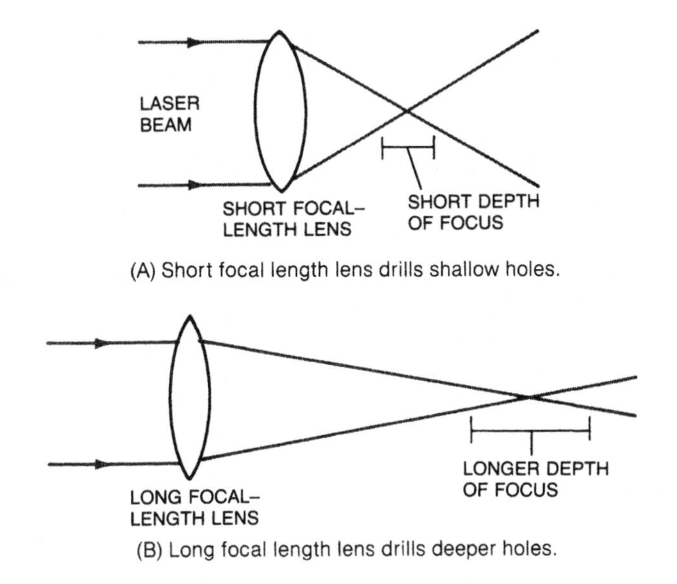

Figure 12-2. Effect of focal length on depth of focus; the longer the depth of focus, the deeper the hole that the beam can drill.

drilling, it may be necessary to change the optical focus during drilling to reach deeper into the drilled hole. The laser beam and/or the object are moved to change the spot being drilled. It also is important to make sure that the vaporized material gets out of the hole.

In the simplest case, a series of laser pulses drills one hole, then the laser is moved to drill another hole with another series of pulses. However, in a few applications many holes are drilled at once, sometimes with a single pulse that is divided into several lower-power pulses by passing through one or more beam splitters. One example is the "drilling" of ventilation holes in cigarette papers. A pulse from a carbon dioxide laser is split into many separate pulses, each of which punches a separate hole in the cigarette paper.

12.3.6 Marking and Scribing

Laser marking and scribing are similar to drilling in that they evaporate material from individual spots on a material, but differ in important ways. In marking, the goal is to write an indelible message, serial number, or trademark on a component by making a series of shallow pits deep enough to be visible. The shallower the pits, the faster the job can be done. Marking often is done by writing a series of dots on the surface to create the desired pattern (e.g., a serial number). Alternatively, the laser beam can be focused through a stencil-like mask to remove material from a larger area. The laser beam can mark a bare surface, but marking efficiency can be enhanced by coating the surface with a layer of black paint or other light-absorbing material that evaporates easily.

Scribing involves drilling a series of deeper holes that form what is essentially a perforated line in a brittle ceramic material. The holes may go through the material, but don't have to penetrate completely. Bending the ceramic breaks it along the weakened line scribed by the laser pulses, a technique used in electronics manufacture.

12.3.7 Laser Cutting

Laser cutting is somewhat like drilling a series of holes that overlap. In cutting, the beam and/or the object move continuously, and the beam itself normally is continuous rather than pulsed. You can think of laser cutting as a process something like slicing with a knife, but in this case the knife is a beam of light.

Cutting typically is done with the assistance of a jet of air, oxygen, or dry nitrogen. For nonmetals, the role of the jet is to blow debris away from the cutting zone, and improve the quality of the cut. Lasers can cut readily through sheets of wood, paper, and plastic, but thick materials are more difficult, and foods tend to char when cut, leaving an unappetizing black layer. Efforts to slice bread with lasers have produced burnt toast.

Metal cutting works differently, and uses a different type of jet. The laser beam heats the metal to a temperature hot enough that it burns as oxygen in the jet passes over it. This process is properly called "laser-assisted cutting," because the oxygen in the jet actually does the cutting.

12.3.8 Laser Welding

Welding may seem similar to cutting, but the two processes differ in fundamental ways. Cutting separates one object into two (or more) pieces. Welding joins two (or more) pieces into a single unit. Both require heating the entire thickness of the material, but the conditions differ.

Figure 12-3 shows how two pieces of metal are welded together. To form a solid bond, they must be fitted together precisely, leaving very little room between them. The laser beam heats the edges of the two plates to their melting points, causing them to fuse together where they are in contact. If the pieces are not in contact along the entire junction, the weld will be flawed. The laser beam must penetrate the entire depth of the weld (the thickness of the material) to make a proper joint. The heat of welding transforms the metal through the entire depth of the weld, and in a zone extending on both sides of the actual junction. The width of that weld zone depends on the nature of the materials being welded, on the power delivered by the laser, and on how fast the laser beam moves along the joint. The further from the mid-point of the weld, the less the heat affects the metal.

The compositions of the pieces being joined together are very important in welding. The materials do not have to be identical, but metallurgists have learned that some compositions do not bond together well, no matter how thoroughly they are heated or what technique is used for welding.

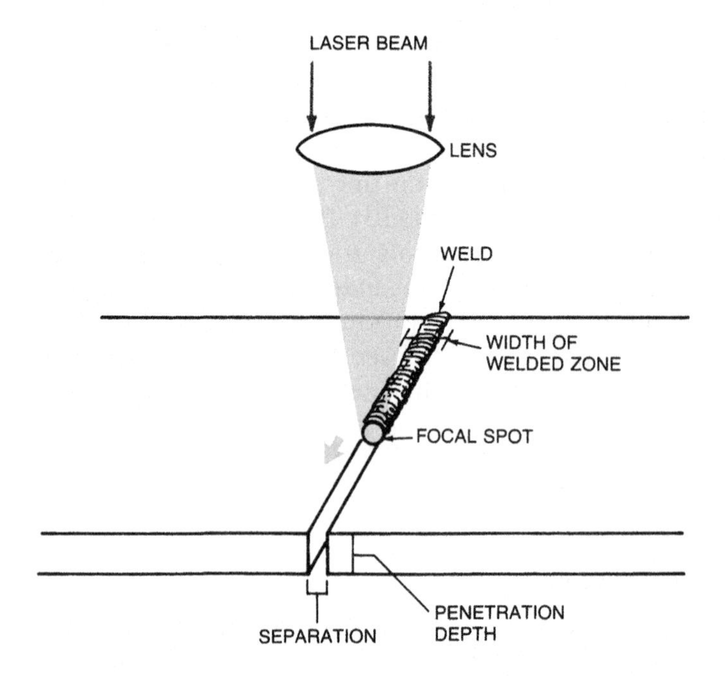

Figure 12-3. Two metal plates being welded with a laser.

Like cutting and drilling, laser welding is limited in the thickness of material it can handle. However, it has gained wide acceptance in manufacturing high-technology products. The cases of heart pacemakers and the blades of Gillette Sensor razors are among products welded by lasers.

12.3.9 Heat Treating

High-power lasers also are used in heat treating, which raises the temperature of the surface of a metal to convert it to another crystalline state. Typically, the goal of the process is to produce a state that is harder and more resistant to wear than the rest of the material. That hard crystalline form of the metal may be too brittle to use in bulk form, but strong enough if it merely forms a layer on another type of metal crystal.

In laser heat treating, the beam—usually from a carbon dioxide laser—scans across the surface of the metal. The surface may be covered with a coating to help it absorb the laser energy more efficiently. Metal surfaces can be heat-treated in other ways, but a laser beam can reach into areas that other processes cannot, such as the insides of engine cylinders. Typically, heat treating is done with a high-power continuous laser, with the beam spread out to cover a large area.

A related application is the annealing of silicon wafers used in semiconductor fabrication to change them from amorphous to crystalline states. However, silicon annealing is done with ultraviolet pulses from excimer lasers.

12.3.10 Types of Lasers Used in Materials Working

Laser Focus World reported that more than $1.68 billion worth of lasers were sold for materials-working applications in 2006. The sales are the largest for any laser application. The major types of lasers sold were:

- Carbon dioxide gas lasers, used in a range of applications, largely for nonmetals—$695 million
- Solid-state lasers based on rods, largely doped with neodymium—$465 million for a range of applications, including metalworking
- Excimer lasers, largely used in semiconductor photolithography—$349 million
- Fiber lasers, largely doped with ytterbium—$144 million, used in a range of applications

12.4 ELECTRONICS MANUFACTURING

So far, we have mentioned two laser techniques used in electronics manufacture: drilling holes to scribe ceramic or semiconductor wafers, and annealing the surface of silicon devices. They are among a number of laser materials-working processes widely used in the electronics industry.

12.4.1 Semiconductor Photolithography

The most important use of lasers in semiconductor production is in *photolithography,* the process used to fabricate the patterned layers of semiconductor that determine the function of integrated circuits. The process starts with coating the surface of a semiconductor wafer with a light-sensitive material, called a "photoresist." Then a light source illuminates the photoresist through a pattern called a mask, exposing patterns in the photoresist. Depending on the type of photoresist, either the exposed or unexposed areas are etched away chemically. If additional circuit layers are needed, another layer of photoresist is deposited, and the process is repeated to create another layer of patterns on the wafer.

Today's photolithography systems depend on ultraviolet light, and excimer lasers have become the preferred light sources for state-of-the-art chips. The driving factor in increasing the power and complexity of integrated circuits is the number of circuit elements that can be integrated on a single chip, and this in turn depends on the minimum feature size that can be fabricated. To shrink feature size, photolithography has moved to shorter and shorter wavelengths, starting from the blue and now in the ultraviolet. The industry is now moving from 90-nm features to 60-nm geometry, with the next target 45 nm.

Shrinking those features requires shorter laser wavelengths and optical systems that focus the light tightly, so it can expose features only a fraction of a wavelength. The 248-nm wavelength of krypton–fluoride excimer lasers was used in the past, but the state of the art is now the 193-nm wavelength of argon–fluoride (ArF) excimers. That technology has been extended by immersing the optics in liquids with high refractive index, which has the effect of increasing the focusing power of the lens to make tighter focal spots.

The industry is working on new techniques to generate even smaller features. The next logical type of laser to try is the 157-nm molecular fluorine laser, but that system has proved difficult to work with. Other efforts are focusing on improving the optical systems to stretch ArF laser technology to produce finer details. Semiconductor fabrication is by far the largest single application for excimer lasers, so extending ArF technology is important to the laser industry.

12.4.2 Scribing, Mask Repair, and Circuit Changes

Laser systems are also used in other electronics fabrication operations. One example is the scribing systems described in Section 12.3.6.

Resistor trimmers are laser systems designed to modify thin-film resistors deposited on chips or hybrid circuits. Their resistance depends on their surface area, which can be reduced by vaporizing some material with laser pulses, fine-tuning the resistor to the desired value.

Other laser systems repair the masks used to expose integrated circuits. Masks are costly devices, and a laser pulse can vaporize dirt or other excess material, repairing the mask so it can remain in manufacturing use.

Laser pulses can customize certain types of integrated circuits by making or breaking connections. These special circuits are designed with laser customization in mind, and include extra connections that can be broken or made with a laser beam. If the chip is used for one application, the manufacturer can break one connection; if it is used for another application, the company can break another connection or set of connections. Changing these connections changes the behavior of the integrated circuit. Although certain chips have to be altered, the economics of integrated-circuit production favors mass production of many identical chips over small runs of customized chips.

12.5 THREE-DIMENSIONAL MODELING

Lasers also can be used to "build" three-dimensional objects, using techniques very different than other types of laser materials working. The laser systems are among several competing technologies for *rapid prototyping,* which produces three-dimensional objects in shapes specified by computers. The parts typically are used in visualization or testing, but, eventually, they may be made on demand for general use.

One laser technique, called *stereolithography,* relies on the fact that some liquids solidify when exposed to ultraviolet light. In a stereolithography system, an ultraviolet laser beam is focused onto a platform slightly below the surface in a bath of liquid that solidifies when exposed to ultraviolet light. The computer controls scanning of the beam across the surface of the liquid, forming a thin layer of solid on the platform. Then the platform is lowered further into the liquid, and the laser beam scans the surface again, following another pattern to build the next layer of the model. After that layer is formed, the platform is lowered again and the cycle repeats layer by layer to build a three-dimensional model.

A second technique is *selective laser sintering,* which uses the scanned beam from a high-power laser to melt small particles in a powder that covers a small platform. Then the platform is lowered and the process repeated to solidify another layer. Other than in its use of a powder and longer-wavelength lasers, the technique is similar to stereolithography.

Several other rapid prototyping techniques have been developed using nonlaser technologies such as inkjet-like printers to build up structures layer by layer.

12.6 LASER MEDICAL TREATMENT

Physicians started experimenting with lasers soon after the first lasers were demonstrated. The first medical specialists to use lasers were those who were already familiar with light, notably eye specialists (ophthalmologists) and skin specialists (dermatologists). Other specialties followed, and laser medicine has become a well-developed field.

12.6.1 Laser Interactions with Tissue

Laser medicine is based on an understanding of the interaction of laser light with tissue. Inevitably, the interaction depends on the wavelength of light and the nature of the tissue. As shown in Table 12-1, dark skin absorbs more than light skin at visible wavelengths, but there is no difference between the two at the 10.6-μm wavelength of CO_2 lasers. Dark skin absorbs visible light because it is rich in the dark pigment melanin. The infrared absorption depends on the composition of the tissue.

Water is the most important component of tissue, and it absorbs strongly in much of the infrared. In fact, water absorbs infrared light so strongly and accounts for so much of the bulk of soft tissue that it is not a bad approximation to consider tissue as absorbing light as if it was water. Water absorbs about 80% of the incident 10.6-μm wavelength of a CO_2 laser in the first 20 μm, corresponding to the surface of exposed skin or tissue. Absorption is even stronger at 3 and 6 μm, although it is not uniform through the infrared.

That high absorption and the ready availability of CO_2 lasers has made the 10.6-μm line a favorite for laser surgery. Focus a CO_2

laser beam onto tissue at high enough intensity, and the laser energy will vaporize cells. The absorption is so strong that only the upper layer of cells are killed; they absorb virtually all of the light, so the lower level of cells survive with little damage. The lower layer suffers even less damage from a laser emitting at 3 μm because water absorbs that wavelength even more strongly.

Another advantage of the 10.6-μm CO_2 laser wavelength is that it penetrates just deep enough into tissue to seal small blood vessels and stop bleeding. Figure 12-4 shows how a CO_2 laser can remove the upper layer of tissue. When a surgeon scans the beam across the tissue, water in the upper layer of cells absorbs the 10.6-μm light. The light energy vaporizes the uppermost cells, but little reaches the underlying tissues to cause damage. The process also cauterizes the tissue, sealing blood vessels smaller than about a millimeter in diameter, which effectively stops bleeding. This cauterization effect makes the CO_2 laser especially valuable for surgery in regions rich in blood vessels, such as the gums and the female reproductive tract, by giving the surgeon a tool to remove thin layers of blood-rich tissue. This is particularly useful in the treatment of gum disease and of endometriosis, a condition affecting several million American women.

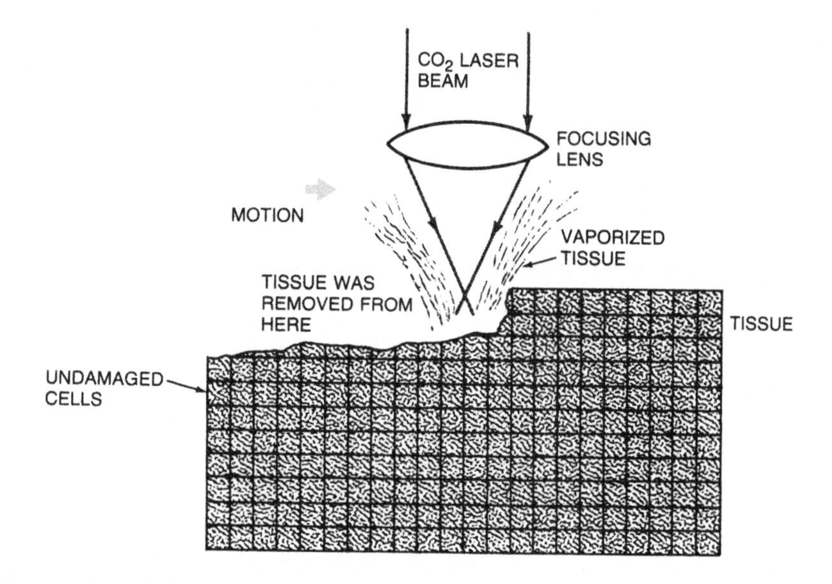

Figure 12-4. A CO_2 laser removes the top layer of cells with minimal damage to those underneath.

Carbon dioxide lasers also are used in a type of heart surgery that creates new paths for blood vessels in the heart, called transmyocardial revascularization.

Bone contains less water than other tissue, so CO_2 lasers cannot cut it. Lasers emitting at 3 μm can cut bone, because it contains some water that strongly absorbs that wavelength. However, surgeons generally use other tools for cutting bone, and for many other types of surgery, particularly general surgery on internal organs, where lasers have no particular advantages over conventional scalpels.

Other laser wavelengths in the ultraviolet, visible, and near infrared have other advantages for certain types of medical treatment. For example, skin blemishes from dark port wine stains to tattoos can be bleached by illuminating them with lasers selected to match the peak absorption wavelength of the blemish. In the case of port wine stains, the dark color of the birthmark comes from defective blood vessels near the surface of the skin, so the laser is matched to the absorption of the blood vessels. The 193-nm wavelength of the argon–fluoride excimer laser is strongly absorbed by the lens of the human eye, making it ideal for refractive surgery, as described in Section 12.6.2. Recent advances in laser medicine have been made possible by matching the laser wavelength, power level, and pulse duration to specific treatment needs.

In the decades since lasers were first used in medical treatment, interest has shifted away from general surgery to more specialized treatments. Let us look at some important procedures to understand how they work.

12.6.2 LASIK and Other Refractive Surgery

The best known and most intensely promoted laser surgery today uses an argon–fluoride excimer laser to reshape the lens of the eye to correct refractive defects that otherwise would force people to wear eyeglasses or contact lenses. The idea of refractive surgery emerged after researchers found in the early 1980s that 193-nm pulses from argon–fluoride excimer lasers could ablate tissue very efficiently from the cornea, the transparent front layer of the eye, shown in Figure 12-5. The cornea provides much of the eye's refractive power, so that opened the possibility of surgery to correct vision by reshaping the cornea.

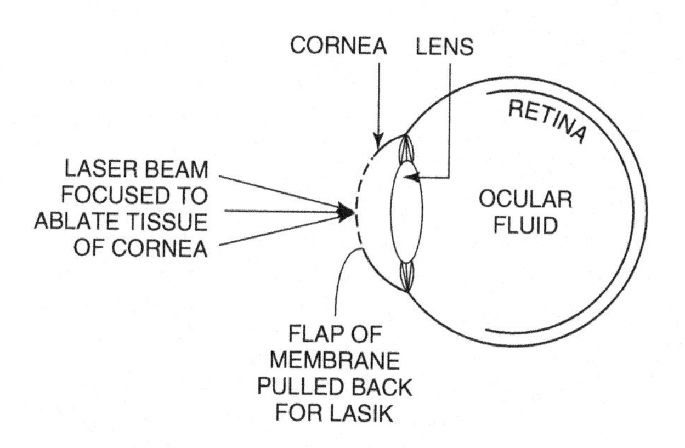

Figure 12-5. The anatomy of the eye. Refractive surgery removes tissue to change the shape of the cornea, the outer layer that provides much of the eye's refractive power.

The first type of laser refractive surgery developed was photorefractive keratectomy (PRK), in which an excimer layer ablated material from the cornea after either scraping away or ablating the top layer of the cornea, known as the epithelium. Concern over the effects of removing the epithelium led to the development of LASIK (laser-assisted in-situ keratomileusis), in which the epithelium is peeled back before the excimer laser ablates tissue from the underlying layer, then put back in place afterwards. LASIK has become the most common form of refractive surgery, and advocates say it has a lower rate of complications, although PRK is still in use and new variations are being developed.

Equipment for laser refractive surgery has steadily improved. Current systems can measure the refractive profile of the eye to compute the precise corrections required to compensate for the eye's defects. However, some complications remain, and accuracy of the correction depends on the healing response of the eye. Some patients require a second surgery to complete correction.

12.6.3 Laser Surgery Inside the Eye

The first major successes of medical lasers were in other types of eye surgery. These treatments focused the laser beam through the eye to treat damaged or diseased retinas, instead of performing risky open-eye surgery, and remain important tools for ophthalmologists. A newer treatment also focuses light inside the eye, but to help clear up a common complication of cataract surgery.

The simplest treatment to explain is detachment of the retina, the light-sensitive layer at the back of the eye, from the eyeball. The damage can spread, and without treatment the entire retina can come loose from the back of the eye, causing blindness. The detachment can be stopped by focusing a laser pulse so it causes a small burn on the retina that forms scar tissue. Although this damages a small area of the retina, the scar tissue "welds" the retina down to the back of the eyeball, so it cannot break free. That was a huge advance in treating detached retinas, which otherwise could require dangerous surgery or cause blindness.

A more common laser procedure is treatment of the abnormal proliferation of blood vessels that spread across the surface of the retina in many people with diabetes, a condition called "diabetic retinopathy." These blood vessels are very fragile, so they can leak blood into the normally clear liquid of the eye, gradually dimming vision and leading to blindness. Ophthalmologists found that illuminating the diseased retina with less than a watt of continuous-wave visible laser light can close off the abnormal blood vessels before they spread and do serious damage. The laser treatment does not always prevent blindness, but it slows the progress of the disease and helps victims retain their vision longer. It has been used on tens of millions of eyes around the world.

Cataracts occur when the natural lens of the eye becomes cloudy, and are treated by replacing the natural lens with an plastic one. The surgery leaves behind a membrane that holds the natural lens in place, and in about one-third of all cases, the natural membrane becomes cloudy after the surgery, again obstructing vision. That can be treated by firing a short, bright pulse from a neodymium laser through the implanted lens to break the cloudy membrane. This requires firing a series of laser pulses into the eye through a lens with a short focal length, so the energy is focused onto the thin membrane but spreads out over a large area inside the eye, so it does not harm the retina or other parts of the eye. The operation is done in a physician's office, and the membrane opens up almost instantaneously so the patient can see. Some patients almost jump for joy. Like other treatments that focus light into the eye, it avoids the time and expense of hospitalization, and the risk of surgically opening the eye.

12.6.4 Laser Dermatology

Dermatologists were early laser enthusiasts, but development of standard laser treatments for skin conditions took a while. Today, lasers are used to bleach skin blemishes called port wine stains and tattoos, although both are stubborn targets. Over the past several years, lasers have also been widely adopted for cosmetic skin treatments to remove wrinkles, improve skin tone, and remove unwanted hair.

The first major success of laser dermatology was in treating dark-red birthmarks called "port wine stains," which often appear on the face or neck. The discoloration comes from blood in networks of abnormal blood vessels just under the surface of the skin. Because port wine stains are spread over the surface of the skin, they couldn't be treated by conventional surgery, and many people tried to cover them with cosmetics. Early laser treatments tried to use blue-green light from argon lasers, but these were painful, worked only with dark blemishes and could cause scarring.

These problems were reduced by changing to pulsed dye lasers tuned to emit near the peak absorption of hemoglobin in blood at about 580 nm. This concentrates light absorption in blood vessels, so it avoids skin blisters and can be used with lighter birthmarks or on children. Lasers also are used for treating other skin blemishes, such as acne scars.

Laser bleaching of tattoos has received considerable attention, but remains a mixed success. The idea is to illuminate tattoos with laser light at a wavelength absorbed by the tattoo pigment, breaking down the pigment molecules to bleach the tattoo. How well it works depends on the ink; some almost disappear, but others fade only slightly. If you are lucky, laser treatment will leave a slightly discolored spot where a traditional tattoo used to be; if you are not, you will still wear your pledge of undying love for Annie to your wedding to Zelda.

A new family of encapsulated tattoo inks designed for people who might have second thoughts could make laser tattoo removal much easier. Tiny transparent plastic spheres contain the dyes, and protect them from degrading in the body. But a single laser treatment splits open the spheres, dumping the dye into cells which break down the molecules to bleach the color.

Other types of laser skin treatment are largely cosmetic. Laser skin resurfacing scans a carbon dioxide laser across aged areas, particularly the face, neck, and hands, to remove the surface layer. This removes surface wrinkles and blemishes, exposing a fresh layer of skin, which requires a few weeks to heal and return to normal color.

Laser hair removal aims a pulsed near-infrared laser emitting in the near infrared at hair follicles. Melanin in the hair and follicle absorbs the light, heating and killing the follicle, and thus stopping hair growth in areas where hair is undesired. The effectiveness of the treatment depends on the contrast between skin and hair color; the darker the hair and lighter the skin, the more effective the treatment. Different lasers may be used for different skin and hair combinations. The 755-nm output of a pulsed alexandrite laser works well, but only on light skin. Arrays of diode lasers emitting pulses at 810 nm are used for light to medium skin, and neodymium lasers emitting at 1064 nm are used for darker skin. Removing light-colored hair is difficult.

12.6.5 Laser Dentistry

It is hard to miss the placards heralding "laser dentistry," but from outside the office it is not clear what that entails. There are three main applications.

The obvious one is "drilling" teeth, using solid-state erbium lasers emitting at 3 μm. These lasers cannot remove solid enamel well, but they can remove decayed areas and prepare cavities for repair. More important to both dentist and patient, laser pulses are less threatening than the traditional whirring mechanical drill, and can be less painful as well.

As mentioned earlier, carbon dioxide lasers can treat gum disease by removing swollen tissue.

Visible argon and neodymium lasers are used to activate solutions that whiten teeth, a popular cosmetic treatment.

12.6.6 Fiber-Optic Laser Surgery

Conventional laser surgery works well for parts of the body that are exposed to the laser beam, but delivering the beam to the inside of the body can be a problem. That problem can be solved for some treatments by delivering the laser energy through an optical fiber that can be threaded into the body without major surgery.

An important example is treatment of kidney stones by threading a fiber-optic instrument through the urethra into the kidney or bladder. The fiber is pointed directly at the stone, then laser pulses are fired through the fiber to shatter the stone into pieces small enough to pass through the urethra without pain.

A similar treatment can be used to treat benign enlargement of the prostate gland by threading an optical fiber into the urethra. In this case, green pulses from a frequency-doubled neodymium laser are fired through the fiber to remove tissue from the enlarged prostate, allowing urine to flow more freely through the urethra.

Another fiber-optic laser therapy that has found less success is laser angioplasty. The goal was to deliver laser energy to remove plaque from the inside of clogged arteries, reducing the risks of strokes and heart attacks. The idea of a laser Roto Rooter sounded appealing, but clinical trials found serious potential complications, including a high risk that the artery would close again, increased probability of a heart attack, and perforation of the artery. Today the technique is rarely used.

12.6.7 Laser Cancer Treatment—Photodynamic Therapy

Early attempts to use lasers in cancer surgery failed because laser ablation splattered cancer cells around the surgery site, making the procedure counterproductive. However, lasers now treat cancer in a different way, called photodynamic therapy.

The treatment requires two steps. First, the patient is injected with a compound that strongly absorbs red light and is strongly absorbed by cells that are growing rapidly, particularly cancer cells. After a waiting period, which may last a couple of days, a red laser illuminates the affected zone, where the light-absorbing compound absorbs the light and releases reactive compounds that destroy the cancer cells. The technique is used against some hard-to-treat cancers of the esophagus and a few other cancers. One significant drawback is that the treatment leaves the patient hypersensitive to light for about four weeks, until the compound has dissipated from the patient's body.

12.6.8 Cold Laser Treatment and Biostimulation

All the laser treatments described so far produce observable changes in some tissue. However, low-power lasers emitting only a few milliwatts are used for biostimulation and other treatments that do not produce readily observable changes. Many of these lasers are used in alternative medicine, and the orthodox medical establishment is generally skeptical about their effectiveness. Examples include:

- Laser "acupuncture," in which a low-power laser beam illuminates the acupuncture points originally defined for classical Chinese acupuncture with needles. The effects are said to be similar to needle acupuncture.
- Alleviation of chronic pain by illuminating affected areas with a low-power laser.
- Speeding of wound healing by illumination with a low-power laser beam.

12.7 PHOTOCHEMISTRY AND ISOTOPE SEPARATION

The narrow range of wavelengths emitted by a laser make it possible to selectively trigger chemical reactions by illuminating certain materials with specific wavelengths of light. The light may remove an electron from the atom to ionize it, excite an atom or molecule to a highly reactive state, or split apart a molecule. These are called *photochemical* reactions, and their selectivity makes them attractive for separating materials that otherwise behave very similarly, such as different isotopes of the same element, which absorb light at slightly different wavelengths because they have different atomic weights.

In principle, laser photochemistry can be used in many ways but, in practice, the equipment required is so expensive that it makes sense only if the products are very expensive and are hard to produce in other ways. That led to a major U.S. government research program to separate uranium and plutonium isotopes for use in nuclear reactors and nuclear weapons, respectively. The larger uranium program began in the 1970s, was turned over to private industry in the 1990s, and was largely shelved in 1999. Plutonium enrichment is no longer pursued actively in the United States.

12.7.1 The Isotope Problem

Isotopes are atoms of the same element that contain different numbers of neutrons in their nucleus but the same number of protons. The number of protons in the nucleus (which equals the number of electrons in a neutral atom) defines the atomic number and, hence, the identity of an element. For example, all atoms with 92 protons in their nucleus are uranium atoms. However, uranium atoms can have different numbers of neutrons in their nucleus. The two most common uranium isotopes are uranium-238 (with 146 neutrons) and uranium-235 (with 143 neutrons). Uranium-238 is the most common in nature, but it cannot sustain the fission chain reaction needed to drive a nuclear reactor. To sustain a fission reaction, you need the less-common U-235. Conventional light-water fission reactors need a few percent of U-235 to sustain a chain reaction, but the natural U-235 concentration is only about 0.7%. The only way to get those reactors to work is to enrich the U-235 concentration above the natural level.

The problem is finding a way to separate the isotopes. The differences in nuclear mass shift electronic energy levels slightly, but not enough to make their chemical properties significantly different. That rules out chemical separation. Two conventional isotope-enrichment techniques, gaseous diffusion and centrifuges, rely on small differences in the mass of different isotopes or molecules that contain them, but those differences are small and separation is difficult.

Lasers offer a way around these limitations by generating light in a narrow range of wavelengths that can excite one isotope but not the other.

12.7.2 Atomic Vapor Laser Isotope Enrichment

The most important of the U.S. programs was the atomic vapor laser isotope enrichment process for uranium developed at the Lawrence Livermore National Laboratory in California. As shown in Figure 12-6, a copper-vapor laser pumped a dye laser that was tuned to a narrow range of wavelengths that excited U-235 atoms but not U-238 in uranium vapor. The excitation process ionized the U-235 atom, leaving it positively charged so it was attracted to a negatively charged electrode. In contrast, U-238 atoms retained all their electrons and were not collected by the negative electrode.

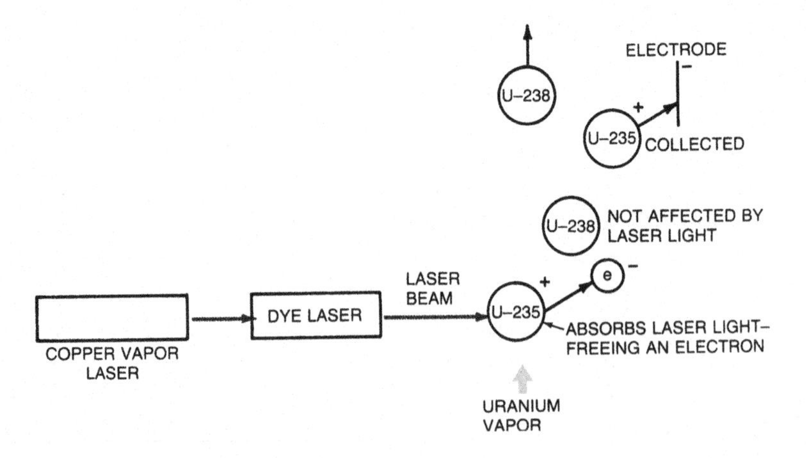

Figure 12-6. To enrich concentration of the fissionable U-235 isotope, a dye laser is tuned to a wavelength that ionizes U-235 but not the more abundant U-238.

The figure shows the negative electrode collecting only U-235 but, fortunately, the process is not that simple, because U-235 is the fissionable isotope that can be used in nuclear bombs as well as nuclear reactors. Each pass through the laser system increases the concentration of U-235 somewhat, but does not produce pure U-235. The process also poses other problems, such as handling uranium vapor. In the end, those complications combined with low demand for enriched uranium brought the program to a halt.

Livermore also developed a similar process for plutonium isotopes. In that case, the goal was to reduce levels of the heavier, nonfissionable plutonium-240 and -241 isotopes, which absorb neutrons, slowing down the nuclear chain reaction in a plutonium-239 bomb. That program was abandoned after the end of the Cold War left the United States with a surplus of weapon-grade plutonium.

12.7.3 Molecular Laser Isotope Enrichment Processes

Another approach to laser uranium enrichment starts with uranium hexafluoride (UF_6), which is a gas at room temperature. The United States developed a technique based on selectively exciting UF_6 containing U-235 at a wavelength of 16 μm, but that project was abandoned. An Australian company called Silex Systems Ltd. later developed a different process that reportedly selectively ionizes molecules containing U-235, then traps and collects them.

Silex has since licensed the technology for commercial development, but it remains unclear if it will prove practical.

12.7.4 Laser Separation of Other Isotopes

Laser techniques can separate the isotopes of other elements, for applications in medicine, research, or industry. The need for these isotopes is much smaller than the need for uranium or plutonium isotopes, but laser techniques are promising for production of relatively small quantities of isotopes including calcium, gadolinium, and erbium.

12.8 LASER-DRIVEN NUCLEAR FUSION

Government research on laser initiation of nuclear fusion began in the 1960s and has continued since then on two parallel lines, one devoted to development of civilian fusion reactors, the other to simulation of the explosion of hydrogen bombs. The civilian side of the project was long stressed in public, but the military program has long paid for most of the research.

The idea behind laser fusion is to heat and compress hydrogen isotopes to the high temperatures and pressures needed for their nuclei to fuse together, as they do in stars and hydrogen bombs. This would be done by focusing very short and very high-energy laser pulses onto a target made of fusion fuel, to simultaneously heat and compress the target to very high pressures and temperatures for a fleeting instant. Inertial forces implode the fusion targets, so this approach is called *inertial confinement fusion,* in contrast to the older concept of *magnetic confinement fusion,* in which fusion fuel is heated to high temperatures while confined for a long time in a magnetic field. Today, magnetic-confinement fusion is a civilian research project; inertial confinement fusion is largely military funded to study bomb physics, although civilian power is a long-term goal.

Laser fusion is based on illuminating pellets containing a mixture of the heavy hydrogen isotopes deuterium and tritium, which fuse together when heated to high temperatures and compressed to high densities. Laser energy is concentrated in short, intense pulses and the light is distributed uniformly across the surface of the target. This produces a symmetrical explosion that forces the target material inward while heating it to high temperatures. Fusion occurs at the peak of the implosion.

The details of inertial confinement fusion are closely related to the workings of a hydrogen bomb. In both, the rapid deposition of energy vaporizes material to form a plasma that emits intense X-rays. The powerful burst of X-rays then compresses the target to produce fusion, which releases another powerful burst of energy. The small-scale processes in laser fusion model the larger-scale processes that occur in a nuclear bomb, allowing much more detailed observations than otherwise possible of the operation of a bomb.

Government scientists have built a series of increasingly larger and more powerful lasers, which have demonstrated that laser fusion is possible but not easy. Each step paved the way for a bigger and more powerful laser. The U.S. Department of Energy began construction of the latest in the sequence, the National Ignition Facility (NIF), in 1997 at the Lawrence Livermore National Laboratory in California. (The name comes from the goal of crossing a fusion threshold called ignition.) However, it is not scheduled to begin full-scale fusion experiments until 2010.

NIF is the biggest laser ever built, occupying three connected buildings that together are 704 feet long, 403 feet wide, and 85 feet tall, shown in Figure 12-7. The laser is a massive laser oscillator–amplifier. The seed pulse comes from a single one-nanojoule oscillator, which is amplified then divided among parallel chains of amplifiers. All told, there are 192 beam lines, all converging on a single target chamber. At the fundamental wavelength of 1053 nm, the neodymium–glass laser will deliver energy of 4.2 megajoules per pulse, which the final optics will convert to the third harmonic at 351 nm at the cost of about half the energy, leaving 1.8 megajoules to be delivered to the target. The peak power reaching the target will be 500 terawatts, roughly a thousand times the electric power generation capacity of the whole United States, but the pulse will last only 3 to 20 nanoseconds.

As you might expect from the dozen years needed to complete the project, it has suffered a series of problems and delays. Nonetheless, construction has continued because it is a key part of the U.S. Stockpile Stewardship Program, which seeks to maintain the country's nuclear weapon capability without a continuing series of nuclear tests. Researchers also hope it will help point the way to progress in civilian fusion-energy research.

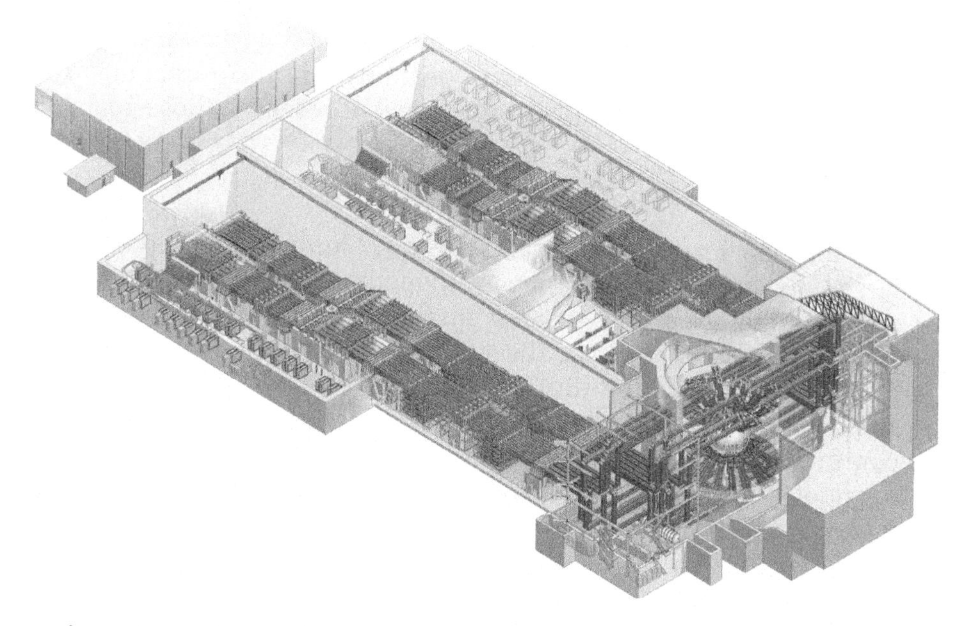

Figure 12-7. View of the National Ignition Facility, with the surrounding building cut away to expose the laser equipment. The beam lines feed into the target chamber at lower right. (Lawrence Livermore National Laboratory illustration; government work not subject to copyright.)

France is building a similar system called Laser Megajoule, designed to produce similar power and energy levels with a total of 240 parallel beams. Plans call for it to be completed about the same time as NIF.

12.9 HIGH-ENERGY LASER WEAPONS

The idea of laser weapons goes back to the dawn of the laser age in 1959, when the Pentagon's Advanced Projects Research Agency issued a $1 million contract to try to build a laser based on ideas proposed by Gordon Gould. Since then, the Pentagon has spent billions of dollars trying to develop high-energy lasers for use on the ground, at sea, in the air, and in space. As of this writing, none of those weapons have ever been deployed, much less used in combat.

The long history of laser weapon development testifies that the idea of a weapon that strikes at the speed of light is attractive.

However, the long history also shows that making laser weapons is a formidable problem. To understand this long history, let us look at the missions for laser weapons, the requirements to achieve those goals, and programs aimed at reaching those goals.

12.9.1 Missions for Laser Weapons

The mission for laser weapons is to disable or destroy an enemy target. Exactly what that means depends on the target. If the enemy target is a heat-seeking missile shot at one of your airplanes, the target could be disabled by shining a moderate-power laser at the missile to blind the sensors needed to guide the missile to hit the plane it was aimed at. If the target is an enemy intercontinental ballistic missile launched toward your city, you want to shoot it down before it reaches your territory. Targets could be rockets, artillery shells, helicopters, airplanes, missiles, or satellites.

Sometimes, it suffices to blind a sensor or heat the enemy target to a temperature that cooks its electronic guidance system, but it is easiest to think of a laser weapon as performing materials working on unfriendly objects. The beam is intended to punch holes in the target, cut it open, or make it blow up. From the standpoint of battle management, a laser weapon works best if its impact on the target can be detected by the side firing the laser, so they know the target is no longer a threat and can aim the laser at something else. In general, reaching these goals means the laser must deliver high power to the target, and keep that power focused on the target long enough to do lethal damage.

The immediate goal may be to win the battle at hand, or to win the war. Tactical weapons are built to use to win a battle. Strategic weapons generally exist to discourage anyone else from waging war against you. Laser weapons can be either type.

Ronald Reagan called his "Star Wars" missile defense system the Strategic Defense Initiative because it was a strategic weapon system. Its purpose was to discourage an attack by the Soviet Union, by making it impossible for the Soviets to win a war. The orbiting laser battle stations he hoped might provide a defensive shield were intended to be strategic weapons—ready to be used if necessary, but hopefully not to be needed.

The Advanced Tactical Laser being built and tested by the Air Force is intended to study how lasers could be used on a battlefield. The goal is to see how effectively the laser can destroy enemy targets on the ground or in the air. The experimental Tactical High-Energy Laser (THEL) developed by the U.S. Army and the Israeli Ministry of Defense was intended to shoot down short-range missiles, including mortar shells and inexpensive rockets often used against Israel. In the end, THEL technology was not ready to deploy, but the project's sponsors hope that more advanced lasers will be up to the job.

12.9.2 Requirements for Laser Weapons

A high-power laser alone does not make a laser weapon. Practical laser weapons require four key subsystems:

1. A target acquisition and tracking system, which identifies the target, tracks it, and gathers data on its location and track to aim the laser at the target.
2. The laser weapon itself, including the laser, the power supply for the laser, and other components.
3. Optics for beam control and pointing, to deliver the laser beam to the target.
4. The "platform" that houses and carries the laser and associated subsystems, which may be a plane, ship, ground vehicle, satellite, or a building.

Each of those subsystems must meet stringent requirements. The targeting system must find and follow the right target. The laser weapon platform must carry the laser and power supply, stabilize the laser so it can operate properly (vibrations can cause problems), and keep the laser running. The beam control and pointing optics must deliver a lethal amount of laser energy to the target through the atmosphere. They also must not be too bulky, putting a premium on lasers that operate on shorter wavelengths that are easier to focus in tight beams with reasonable-sized optics.

Focusing lethal laser energy onto a target that may be tens, hundreds, or (in the case of strategic defense) thousands of kilometers away poses tremendous challenges. Air looks clear, but it always absorbs a small fraction of the light it transmits. Absorption of only 0.1% of the energy in the beam may seem a small loss, but if the beam is powerful, that tiny fraction can heat the atmosphere significantly. For a 1-MW laser beam, 0.1% is a kilowatt, and that's enough to heat the atmosphere and make it turbulent so it bends the laser beam in unpredictable directions. This turbulence makes it hard both to track the target and to focus the laser beam onto it effectively, as if you were looking over the hood of a car on a hot sunny day.

Laser weapons require power levels of 100 kilowatts and up for use on the battlefield, and in the megawatt range for use over distances of tens or hundreds of kilometers. That may sound like a lot compared to a kilowatt-class industrial laser that can cut almost anything, but remember that the industrial laser sits right next to the target. A laser weapon is far away from the target.

12.9.3 Laser Weapon Physics

The basic physics of high-energy laser weapons resemble that of laser materials working, but there are some important differences that are not always obvious:

- The beam must be focused onto the target from a distance. When a laser is close to the object, a lens can focus the beam down to a small spot on the object. When the laser is far away, a diffraction-limited beam forms a much larger spot. If beam divergence is one milliradian, the spot size is roughly one meter at a distance of one kilometer. The longer the distance, the more the air is likely to perturb and defocus the beam.
- The beam from a laser weapon cannot be precisely controlled to ablate material from the target surface with pinpoint precision. The beam is spread over an appreciable area on the target surface, so the goal is to deposit enough energy so the surface or a critical internal component fails. Short intense pulses of laser energy can shock the surface, causing cracks that can make a missile go off course. Depositing laser energy can heat a missile enough to ignite rocket fuel, or damage the warhead.
- Laser beams do not have appreciable momentum, so they have to deposit energy to damage the surface. It is not enough just to hit the target; the photons have to be absorbed. In contrast, bullets and other projectile weapons do lethal damage because their momentum carries them into the target, although they do not carry a huge amount of energy. Getting hit by an intense laser pulse might burn a hole in your shirt or burn your skin,

but a bullet would penetrate your skin and might punch a hole all the way through your body due to its momentum.

12.9.4 Lasers Developed for Weapon Applications

Several types of lasers have been tested for potential use as weapons, but some have been abandoned because of limited power, difficulty of operation, or problems with atmospheric transmission at their output wavelength. The search continues for new types of high-power lasers, but a few types dominate recent laser-weapon programs:

- *Hydrogen fluoride chemical lasers* emitting at 2.6 to 3.0 μm are the leading candidates for space-based lasers, but cannot be used in the atmosphere because of high absorption. However, currently there is little interest in space-based laser weapons.
- *Deuterium fluoride chemical lasers* emit at 3.6 to 4.0 μm using hydrogen's heavy isotope deuterium to produce a wavelength readily transmitted by the atmosphere. They have been tested on the ground and are being tested in the air.
- *Chemical oxygen–iodine lasers* emitting at 1.3 μm are used in the Airborne Laser and the Advanced Tactical Laser.
- *Diode-pumped solid-state lasers* emitting near 1 μm are being developed for possible tactical applications on the ground or in the air. Diode-pumped neodymium lasers are the leading candidates, but arrays of diode-pumped fiber lasers are also being considered.

12.9.5 Laser Weapon Programs

The Pentagon has run a long series of laser weapon programs. We cannot cover them all, but the most important ones of the past decades deserve mention.

The *Strategic Defense Initiative* proposed building orbital chemical laser battle stations, generating about five megawatts each with 4-m focusing mirrors to aim the beams up to 5000 kilometers away. This was far beyond the state of the art, and launching the laser stations and their fuel would have been challenging at best. A test bed called *Alpha* reportedly demonstrated megawatt-level output in ground tests, presumably for brief intervals. The *Space-Based Laser* was another plan, but that was cut after the end of the Cold War.

The *Tactical High-Energy Laser* (THEL) was built by Northrop-Grumman for the Army and the Israeli ministry of defense to test the feasibility of ground-based defense against rockets and mortars. It used a high-power deuterium fluoride chemical laser and the testbed filled several trailers. The laser was able to shoot down mortar shells and short-range rockets on test ranges, but a planned mobile version based on a chemical laser was never built, largely because military planners did not want to field a system that required special chemical fuels.

The *Advanced Tactical Laser* is a high-power chemical oxygen/iodine laser mounted in a C-130 aircraft for tests of how lasers might be used on the battlefield.

The *Airborne Laser* (Figure 12-8) a megawatt-class chemical oxygen/iodine laser mounted in a customized Boeing 747, intended to fly a missile-defense patrol within a few hundred kilometers of rogue states ready to shoot down any long-range missiles launched toward the United States.

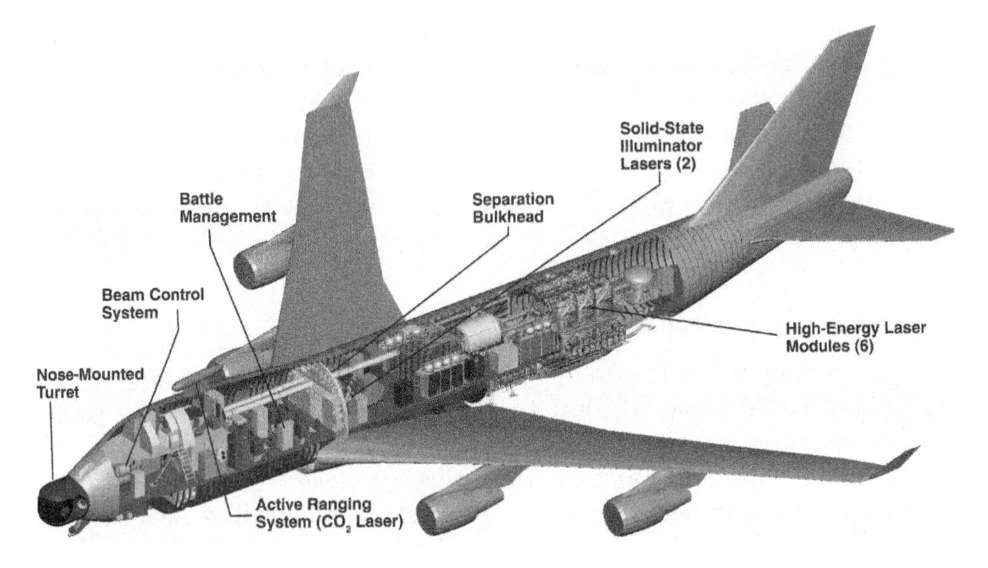

Figure 12-8. Cutaway view of the Airborne Laser, showing the chemical oxygen iodine laser in the back of the plane, with crew and control systems in front. The beam emerges from the nose turret. (Courtesy of Boeing Corporation.)

The *Joint High-Power Solid-State Laser* is a demonstration diode-pumped solid-state laser being built to assess feasibility of a 100-kilowatt solid-state laser weapon. If the technology meets expectations, it could be adapted for other uses, such as replacing the chemical lasers in the Advanced Tactical Laser and THEL. Military planners prefer solid-state lasers for battlefield use because they would run off electricity that could be produced by diesel generators, without requiring special fuels.

At this writing, the last three programs are still in the testing or development stages, with the Advanced Tactical Laser furthest along, followed by the Airborne Laser; the feasibility of 100-kW solid-state lasers is still being assessed.

12.9.6 Antipersonnel Lasers

Because laser weapons can only cause damage by depositing energy, they are not as lethal against enemy soldiers as guns that fire bullets, which carry deadly momentum. However, the eye is extremely sensitive to laser light, and it is technologically feasible to zap people with enough laser light to seriously damage their retinas and permanently impair their vision. They probably would retain some vision, but not enough for most types of work. This is considered inhumane, and international treaties ban antipersonnel laser weapons.

12.10 FUTURISTIC HIGH-POWER LASER IDEAS

What about peaceful uses for lasers of weapon powers or higher? A number of ideas have been proposed but, curiously, many of them are leftovers from the early days of laser research.

Power beaming from orbiting solar power satellites was among the proposals to deal with the energy crisis of the 1970s. However, advocates of solar power satellites were wary of using high-energy lasers to beam power down to the ground, worried that the lasers might be deliberately misdirected as weapons.

Powerful laser beams might propel spacecraft by evaporating fuel kept on the satellite. One idea is to nudge satellites into different orbits by evaporating ice they carry, so the force from the evaporating gas would push the satellite in the right direction. People have even proposed using lasers to launch spacecraft from the ground, but extremely high powers might be needed.

Lasers might even save the world someday if it was threatened with the possible impact of a wayward asteroid. If the asteroid's path was known to be dangerous many decades in advance, a laser beam could be aimed at the asteroid to transfer a small bit of energy and the little momentum that photons carry to the asteroid. Over many years, that tiny push could deflect the asteroid safely away from an impact.

12.11 WHAT HAVE WE LEARNED?

- High-power lasers cause significant changes to materials not intended to be sensitive to light.
- Short laser pulses concentrate energy to produce very high peak power for short intervals, affecting materials more than exposure to the same power over longer time.
- Lasers do not move or distort objects, allowing noncontact processing.
- Materials working includes cutting, welding, drilling, and heat treating both metals and nonmetals.
- Laser pulses can drill holes in diamond, the hardest known material.
- Laser pulses can punch holes in very soft materials without distorting them.
- Laser pulses do not get dull or bend.
- Energy deposition depends on wavelengths at which materials absorb.
- Plasma released from an illuminated object's surface can block the laser beam.
- Drilling requires pulsed lasers and depends on peak power.
- Lasers can mark or scribe materials without drilling holes all the way through.
- Laser cutting of nonmetals is aided by jets of air or dry nitrogen, to blow away debris.
- Laser cutting of metals is aided by oxygen, which reacts with the hot metal.
- Welding joins two pieces into a single unit.
- Heat treating transforms the surface of a metal to a harder form.
- Materials-working applications are the largest market for lasers in terms of dollar sales.
- Semiconductor photolithography uses ultraviolet light from ex-

cimer lasers to fabricate integrated-circuit patterns on semiconductor wafers.

- The shorter the wavelength used for photolithography, the finer the details that can be produced, and the more electronic components that can fit on a chip.
- Laser stereolithography is a rapid prototyping technique that uses ultraviolet light to produce solid three-dimensional models.
- Laser medicine is based on understanding the interactions of laser light with tissue.
- Most surgical lasers emit at wavelengths strongly absorbed by water, which is the most important component of tissue.
- The 10.6-μm CO_2 laser can vaporize tissue and prevent bleeding.
- LASIK and other refractive surgeries uses 193-nm ArF lasers to change the refractive power of the cornea.
- Lasers can treat detached retinas and diabetes-related blindness.
- Pulsed dye lasers can bleach dark birthmarks called port wine stains.
- Laser success in erasing tattoos depends on the ink used. Many traditional inks are hard to treat.
- Laser skin resurfacing removes surface wrinkles and blemishes, exposing a fresh layer of skin.
- Lasers can stop growth of unwanted hair by killing hair follicles.
- 3-μm erbium lasers can remove decayed regions of teeth and prepare cavities for filling, but do not remove solid enamel well.
- CO_2 lasers can treat gum disease.
- Laser energy delivered through optical fibers can shatter kidney stones and treat benign enlargement of the prostate.
- Cold laser treatment includes acupuncture, pain treatment, and wound healing.
- Lasers can selectively excite isotopes of uranium and plutonium to produce mixtures of isotopes better suited for nuclear reactors or bombs.
- Inertial confinement fusion uses high-energy laser pulses to heat and compress fusion fuel to the high densities and temperatures needed for nuclear fusion.
- Laser fusion experiments simulate the explosions of hydrogen bombs.

- Laser light carries very little momentum, so laser weapons must damage targets by transferring energy to them.
- Laser weapons require at least 100 kW on the battlefield, and megawatt-class powers for destroying targets tens or hundreds of kilometers away.
- Chemical oxygen–iodine lasers are used in laser weapon test beds such as the Airborne Laser.
- Diode-pumped solid-state lasers are being developed for use as battlefield laser weapons.

QUIZ FOR CHAPTER 12

1. Which of the following is not an attraction of high-power lasers?
   a. Ability to generate extremely high powers in very short pulses
   b. Amenable to robotic control
   c. Do not apply physical force to objects
   d. Are extremely efficient
   e. Can focus light energy tightly onto a small spot to generate very high powers
2. Which of the following is most important for laser drilling of baby-bottle nipples?
   a. Ability to generate extremely high powers in very short pulses
   b. Amenable to robotic control
   c. Do not apply physical force to objects
   d. Are extremely efficient
   e. Can focus light energy tightly onto a small spot to generate very high powers
3. A 1000-watt continuous-wave carbon-dioxide laser illuminates a titanium sheet for one second. How much energy does the titanium absorb (assuming that absorption does not change with heating)?
   a. 80 watts
   b. 80 joules
   c. 120 joules
   d. 1000 joules
   e. None of the above
4. Which of the following lasers emits a wavelength that is absorbed most efficiently by human skin?

a. Argon-ion
b. Neodymium–YAG
c. Ruby
d. Carbon dioxide
e. Semiconductor diode (GaAlAs)

5. What type of laser machining is assisted by a jet of air or oxygen?
   a. Cutting
   b. Drilling
   c. Heat treatment
   d. Scribing
   e. Resistor trimming
6. The biggest single application for excimer lasers is
   a. Refractive surgery
   b. Research
   c. Isotope separation
   d. Laser fusion
   e. Semiconductor photolithography
7. What material in skin and tissue absorbs most strongly at 10 μm?
   a. Melanin
   b. Bone
   c. Water
   d. DNA
   e. Fat
8. Laser hair removal works best for people with
   a. Light hair and dark skin
   b. Light hair and light skin
   c. Dark hair and dark skin
   d. Dark hair and light skin
   e. Patients whose skin absorbs a dye before treatment
9. How does laser-induced nuclear fusion work?
   a. Intense laser pulses compress tiny pellets containing a mixture of deuterium and tritium until their nuclei fuse.
   b. Intense laser pulses separate uranium-235 from uranium-238.
   c. Intense laser pulses ignite combustion of hydrogen in an oxygen atmosphere.
   d. Intense laser pulses vaporize a frozen mixture of deuterium and hydrogen.
   e. The technique does not work.

10. Which of the following high-energy lasers could only be used in space?
    a. Deuterium fluoride chemical laser
    b. Hydrogen fluoride chemical laser
    c. Chemical oxygen–iodine laser
    d. Diode-pumped solid-state laser
    e. Flashlamp-pumped solid-state laser
11. A laser in an orbiting battle station must be able to focus a diffraction-limited beam onto a 3-meter spot on a missile 5000 km away. If the laser emits at 3 μm, how big must the beam-directing optics be?
    a. 10 cm
    b. 1 m
    c. 3 m
    d. 5 m
    e. 1 km
12. What target is most vulnerable to a laser weapon?
    a. Ballistic missiles
    b. Mortar rounds
    c. Helicopters
    d. Tanks
    e. The retina of the human eye

CHAPTER 13

LASERS IN RESEARCH

ABOUT THIS CHAPTER

Scientists pick lasers for their experiments for the same reason that engineers pick lasers for the equipment they design—the characteristics of laser light offer some special advantages. Lasers have been used in many types of scientific research since the first laser was demonstrated. It is impossible to cover all the ways that lasers are used in research, so this chapter only samples the high points.

13.1 LASERS OPEN NEW OPPORTUNITIES

The coherent, highly directional, monochromatic, and intense beam from the laser opened new opportunities for scientists and, like engineers and physicians, soon after the first laser was demonstrated, they lined up to borrow, buy, or build their own lasers and test them in their laboratories. They soon began getting interesting results.

One important example was harmonic generation. Physicists had known that the nonlinear interaction of electromagnetic waves with matter could generate harmonics of the fundamental frequency. However, the effect was very small at low intensities, so it had been impossible to observe at the high frequencies of light waves, until the laser came along. A year after Maiman demonstrated the ruby laser, Peter Franken focused ruby pulses into quartz and observed a feeble emission at 347 nm in the ultraviolet, the second harmonic of the ruby 694-nm wavelength. The second-harmonic spot looked so faint on the photograph that a production editor at the journal that published Franken's paper thought it was a flaw and removed it, but Franken had nonetheless made a breakthrough. Today, nonlinear optics is a powerful tool, and harmonic generation is routinely used to generate light at many wavelengths, like the green of green laser pointers.

Many other research advances followed, some important enough to be rewarded with Nobel Prizes. Charles Townes, Nikolai Basov, and Alexander Prokhorov were the first to share one, the 1964 physics prize for developing the principles behind the maser and the laser. (Basov and Prokhorov were recognized for pioneering theoretical work on microwave masers.) Four of the ten physics prizes awarded between 1997 and 2006 were for laser-related work. Table 13-1 lists all laser-related Nobel Prizes through 2007.

Table 13-1. Laser-related Nobel Prizes through 2007

Year and prize	Recipients	Research
1964, Physics	Charles Townes, Nikolai Basov, Alexander Prokhorov	Fundamental research leading to the maser and laser
1971, Physics	Dennis Gabor	Holography (made practical by the laser)
1981, Physics	Nicolaas Bloembergen, Arthur Schawlow	Development of laser spectroscopy
1997, Physics	Steven Chu, Claude Cohen-Tannoudji, William Phillips	Laser trapping and cooling of atoms
1999, Chemistry	Ahmed Zewail	Studies of chemical reaction dynamics on femtosecond time scales
2000, Physics	Zhores Alferov, Herbert Kroemer	Invention of heterostructures, essential for high-speed optoelectronics
2001, Physics	Eric Cornell, Carl Wieman, Wolfgang Ketterle	Producing Bose–Einstein condensates, sometimes called "atom lasers"
2005, Physics (separate citations)	Roy Glauber	Quantum theory of optical coherence
2005, Physics (separate citations)	John Hall, Theodor Hänsch	Ultraprecise laser spectroscopy and frequency-comb generation

Looking at the list of Nobels shows a clear pattern—elegant measurements and experiments using lasers have been richly rewarded, showing their importance to both physics and chemistry. An exception was the 2000 prize, shared with Jack Kirby for his development of the integrated electronic circuit, which recognized the practical importance of diode lasers and was the first Nobel prize citation to mention the Internet.

13.2 LASER SPECTROSCOPY

The breakthroughs that launched the era of laser spectroscopy came in the late 1960s—first the dye laser, then ways to tune the wavelength of the dye laser. Early dye lasers were complex and messy devices, but they could provide something available from no other light source—a large number of photons concentrated in a very narrow range of wavelengths. Any source of white light could produce a few photons at a narrow range of wavelengths if the light was passed though optics that first spread them into a broad spectrum; a narrow slice of wavelengths was then selected. But only lasers could deliver a lot of photons in a narrow range. That opened the doors to more spectroscopic experiments.

Spectroscopy has achieved fascinating things, but there is only room here to describe a few notable achievements.

13.2.1 Doppler-Free Spectroscopy

A key early advance was the development of a way to study the spectra of gases at higher resolution than previously possible. Normally, the atoms and molecules in gases move at random thermal velocities, spreading their observed spectral lines across a wider range of wavelengths than if the atoms were at rest. The effect is called *Doppler spreading* because the wavelength shifts caused by motion are known as Doppler shifts, and it previously limited the resolution possible when studying gases.

The high intensity of laser beams led to an ingenious scheme called Doppler-free saturation spectroscopy, devised by Theodor Hänsch and Arthur Schawlow at Stanford University, both of whom later received Nobel Prizes. They recognized that a dye laser could generate a beam with spectral width much smaller than the Doppler width of a gas transition, so they could use the laser beam to saturate absorption in a narrow slice of the spectrum. The ingenious trick was to split the laser beam into two unequal parts, passing in opposite directions through a tube containing the gas, as shown in Figure 13-1. The brighter portion of the beam saturated absorption in the gas, exciting all the atoms or molecules to a higher state so they could not absorb any more light at that wavelength. The weaker beam passed in the other direction through the gas and was detected. By inserting a chopper into the high-power beam, they could switch the absorption off and on. Measuring the portion of the probe beam transmitted when the saturation beam was off gave the actual absorption in that narrow slice of spectrum that saw both laser beams at the same wavelength because they were not moving along the length of the tube. Thus, they were able to cancel the natural Doppler spreading to get Doppler-free gas spectra, an impressive achievement in 1970 that laid the groundwork for a long series of spectroscopic advances.

13.2.2 Femtosecond Spectroscopy and Chemical Reactions

Instead of studying very narrow slices of the spectrum of light, Ahmed Zewail decided to study narrow slices of time in his Caltech chemistry lab. His goal was to probe how chemical reactions occur.

Chemical reactions are fast; chemical bonds are made and broken on a time scale of picoseconds to femtoseconds (10^{-12} to

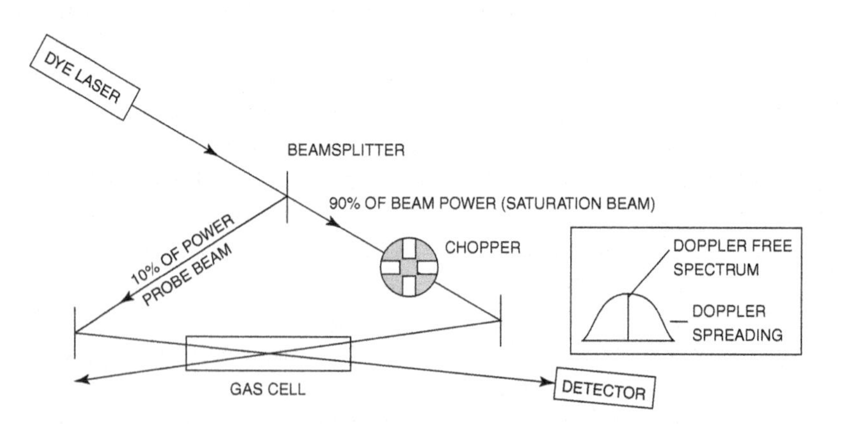

Figure 13-1. Doppler-free saturation spectroscopy.

10^{-15} second). To freeze those events, we need lasers that flash on a femtosecond time scale. Zewail did just that, illuminating reacting atoms and molecules with femtosecond pulses and watching events unfold. The experiments were not as simple as they might sound, because making precise measurements required carefully timing the pulses and the course of the chemical reactions in order to determine the exact sequence of events. Their success gave tremendous new insight into chemical reactions.

13.2.3 Frequency Combs and Atomic Clocks

We tend to think of femtosecond pulses as isolated extremely short events, and that view works very well for studying the dynamics of chemical reactions. However, mode-locked lasers naturally generate a series of picosecond or femtosecond pulses separated by regular intervals, which offers additional possibilities if you look at the pulses on a frequency scale rather than on a time scale.

Understanding what that means requires a brief look at the ways to describe the variation of laser power, or any signal, over a period of time. Normally, we think of the variation as a function over time, such as a sine wave that varies regularly, or a signal turning off and on regularly, forming a series of square waves. However, that variation also can be described as the sum of a number of sine waves having different frequencies and amplitudes, using a technique called *Fourier analysis*. The mathematical details are complex, but the point is that any repetitive wave can be described as the sum of sine and/or cosine waves at a number of different frequencies. For a sine wave, that description is just a single wave at the wave's own frequency. For a square wave, it is the sum of sine waves at odd harmonics of the square wave frequency, with the intensity decreasing at higher harmonics.

If you have a regular series of very short pulses in time, as shown in Figure 13.2, transforming the signal into the frequency domain has a very interesting result—a series of signals spaced uniformly in frequency, separated by an increment equal to the repetition rate of the laser pulses, also shown in Figure 13.2. This is called an optical frequency comb, and the shorter the pulses are, the wider the range of frequencies they span. Each frequency component is a continuous sine wave at precisely that frequency, with an amplitude that depends on the original pulse pattern. For a se-

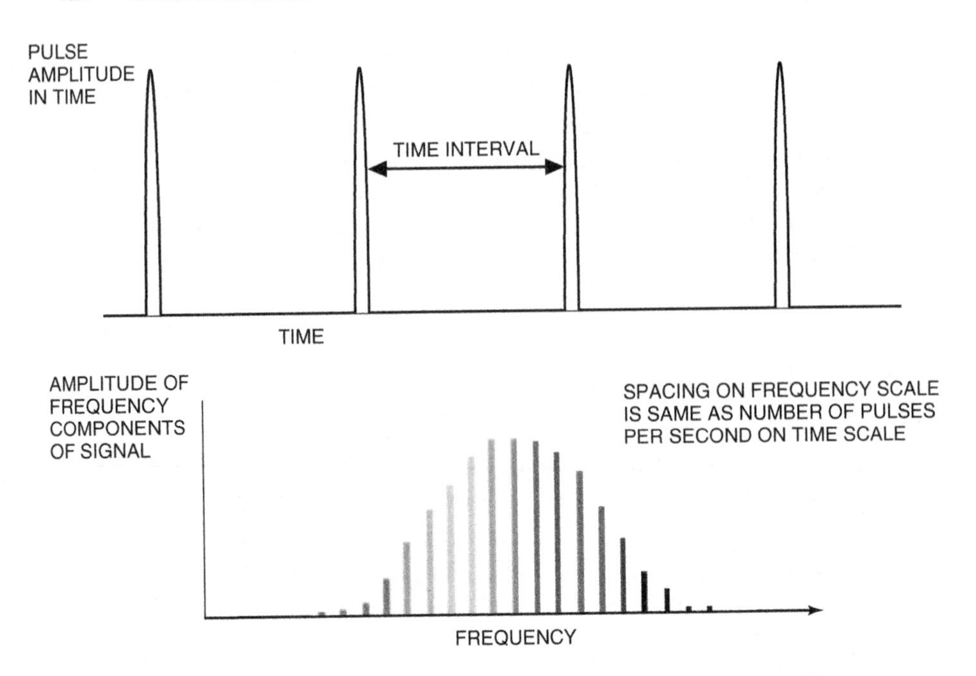

Figure 13-2. A series of ultrashort short pulses separated uniformly in time (top) corresponds to a series of uniformly spaced frequencies that are graded in amplitude (bottom).

ries of very short pulses, the variation in amplitude is roughly as shown.

Hänsch, working at the Max Planck Institute for Quantum Physics in Germany, and John Hall at the National Institute of Standards and Technology transformed this interesting bit of esoteric physics into a powerful measurement technique by using another elegant optical trick. When intense light passes through a highly nonlinear medium, the wavelengths are distorted, spreading across a wider range of wavelengths. The result is called a *white-light continuum* or, sometimes, a *supercontinuum,* to stress how much the wavelengths are being spread. Stretching the range of wavelengths makes it possible to squeeze the pulses so they last for shorter intervals. This pulse squeezing is possible because of a variation on the uncertainty principle—the wider the range of frequencies in a pulse, the better it can be defined in time and, correspondingly, the shorter the pulses, the wider the frequency range. With short enough pulses, the frequency comb can span a whole octave—a factor of two—in frequency.

Once a frequency comb spreads across an octave of frequencies, it is possible to double one of the lowest frequencies in the comb and find the doubled frequency on the high-frequency end. Knowing the pulse repetition rate, it is possible to calibrate the whole spectrum of millions of precisely spaced optical frequencies, and use them to make extremely precise measurements. That marked a major advance, because precise calibration of optical frequencies had been extremely difficult, and Hänsch and Hall shared a well-deserved Nobel Prize for developing the elegant technique. The technique is now being used in extremely accurate tests of fundamental theories of physics.

It also may open the door to a new generation of atomic clocks based on light rather than microwaves. For the past 40 years, the fundamental standard of time has been the oscillation of a cesium atom at 9,192,631,770 hertz, but the accuracy is inevitably limited by the microwave frequency. Optical oscillations would slice time into much shorter intervals, promising much higher accuracy, and several types of optical clock are in development.

13.3 MANIPULATING TINY OBJECTS

Another fascinating research use of lasers is in manipulating tiny objects using laser energy. There are two important variations: optical trapping, which physically manipulates a tiny object in space, and laser cooling, which slows the thermal motion of an atom or molecule so it reaches extremely low temperatures.

13.3.1 Optical Trapping and Tweezers

The radiation pressure of light—the net force that light applies to an object it encounters—can be used to trap and manipulate tiny objects, an effect called *optical tweezers.*

First demonstrated in 1970, optical trapping uses a laser beam to manipulate nanometer- or micrometer-sized objects, which must be made of nonconductive materials, such as tiny glass spheres. A lens or microscope objective focuses the laser beam to a small spot, creating a narrow waist in the beam; beyond that point, the focused beam expands to larger size, and usually is focused through another lens. Radiation pressure from the light in the beam pushes an object toward the most intense zone of the beam—the sharply focused waist—and holds it close to that point. The force of photons passing through that zone tends to push the object slightly away from the point of peak laser intensity.

This trapping technique holds the tiny object at a point where its properties can be measured. The technique has also been extended to molecules and single biological cells, allowing them to be studied in great detail.

13.3.2 Laser Cooling

Laser cooling works differently than optical tweezers, but also serves to slow objects, in this case atoms, to a virtual halt.

The trick of laser cooling is to slowly drain from atoms the thermal energy that keeps them in constant motion. To do that, a laser is tuned to a wavelength a little longer than that of an electronic transition in the atom, so the photons it emits have a little less energy than the atomic transition. When the laser photons illuminate the atoms, they will not have quite enough energy to excite atoms standing still. However, atoms moving toward the laser see the laser photons Doppler-shifted to the blue, a slightly longer wavelength, because of the atoms' motion. This makes them more likely to absorb those photons.

Atoms in a gas naturally absorb and reemit photons on the transition so, eventually, the atom that has absorbed the blue-shifted photon will emit another photon with the same amount of energy. But because the atom did not pick up quite as much energy from the absorbed photon as it needed for the transition, it will lose a little bit of its thermal energy. The more times the atom absorbs and emits a photon in that way, the more thermal energy it loses, cooling it to lower temperatures.

Atomic motion is in three dimensions, so a single laser beam is not enough to cool atoms to seriously low temperatures. That requires multiple beams, which serve to slow down atoms going in any direction. It also requires adjusting the laser wavelength, because as the atoms cool down they move more slowly, and their motion does not blue-shift the laser light as much as when they were warmer. Thus, the laser wavelength must be tuned closer to the transition wavelength as the atoms are cooled.

Eventually, laser cooling can slow atoms down enough to produce what is called *optical molasses,* in which cold atoms are moving very slowly and held loosely in place by the laser light. At this point, their temperatures are in the microkelvin range (millionths of a degree above absolute zero), but additional types of trapping such as magnetic fields are needed to hold the cooled atoms longer, even with the cooling lasers remaining in operation. Other techniques are needed to push temperatures even lower.

13.4 ATOM LASERS AND BOSE–EINSTEIN CONDENSATES

Laser cooling did more than set records by creeping ever closer to absolute zero. It also pointed the way to a new state of matter, sometimes called an atom laser.

Albert Einstein and Indian physicist Satyendra Nath Bose predicted in the 1920s that cooling certain particles, now called bosons, below an extremely low temperature should cause them to collectively drop into the lowest possible energy state, a process called condensation. Laser cooling experiments started pushing toward that goal around 1980, but it was not reached until 1995, when Carl Wieman and Eric Cornell of the University of Colorado cooled a few thousand rubidium atoms below about 170 billionths of a degree Kelvin, and the atoms became a virtually motionless cluster for up to 15 seconds at a time. Like photons in a laser beam, all the atoms occupied the same quantum state.

It was an elegant research achievement that took only six years to earn Wieman, Cornell, and Wolfgang Ketterle of the Massachusetts Institute of Technology a Nobel Prize. "This is a completely new state of matter," Wieman told me when I was writing a news article about the experiment. "It never existed before because the universe was way too hot." Nor could it ever occur naturally as the universe cools in the far distant future. It is a state distinct from a solid, liquid, gas, or plasma. Laser cooling was not enough to produce a Bose–Einstein condensate; a magnetic trap and further cooling were needed. But the laser cooling was a vital step in the process.

Later experiments showed that a Bose–Einstein condensate could be made to release individual atoms that briefly remained in the same state. This was called an "atom laser" because the atoms were coherent with each other. This coherence could also be seen in the condensate, where the atoms belonged as a collective whole. (The coherence is most evident when the atoms are considered as "matter waves," which are all coherent with each other.)

Although the matter waves in condensates are similar to light waves in many ways, differences are inevitable because atoms differ from photons. Condensates are superfluids that move without friction. Atoms interact directly with each other; direct interactions between photons are rare except in nonlinear materials. Although experiments in the years since the first demonstrations have revealed interactions among coherent atoms that are the counterparts of familiar optical effects, the details differ in important ways. For example, atom lasers can release some atoms to the outside world like a laser emits a fraction of the light from the optical cavity, but there is no precise analog of stimulated emission to generate new atoms.

13.5 SLOW LIGHT

Another intriguing area of laser research is slowing light down to a veritable snail's pace. As you learned earlier, the speed of light in a material equals the speed of light in vacuum divided by the refractive index, so increasing the refractive index can slow light several times below its normal 300,000 kilometers per second. The refractive index can reach extremely high values, but those peak values are possible only at wavelengths at which the materials have resonant transitions, wavelengths at which the materials strongly absorb light.

Steve Harris at Stanford University found a way around this problem by illuminating the material with so much laser light that he saturated the transition. In 1995, he used this approach to slow down the velocity of light at a resonant wavelength in lead vapor to 1800 km/s, a factor of 165. However, it took much more to capture the public attention. That came in 1999, when Lene Hau at Harvard University slowed light to 17 meters per second by passing it through sodium atoms cooled below the threshold for forming a Bose–Einstein condensate and illuminated by a low-power laser tuned to the same wavelength. That was an impressive achievement, but it required extremely low temperatures and extremely narrow laser linewidths.

Since then, the technique has been extended to other materials and warmer temperatures, using variations on making materials transparent at wavelengths at which their refractive index is extremely high. Slow light has also stimulated practical interest, because slowing down the speed of light could allow construction of optical buffers able to store optical signals for use in fiber-optic communication systems.

13.6 NANOSCALE LASERS

In the era of nanotechnology, it is natural to ask how small a laser can be made, and whether oscillation is possible in nanocavities. Some interesting work has been done, but the field is still very young.

As mentioned in Chapter 9, quantum dot semiconductor lasers have been demonstrated and offer low thresholds, but have yet to reach the high efficiency sought by developers. The quantum dots fabricated in semiconductors are not isolated, but scattered in the active layer of the laser. The quantum dot structure confines electrons but not light. The resonant cavity is defined by the overall diode structure, not by the quantum dot itself, so it is not a true nanoscale laser.

13.6.1 Photonic Crystal Lasers

Laser action has been demonstrated in photonic crystals, which are periodic structures consisting of many very small elements that act together to confine light. Typically, the elements are stacks of very thin layers with different refractive indexes that collectively act to block transmission of light at certain wavelengths in ways similar to thin-film optical coatings. In some cases, the layers are two different compounds; in others, one is solid glass and the other layer is a material containing holes smaller than the wavelength of light, which reduce the effective refractive index of that layer.

Photonic crystals can create cavities with modal volumes smaller than the cube of the emitted wavelength (in the solid) and with high cavity confinement. This allows photonic crystal lasers to have extremely low thresholds, although their output power is low, and single-photon emission is difficult. Researchers at the University of Tokyo have demonstrated continuous room-temperature laser action at 1.3 μm from a photonic crystal nanocavity laser, but the device had to be pumped with an external light source, and was easily damaged if the pump energy remained well above laser threshold.

13.6.2 Single-Atom Lasers

The ultimate in miniaturization is a single-atom laser, which has been demonstrated by placing a single atom inside a resonant cavity. Light can excite a single atom or ion that is either trapped in a resonant optical cavity or one of a series falling though the cavity. Jeff Kimble's group at Caltech in 2003 trapped a single cesium atom at a time in an optical cavity for about 100 nanoseconds, and optically excited it to emit a series of photons resonant with the cavity. Other experiments have used a similar scheme, but observed light emitted by a series of atoms falling through a resonant cavity, where more than one atom may be in the cavity at a time.

13.6.3 Single-Photon Sources

Getting even such a single-atom laser to emit only a single photon at a time is extremely difficult because lasers normally generate a cascade of stimulated emission as light oscillates within the cavity. Light sources that emit no more than one photon per time interval are needed for quantum cryptography and quantum computing, but they generally use a light source that either produces one photon at a time by spontaneous emission or a laser source that is filtered so strongly that only a tiny fraction, perhaps one in a trillion, of the photons get through. These sources also tend to be inefficient, emitting very few photons and generally emitting nothing in most time intervals.

13.7 PETAWATT LASERS

On the other end of the power scale, researchers are pushing to higher and higher peak powers. Today's most powerful lasers can exceed one petawatt (10^{15} watts), and European researchers plan to build a laser called the "Extreme Light Infrastructure" with peak power of an exowatt (10^{18} W) and pulses lasting only 10 femtoseconds.

Although their peak powers are more than double the 500 terawatts to be delivered by the National Ignition Facility, most petawatt lasers are much smaller than the giant fusion laser. They manage to achieve similarly high peak powers by squeezing about a thousandth the energy of NIF or the French Laser Megajoule into a pulse lasting around a picosecond, a thousand times shorter than the pulses from the two giant fusion lasers. By tightly focusing their short pulses onto targets, they momentarily can produce power densities of 10^{18} to 10^{21} watts per square centimeter, opening new realms for research into the interactions between ultraintense light and matter.

The technology for petawatt lasers was a spin off of the development of high-power fusion lasers at the Lawrence Livermore National Laboratory in California during the mid-1990s. The first petawatt laser was demonstrated using a chain of neodymium–glass amplifiers from the Nova fusion laser, the predecessor of NIF. Previously, the amplification of high-power pulses had been limited by nonlinear effects and optical damage arising from the extremely high power of the amplified pulses. To avoid this problem, Livermore used a technique called *chirped pulse amplification,* shown in Figure 13-3, which relied on the ingenious trick of stretching out the pulse before amplification, then squeezing it to a very short pulse after amplification was completed.

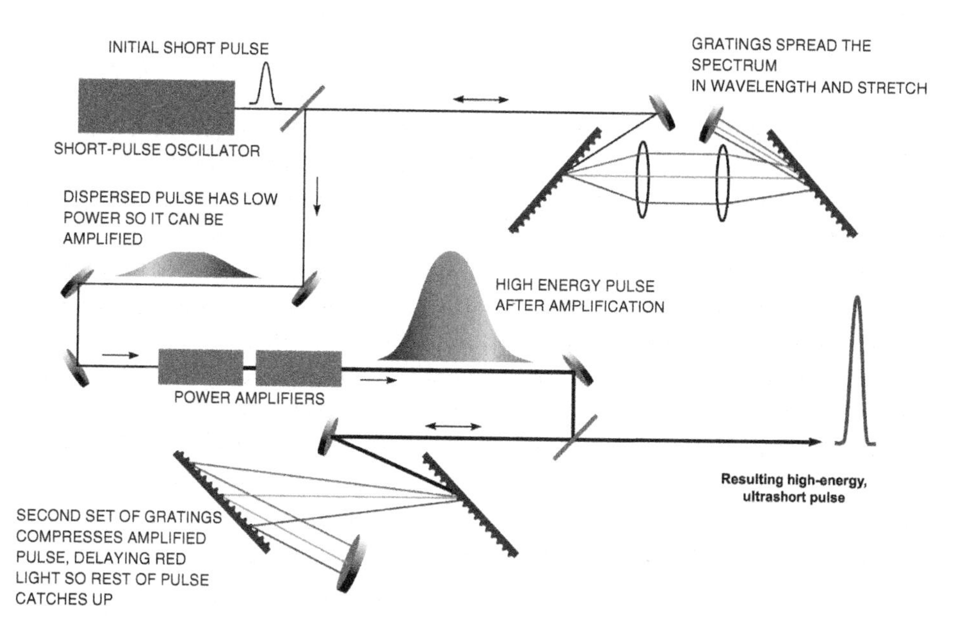

Figure 13-3. Chirped-pulse amplification permits pulses to be amplified to very high peak powers without damaging amplifier optics. (Adapted from Government work, not subject to copyright, Lawrence Livermore National Laboratory.)

Instead of going directly into an amplifier, the short pulse from an oscillator first passed through a pair of diffraction gratings arranged so shorter wavelengths traveled a longer distance than longer wavelengths. (The short wavelengths normally are called "blue" and the long ones called "red" in describing optical systems, although both are really in the infrared.) The pulse that emerged was much longer in duration, with its energy spread out over a longer interval, so the peak power was much lower than when it originally left the oscillator. That left the peak power low enough that the pulse could safely be amplified without the risk of nonlinear effects or optical damage. After that final amplification, the light made a final pass through a second pair of gratings, arranged to delay the red light so the blue light could catch up, recreating the original short pulse, but at a much higher peak power, without damaging any optics.

The new generation of petawatt lasers starting to come on line are based on this approach. They fall into two families. Those using neodymium-doped glass like the original Livermore system can generate pulse energies of hundreds of joules, but the bandwidth of neodymium is limited, so the pulse cannot be compressed to less than 500 femtoseconds. A second family has much broader bandwidth because they use titanium–sapphire amplifiers, so they can generate pulses as short as 20 femtoseconds, but the limited crystal size available restricts pulse energy to a few joules. Some intermediate configurations also exist.

These impressive new lasers will let laser researchers concentrate laser power to higher densities than ever before in femtosecond pulses. They expect to reach regimes in which electrons are accelerated to speeds at which relativistic effects multiply their mass by as much as a factor of 10,000. They expect it to yield the hottest and densest matter ever generated in a laboratory, opening new opportunities in physics.

13.8 ATTOSECOND PULSES

We saw earlier that pulses at optical wavelengths can be no shorter than few femtoseconds. At 800 nm, in the center of the tuning range of titanium–sapphire lasers, a single cycle of a light wave lasts only about 2.7 fs, so it is not surprising that the shortest pulses on record from Ti–sapphire lasers are about 5 fs. It is reasonable that a photon should be a couple of wavelengths long.

The fundamental limit arises from the uncertainty principle. As described in Section 4.6.2, the uncertainty in the frequency (the bandwidth) of a pulse times the duration of a pulse must be at least 0.441. That means that generating a 5-fs pulse requires a bandwidth of at least 88 terahertz, or roughly 710 to 910 nm from a Ti–sapphire laser. Using the same formula, you need a bandwidth of 441 THz to produce a 1-fs pulse. That's a little bigger than the visible spectrum, which is about 420 THz wide. To produce even shorter pulses, with duration measured in attoseconds (10^{-18} second), you need to go to higher frequencies in the ultraviolet or shorter wavelengths.

Attosecond pulses are the thinnest slices of time on the research horizon, and promise insight into new physical processes. Ahmed Zewail used femtosecond pulses to probe the mechanisms of molecular chemistry. Attosecond pulses similarly promise a way to study how electrons behave within atoms. Researchers have already crossed the attosecond frontier, using laser techniques that generate light in the extreme ultraviolet and beyond.

The leading way of generating attosecond pulses is by focusing intense femtosecond pulses of laser light into a jet containing atoms of a rare gas such as argon or xenon. Each femtosecond pulse contains a series of peaks in the electric field marking oscillations at the frequency of the laser light. At these peaks, the electric field is powerful enough to pull valence electrons out of the atoms in the gas, but as the field drops back to a lower level the atomic nucleus pulls the electron back into place. The recaptured electron releases the extra energy it collected from the field of the laser pulse in a burst of extreme ultraviolet waves and X-rays lasting a few hundred attoseconds.

From a theoretical standpoint, the electric field of the laser pulse pulls only part of the electron out of the atom, so it essentially collides with itself when the nucleus pulls it back into the atoms. That sounds messy, but it actually makes the electron radiate extreme ultraviolet pulses at the peak of each wave in the laser pulse that excites it. The shortest pulses yet produced lasted only 130 attoseconds, and researchers predict they should be able to go below 100 attoseconds, which would span a frequency range of about 5000 THz. Physicists are beginning to see what they can do with attosecond pulses.

13.9 LASER ACCELERATION

Another application for the extremely high electric and magnetic fields generated by femtosecond pulses is in accelerating charged particles to high energies over very short distances. Theoreticians Toshi Tajima and John M. Dawson proposed the idea in 1979, but lasers then available were not powerful enough for the job. That changed with the development of chirped pulse amplification and petawatt lasers, and recent years have seen a series of experiments with electrons, protons, and heavier ions.

Traditional charged particle accelerators use arrays of massive electromagnets in long tubes, with power through the electromagnets switched to keep the fields pulling the particles along. Small accelerators for laboratory use have meter-long tubes, but the big ones used in major research programs are miles long. At best, they can increase electron energy by tens of millions of volts per meter. In contrast, pulses from ultrashort lasers fired into a plasma can reach electric fields of teravolts (trillions of volts) per meter at their peak, and accelerate electrons to 100 million electron volts in only one millimeter.

Laser accelerators will not replace the giant accelerators used in particle-physics research. However, they can fit in ordinary laboratories to allow smaller-scale research. And laser accelerators also can produce more powerful beams, making them better-suited for research in fields such as medical treatment with particle beams.

13.10 OTHER EMERGING RESEARCH

There is no shortage of ideas for laser research, and not enough room to describe them all. In many cases, lasers play only a small part in the research effort; in others the laser is central. Let us close by looking at a few of these ideas.

13.10.1 New Optical Materials

One of the most exciting developments in optics as a whole over the past several years has been the emergence of a new class of optical metamaterials. The concept was first demonstrated at microwave frequencies; the materials were built from macroscopic components that collectively reacted in peculiar ways to electromagnetic fields at microwave frequencies. It has now been extended to microstructured metamaterials that work at optical frequencies.

Optical metamaterials have properties that traditional optics taught was impossible. A vivid example is a material with refractive index of –1, meaning that light bends backward as it enters the material. Such materials have been used to demonstrate the science-fictional concept of an "invisibility cloak," but don't hold your breath waiting for a high-tech Harry Potter to appear from nowhere. The optical metamaterials in the real world work only at a limited range of wavelengths, so the cloak would only hide a person if the room were illuminated only with a single laser wavelength.

13.10.2 Quantum Computing and Quantum Cryptography

Quantum optics also promise weird and wonderful capabilities in cryptography and computing. The laws of quantum mechanics mean that observing a quantum state such as photon polarization immediately fixes at a particular value. If that photon shares a common origin with another, and if the two have been produced in certain ways, their properties are connected or "entangled," so observing the properties of one photon can simultaneously change the value of that property for the second photon.

Quantum computing involves performing operations on the quantum states of objects. These quantum states have peculiar properties, including that they nominally are superpositions of a number of possible states, which are not resolved until someone observes them. In principle, quantum computing could be very powerful for a few particular applications, notably the factoring of large numbers, which is now used to assure the security of computer transactions. Researchers are still working on ways to make the concept practical. One requirement is a source of single photons.

13.10.3 Photonic Integration

Chapters 9 and 10 mentioned the possibility of combining lasers with semiconductor electronics for photonic integration on a single substrate. The idea dates back decades, but the reality has long been challenging. Problems remain in combining active light emitters made of III–V compound semiconductors with electronics, largely silicon. But integrated versions of many optical components have been integrated, including lasers. This book has not gone into the details because they do not directly relate to laser physics, but the emerging technology is important.

13.10.4 Coherent Beam Combination

Fiber lasers have made stunning advances in power levels, but fundamental limits remain on the power available from any single fiber. The next big challenge in raising fiber laser power is to find ways to avoid destructive interference when combining their output beams coherently into a single beam, whether for materials working or laser weapons.

13.10.5 Laser Propulsion

The radiation pressure of light, used to move tiny objects in laser tweezers, in principle could be extended to propelling spacecraft in flight. Lasers could supplement solar power to push solar sails across the solar system. Lasers might be used in other kinds of propulsion systems, such as evaporating material from the back of a spacecraft to produce a rocket-like exhaust push.

Much lower laser powers might also play a very different role in propulsion. Laser light might illuminate arrays of solar cells, which could turn the laser energy into electricity. That is one mechanism that has been proposed to power "space elevators," which might someday climb filamentary cables running from the equator to satellites orbiting in geostationary orbit.

13.10.6 Deep-Space Communications and Optical SETI

As you read in Section 11.5.5, the narrow diameter of a laser beam makes it difficult to use lasers to communicate between satellites in Earth orbit. However, over longer distances the laser beam spreads out to be wide enough so that it is less of a problem. NASA has designed a laser communication system to beam signals from Mars back to the Earth at high data rates, although budget cuts forced cancellation of the system.

Powerful enough lasers might even transmit signals across interstellar space. In 1962, Charles Townes calculated that a 10-kW laser beam focused by a 5-meter mirror in space would look brighter than the sun at the laser line when viewed from deep space, and visible to the naked eye up to 0.1 light year away. A 5-meter telescope could detect the signal up to 100 light years away. A handful of astronomers are scanning the sky with optical telescopes, searching for extraterrestrial intelligence transmitting signals to us on laser lines. They have not spotted ET yet, but neither have astronomers searching at microwave frequencies that many had thought were better candidates for SETI.

13.10.7 Lasers to Save the World—Asteroid Defense

Someday, lasers might even save the world in a very different way, by deflecting a wayward asteroid away from the Earth. Radiation pressure is a weak force, but if we knew decades in advance that the asteroid might be a threat, we could point a powerful laser beam toward it. The laser light would push just a tiny bit on the asteroid, but if the asteroid were small and the lead time long, that tiny push could, over many years, nudge the asteroid's orbit far enough to one side that it missed the Earth. It may sound like science fiction, but it is a much easier, if less dramatic, way to save the world from an asteroid than sending Bruce Willis and a crew of astronauts at the last minute with a payload of nuclear weapons.

13.11 WHAT WE HAVE LEARNED?

- Research has been a major laser application since the laser was invented.
- 18 people have received laser-related Nobel Prizes.
- Four of the 10 Nobel physics prizes from 1997 to 2006 were for laser-related work.
- Laser spectroscopy can be used to eliminate the Doppler spreading caused by the random motion of atoms or molecules, giving very sharp spectra as if the particles were not moving.
- Femtosecond spectroscopy can show the progress of a chemical reaction.
- Trains of femtosecond pulses contain a "comb" of uniformly spaced laser frequencies, which can measure frequencies precisely across the optical spectrum.

- Lasers can trap and manipulate tiny objects, a technique called optical tweezers.
- Laser cooling can slow atoms to a virtual halt by draining away their thermal energy.
- Atoms cooled to extremely low temperatures can all occupy the same low energy state in a Bose–Einstein condensate,
- Atom lasers are Bose–Einstein condensates, which can release atoms that are coherent with each other.
- Light can be slowed to a virtual crawl by passing through materials with very high refractive index.
- Laser action has been demonstrated in nano cavities, and from single atoms.
- Single-photon sources are hard to make, particularly from lasers.
- Chirped-pulse amplification made it possible to amplify very short pulses to peak powers in the petawatt region.
- Pulses have to have extremely wide bandwidth to be made extremely short. Attosecond pulses are possible only at the high frequencies of the extreme ultraviolet.
- The strong electromagnetic fields of short laser pulses can accelerate charged particles over short distances.
- Coherent combination of light from an array of fiber lasers could deliver a high-power beam.
- Laser beams could transmit signals between planets or perhaps between stars.
- High-power lasers might someday be used for propulsion in space.

QUIZ FOR CHAPTER 13

1. What does Doppler-free spectroscopy do?
   a. Stops gas atoms from moving
   b. Measures the motion of atoms in an absolute rest frame
   c. Uses two laser beams pointed in opposite directions to cancel the Doppler spreading caused by the random motion of gas atoms
   d. Uses a powerful beam to trap atoms so they cannot move and a weak probe beam to measure their spectra
   e. Produces a Bose–Einstein condensate
2. What do you need to generate a frequency comb?

a. A series of very short pulses repeated at regular intervals
b. A single pulse stretched across a wide range of frequencies
c. A set of narrowband optical filters, each selecting one frequency
d. A set of harmonic generators that multiply frequency
e. An atomic clock

3. How does laser cooling reduce the temperature of atoms?
   a. Atoms absorb and reemit photons repeatedly, each time losing a little bit of energy.
   b. Atoms absorb and reemit photons repeatedly, each time gaining a little bit of energy.
   c. Two laser beams pointed in opposite directions trap the atoms so they cannot move.
   d. A single powerful beam traps the atoms so they cannot move.
   e. Photon pressure pushes the atoms into a refrigerator.
4. What is the essential property that defines a Bose–Einstein condensate?
   a. Atoms are trapped by a laser beam.
   b. Atoms are stimulated to emit other atoms.
   c. Atoms can never escape from it.
   d. Atoms occupy the same quantum state.
   e. Atoms freeze into a perfect crystal.
5. A light pulse contains 1 joule of energy. How short must the pulse be to reach peak power of 1 petawatt?
   a. 1 ps
   b. 100 fs
   c. 10 fs
   d. 5 fs
   e. 1 fs
6. A laser produces pulses that last 20 fs. How much energy must the pulse contain to reach peak power of 1 petawatt?
   a. 1 J
   b. 10 J
   c. 20 J
   d. 40 J
   e. 50 J
7. Suppose you could compress the duration of a pulse to 1 attosecond. How much energy would the pulse have to contain to reach peak power of a petawatt during that attosecond?

a. 1 mJ
b. 100 mJ
c. 1 J
d. 2 J
e. 10 J

8. What are single-photon sources needed for?
   a. To make single-atom lasers
   b. Bose–Einstein condensates
   c. Quantum cryptography
   d. Narrow-line lasers
   e. Attosecond pulses
9. A laser in orbit around Mars must direct signals to Earth so they can be detected from any point on the surface of the planet facing Mars. Assume the distance to Mars is 75 million kilometers. What should the beam divergence be?
   a. 0.00107 milliradian
   b. 0.0017 milliradian
   c. 0.017 milliradian
   d. 0.107 milliradian
   e. 0.170 milliradian
10. The laser orbiting Mars emits at 1 μm. How big must its output mirror be to emit a beam with the divergence you calculated in Question 9? (Approximate divergence as μ/D.)
   a. 6 mm
   b. 6 cm
   c. 60 cm
   d. 1.7 m
   e. 6 m

ANSWERS TO QUIZ QUESTIONS

CHAPTER 1

1. c
2. b
3. a
4. c
5. d
6. b
7. a
8. c
9. b
10. No single correct answer

CHAPTER 2

1. b
2. c
3. b
4. a
5. c
6. d
7. a
8. b
9. d
10. b
11. d
12. c

CHAPTER 3

1. b
2. a
3. c
4. e
5. c
6. e
7. a
8. b
9. c
10. a

CHAPTER 4

1. b
2. e
3. a
4. e
5. c
6. e
7. b
8. d
9. d
10. b
11. d
12. a

Understanding Lasers: An Entry-Level Guide, Third Edition. By Jeff Hecht

CHAPTER 5

1. a
2. d
3. d
4. d
5. b
6. b
7. e
8. e
9. c
10. a
11. c
12. e

CHAPTER 6

1. d
2. b
3. c
4. e
5. a
6. d
7. a
8. c
9. b
10. e

CHAPTER 7

1. b
2. b
3. d
4. a
5. c
6. b
7. d
8. e
9. e
10. b
11. a
12. b

CHAPTER 8

1. e
2. b
3. a
4. c
5. a
6. d
7. b
8. c
9. e
10. a
11. d
12. b

CHAPTER 9

1. b
2. a
3. c
4. d
5. b
6. e
7. d
8. a
9. c
10. d
11. e
12. b

CHAPTER 10

1. a
2. e
3. e
4. c
5. c
6. b
7. c
8. a
9. c
10. a

CHAPTER 11

1. b
2. c
3. a
4. b
5. d
6. b
7. c
8. d
9. e
10. b

CHAPTER 12

1. d
2. c
3. b
4. d
5. a
6. e
7. c
8. d
9. a
10. b
11. d
12. e

CHAPTER 13

1. c
2. a
3. a
4. d
5. e
6. c
7. a
8. c
9. e
10. a

APPENDIX A

LASER SAFETY

Lasers present two primary hazards: the beam and the power supply. Laser beams can damage your vision, but high-voltage laser power supplies have killed people. Do not be scared, but do be cautious.

A.1 POWER SUPPLY HAZARDS

Electrically, the only lasers you can consider harmless are battery-powered laser pointers. Anything that plugs into household line current or high-voltage lines is potentially dangerous. Line current can kill you. Gas lasers often include transformers that generate very high voltages (thousands of volts or more) to power a discharge in the laser gas.

Pulsed lasers often include capacitors that store a powerful electric charge to excite the laser medium. Capacitors retain a high-voltage charge for some time after the power is switched off or disconnected. Learn how to handle electricity safely before you poke around inside the laser or the power supply.

Modern commercial lasers are built to comply with electrical safety codes, so you should not encounter exposed high voltages unless you open the case. However, you cannot assume dangerous voltages are safely concealed in a laser laboratory or workshop.

A.2 LASER BEAM HAZARDS

The human eye has evolved to be very sensitive to visible light, and this sensitivity makes it extremely vulnerable to excess light levels. The brightest light in nature is the sun, and we have developed an aversion response that makes our eyes instinctively look away if we happen to glance at the sun. There is good reason for this, because staring directly at the sun can permanently damage the retina at the back of the eye, impairing vision for the rest of your life.

The hazards from a 1-mW visible laser beam are similar to those from looking directly at the sun. Although the laser delivers only a tiny fraction of the power delivered by the sun, the light rays in a laser beam are all parallel to each other, and the beam is small enough to pass through the pupil of your eye, so the lens focuses all that light into a tiny spot on the retina. A momentary accidental glance into a 1-mW laser beam will dazzle your eyes, but it will not cause permanent damage unless you stare into the beam.

The most vulnerable part of your eye is the retina, the layer in the back of your eye that senses light. If you have normal color vision, your eye responds weakly to infrared light at wavelengths beyond 700 nm, with the response declining as the wavelength increases, so you can see a little light at the 780 nm wavelength emitted by the diode lasers in CD players. However, the rest of the eye transmits wavelengths as long as about 1400 nm to the retina, so those invisible wavelengths are as much of a hazard to your retina as the laser light you can see. Those wavelengths are particularly dangerous because you cannot see them.

Ultraviolet wavelengths shorter than 400 nm and infrared wavelengths longer than about 1400 nm do not penetrate deeply into the eye. They can damage the cornea, the surface layer of the eye, and the lens just underneath the cornea, but the power thresholds for damage to the cornea and lens are higher than for damage to the retina.

You do not have to be looking at the laser to suffer an eye injury. One common cause of accidents is reflection of the beam off a shiny object in an unexpected direction, hitting someone in the eye who is looking away from the laser.

The best way to prevent such accidents is to wear special laser safety goggles made of glass or plastic that blocks selected laser wavelengths. The goggles should be matched to the lasers being used.

Let us look briefly at the types of lasers and the recommended precautions.

A.2.1 Lasers Packaged inside Equipment

Lasers are packaged inside a wide range of equipment, and in most of them the beam is always contained safely inside the case. You will never see the beam from a laser printer or a CD, DVD, or Blu-Ray disk player unless you take them apart, so this equipment requires no special warnings or precautions .

Laser scanners used in supermarkets are an exception because the beam must emerge from the scanner in order to read the bar codes on packages. However, the scanners are designed so the beam power is very low and the beam scans so fast that it does not emit enough light into the air that it requires a warning label. Engineers worked very hard to achieve this goal, because they did not think customers would be eager to shop at stores in which the checkout counter bore a "DANGER: LASER RADIATION" sign.

A.2.2 Battery-Powered Laser Pointers

The type of lasers you are most likely to encounter as separate devices are pointers. United States laser safety regulations limit laser pointers to power levels less than 5 mW. Even at that level, they are required to bear warning labels that say "Danger—Laser Radiation, Avoid Direct Eye Exposure," identify the maximum power level, and specify the class of laser product according to rules specified by the Food and Drug Administration.

Red laser pointers use diode lasers, and are unlikely to exceed their 5-mW rating, so they are safe if you use common sense and do not point them at anybody's eyes.

Green laser pointers pose somewhat different issues. They look much brighter to the eye because the human eye is much more sensitive to green light than to red light, although the eye hazard from a green pointer at a given power level is similar to that from a red pointer. However, green pointers may exceed their rated output level, and some with powers in the 100-mW range are sold over the Internet in pointer-like packages, without standard safety labels or equipment. Because green pointers are based on diode-pumped, frequency-doubled neodymium lasers, near-infrared light at 808 and 1064 nm is present inside the laser. Better pointers include filters that block the infrared light from exiting in the beam, but many green pointers lack filters, so their output includes those invisible infrared wavelengths, which pose additional hazards as mentioned above.

Thus, green pointers in the 5-mW range deserve somewhat more care than red pointers; those with higher output power deserve much more care. When a green pointer is used to identify objects in the sky, special care should be taken to avoid shining it at aircraft because the bright light could flash-blind a pilot and get you in serious trouble.

A.2.3 Scientific, Industrial, and Medical Lasers

If you are working with scientific, industrial, or medical lasers, they are likely to pose more serious hazards. These lasers are classified according to potential hazards by safety regulating agencies. In the United States, the agency responsible for equipment safety is the Center for Devices and Radiological Health in the Food and Drug Administration (FDA). Table A.1 shows the FDA classes, the maximum power level, and the warning label requirements for lasers emitting continuous beams at 400 to 1400 nm. The U.S. Occupational Safety and Health Administration is responsible for safe operation of lasers in the workplace.

As you might expect, the hazards become more serious as power levels increase, and operating requirements become more stringent. The need for an emission indicator might seem odd, but remember that many laser beams are not visible, or are not emitted continuously, so people working nearby might otherwise not know the laser was in operation.

You are not likely to encounter Class IIIb or IV lasers other than in an academic, medical, or industrial laser laboratory. These are the places where laser goggles are a must. Goggles are designed to protect eyes from accidental exposure both from the front and from the sides. They block only specific laser wavelengths or bands; for example, the blue–green band that includes the 488 and 514.5 nm argon-ion lines and the 532-nm line of doubled neodymium emitted by green pointers. The goggles should transmit as much light as possible at other wavelengths so the wearer can see well enough to work safely in the laboratory.

Anyone using lasers should be familiar with the basics of laser safety, but if you are going to work with Class IIIb or IV lasers, be sure to spend some time getting to know the details, particularly for the types of lasers you will be working with. The Web links below are starting points.

Table A.1. Laser safety classification and warning requirements in the United States for continuous-wave lasers at 400 to 1400 nm, somewhat simplified

Safety class	Max CW power	Warning class	Label text	Special requirements
1	up to 0.39 mW	None required	None required	None
IIa	0.39–1.0 mW	None required if exposure under 1000 seconds	None required	None
II	up to 1 mW	CAUTION, if potential exposure over 1000 seconds	Laser radiation—do not stare into beam	None
IIIa	up to 5 mW Irradiance under 2.5 mW/cm^2	CAUTION	Laser radiation—do not stare into beam or view directly with optical instruments	
IIIa	up to 5 mW Irradiance over 2.5 mW/cm^2	CAUTION	Laser radiation—avoid direct eye exposure	
IIIb	up to 500 mW	DANGER	Laser radiation—avoid direct exposure to beam	Key interlock, emission indicator
IV	over 500 mW	DANGER	Laser Radiation—avoid eye or skin exposure to beam	Key interlock, emission indicator; can't operate with cover off

British Standards <http://www.hpa.org.uk/radiation/laser/index.htm>

Laser Safety and the Eye <http://www.dermweb.com/laser/eye-safety.html>

Sam's Laser FAQ—Laser Safety <http://www.repairfaq.org/sam/lasersaf.htm>

U.S. Occupational Safety and Health Administration on Laser Hazards <http://www.osha.gov/SLTC/laserhazards/index.html>

APPENDIX B

HANDY NUMBERS AND FORMULAS

PHYSICAL CONSTANTS

Boltzmann constant (k): 1.380658×10^{-23} J/K (joule/Kelvin)
Boltzmann constant (k): 8.617385×10^{-5} eV/K
Electron mass: $9.1093897 \times 10^{-31}$ kilogram
Planck's constant (h): $6.6260755 \times 10^{-34}$ joule-second
Planck's constant (h): $4.1356692 \times 10^{-15}$ electronvolt-second
Proton mass: $1.6726231 \times 10^{-27}$ kilogram
Rydberg constant (R): $1.0973731534 \times 10^{7}$ per meter
Rydberg constant (R): 13.6056981 eV
Speed of light in vacuum (c): 2.99792458×10^{8} meters/sec

CONVERSIONS

Frequency in terms of wavelength and speed of light: $\nu = c/\lambda$
Wavelength in terms of frequency and speed of light: $\lambda = c/\nu$
Electronvolts to joules: 1 eV = $1.60217733 \times 10^{-19}$ joule
Photon energy and frequency: $E = h\nu$
Photon energy in joules: $E = h\nu = 6.63 \times 10^{-34}\ \nu$
A 1-eV photon has frequency: $2.41798836 \times 10^{14}$ hertz
Photon energy and wavelength: $E = \dfrac{hc}{\lambda}$
A 1-eV photon has wavelength: 1.2399 μm

SYMBOLS TO REMEMBER

λ (Greek lambda): wavelength
n refractive index
ν (Greek nu): frequency
ω (Greek omega): angular frequency = $2\pi\nu$
c: speed of light
h: Planck's constant
Δ (Greek delta): change or increment
θ (Greek theta): angle

IMPORTANT FORMULAS

Angle of refraction from Snell's law: $\theta_2 = \arcsin\left(\frac{n_1 \sin \theta_1}{n_2}\right)$

Approximate diffraction limit in radians for aperture D: λ/D

Decibel ratio of two powers: $dB = 10 \times \log_{10}\left(\frac{P_{\text{out}}}{P_{\text{in}}}\right)$

Far-field beam diameter at distance D for beam with divergence θ:
$2D \times \tan \theta$

Frequency as function of speed of light and wavelength: $\nu = c/\lambda$

Fresnel reflection at surface: $R = \left(\frac{n_1 - n_2}{n_1 + n_2}\right)^2$

Ratio of populations in states at energies E_1 and E_2:

$$\frac{N_2}{N_1} = \exp\left[\frac{-(E_2 - E_1)}{kT}\right]$$

Resonance condition in cavity of length L and refractive index n:
$2nL = N\lambda$

Snell's law of refraction: $n_1 \sin \theta_1 = n_2 \sin \theta_2$

Wavelength as function of speed of light and frequency: $\lambda = c/\nu$

Table B-1. Metric unit prefixes and their meanings

Prefix	Symbol	Multiple
exa	E	10^{18} (quintillion)
peta	P	10^{15} (quadrillion)
tera	T	10^{12} (trillion)
giga	G	10^{9} (billion)
mega	M	10^{6} (million)
kilo	k	10^{3} (thousand)
hecto	h	10^{2} (hundred)
deca	da	10^{1} (ten)
deci	d	10^{-1} (tenth)
centi	c	10^{-2} (hundredth)
milli	m	10^{-3} (thousandth)
micro	μ	10^{-6} (millionth)
nano	n	10^{-9} (billionth)
pico	p	10^{-12} (trillionth)
femto	f	10^{-15} (quadrillionth)
atto	a	10^{-18} (quintillionth)

APPENDIX C

RESOURCES AND SUGGESTED READINGS

FURTHER READING

Introductory Books on Lasers and Optics

J. Warren Blaker and Peter Schaeffer, *Optics: An Introduction for Technicians and Technologists* (Prentice-Hall, Upper Saddle River, NJ, 2000)

David Falk, Dieter Brill, and David Stork, *Seeing the Light: Optics in Nature, Photography, Color, Vision, and Holography* (Harper & Row, New York, 1986)

Grant R. Fowles, *Introduction to Modern Optics* (Dover, New York, 1989)

C. Breck Hitz, J. J. Ewing, and Jeff Hecht, *Introduction to Laser Technology,* 3rd ed. (IEEE Press, Piscataway, NJ, 2001)

Orazio Svelto, *Principles of Lasers,* 4th ed. (Springer, New York, 1998)

Advanced Books on Lasers and Optics

Eugene Hecht, *Optics,* 4th ed. (Addison-Wesley, Reading, MA, 2002)

Walter Koechner, *Solid-State Laser Engineering,* 2nd ed. (Springer-Verlag, Berlin, 1988)

Bahaa E. A. Saleh and Malvin Carl Teich, *Fundamentals of Photonics,* 2nd ed. (Wiley-Interscience, Hoboken, NJ, 2007)

Anthony Siegman, *Lasers* (University Science Books, Mill Valley, CA, 1986)

Warren J. Smith, *Modern Optical Engineering,* 3rd ed. (McGraw-Hill, New York, 2000)

Richard N. Zare, Bertrand H. Spencer, Dwight S. Springer, and Matthew P. Jacobson, *Laser Experiments for Beginners* (University Science Books, Sausalito, CA, 1995)

Understanding Lasers: An Entry-Level Guide, Third Edition. By Jeff Hecht

History of Laser Development

Joan Lisa Bromberg, *The Laser in America 1950–1970* (MIT Press, Cambridge, MA, 1991)
Jeff Hecht, *Beam: The Race to Make the Laser* (Oxford University Press, New York, 2005)
Theodore Maiman, *The Laser Odyssey* (Laser Press, Blaine, WA, 2000)
Nick Taylor, *Laser: The Inventor, the Nobel Laureate, and the 30-Year Patent War* (Simon & Schuster, New York, 2000)
Charles H. Townes, *How the Laser Happened: Adventures of a Scientist* (Oxford University Press, New York, 1999)

Laser Applications

Gerhard K. Ackermann and Jürgen Eichler, *Holography: A Practical Approach* (Wiley-VCH, Hoboken, NJ, 2007)
Jean-Claude Diels and Wolfgang Rudolf, *Ultrashort Laser Pulse Phenomena,* 2nd ed. (Academic Press, San Diego, CA 2006)
Wolfgang Demtröder, *Laser Spectroscopy,* 3rd ed. (Springer, New York, 2002)
Jeff Hecht, Beam Weapons: *The Next Arms Race* (Plenum, New York, 1984; reprinted Backinprint.com, 2000)
Jeff Hecht, *Understanding Fiber Optics,* 5th ed. (Pearson/Prentice-Hall, Upper Saddle River, NJ, 2006)
Nigel Hey, *The Star Wars Enigma* (Potomac Books, Dulles, VA, 2006)
National Research Council, *Harnessing Light: Optical Science and Engineering in the 21st Century* (National Academies Press, Washington, DC, 1998)
Markolf H. Niemz, *Laser–Tissue Interactions: Fundamentals and Applications,* 3rd ed. (Springer, New York, 2003)
John F. Ready, *Industrial Applications of Lasers* (Academic Press, San Diego, CA, 1997)
D. Sand, *Diode Lasers* (Taylor & Francis, London 2004)
William M. Steen and Kenneth Watkins, *Laser Material Processing,* 3rd ed. (Springer, New York, 2003)

Laser Safety

Roy Henderson and Karl Schulmeister, *Laser Safety* (Institute of Physics Publishing, Bristol and Philadelphia, 2004)

Magazines and Scholarly Journals

IEEE Journal of Quantum Electronics
IEEE Journal of Selected Topics in Quantum Electronics

IEEE LEOS Newsletter
IEEE Photonics Technology Letters
Journal of Lightwave Technology
Laser Focus World
Optics & Photonics News
Optics Express
Optics Letters
Photonics Spectra

Professional Societies

European Optical Society, http://www.europeanopticalsociety.org/
Institute of Electrical and Electronic Engineers, http://www.ieee.org
IEEE Laser and Electro-Optics Society, http://www.ieee.org/portal/site/leos/
Laser Institute of America, http://www.laserinstitute.org/
Optical Society of America, http://www.osa.org
SPIE, http://www.spie.org

Online Links

Sam's Laser FAQ, http://www.repairfaq.org/sam/lasersaf.htm
Sam's Safety Guidelines for high-voltage and/or line-powered equipment, http://www.repairfaq.org/sam/safety.htm
Sam's Laser FAQ–Laser Safety, http://www.repairfaq.org/sam/lasersaf.htm

GLOSSARY

Acoustooptic Interaction between an acoustic wave and a light wave, used in beam deflectors, modulators, and Q switches.

Active medium The light-emitting material in a laser, sometimes used to identify a specific element in a crystal.

Active species The atoms or molecules producing stimulated emission in a laser.

Alexandrite A synthetic crystal doped with chromium to form a tunable solid-state laser that emits near-infrared light.

Amplifier An optical device that increases the power of an input optical signal by stimulated emission, but which lacks resonator mirrors.

Angstrom (Å) A unit of length, 0.1 nanometer or 10^{-10} meter, abbreviated Å. It is not a standard SI unit, but is often used to measure wavelength in the visible spectrum.

Arc lamp A high-intensity lamp in which an electric discharge continuously produces light.

Attenuation Reduction of light intensity, or loss. It can be measured in optical density or decibels as well as percentage or fraction of light lost.

Attenuator An optical element that transmits only a given fraction of incident light.

Average power The average level of power in a series of pulses. It equals pulse energy times the number of pulses divided by the time interval.

Band gap The gap between valence band and conduction band in a semiconductor.

Beam diameter The distance between the edges of the beam, which are defined as the distance from the center where power drops to a certain level, often $1/e^2$ of the central power.

Beam divergence The angle at which a beam spreads.

Beam splitter A device that divides incident light into two separate beams, one reflected and one transmitted.

Birefringent Has a refractive index that differs for vertically and horizontally polarized light.

Brewster's angle The angle at which a surface does not reflect light of one linear polarization

Chemical laser A laser that is excited by a chemical reaction. The most common type produces hydrogen fluoride or deuterium fluoride.

Coating Material applied in one or more layers to the surface of an optical element to change the way it reflects or transmits light.

Coherence Alignment of the phase and wavelength of light waves with respect to each other. If the waves are perfectly aligned, the light is coherent.

Collimate Make light rays parallel.

Concave Curving inward, so that the central parts are deeper than the outside, like the inside of a bowl.

Conduction band Energy level in a solid in which electrons are not bound to individual atoms but are free to carry current through the solid.

Confocal Having the focal point of two mirrors at the same place. Confocal resonators have concave end mirrors that have the same focal point in the middle of the laser cavity.

Continuous wave Emitting a steady beam.

Convex Curving like the outside of a ball so the outer parts are lower than the center.

Cycles per second Number of oscillations a wave makes; the frequency. 1 hertz = 1 cycle per second.

Decibel A logarithmic comparison of power levels, abbreviated dB and defined as the value 10 $\log(P_2/P_1)$, or 10 times the base-10 logarithm of the ratio of the two power levels.

Detector A device that generates an electric signal when illuminated by light.

Dielectric Electrically nonconductive or insulating.

Difference frequency An output wave with frequency the difference of the frequencies of two input waves.

Diffraction Scattering or spreading of light waves when they pass an edge.

Diffraction limit The minimum possible spreading of light, a function of the wavelength divided by emitting area (λ/D).

Diode An electronic device that preferentially conducts current in one direction but not in the other. Semiconductor diodes contain a *p–n* junction between regions of different doping, which lets current to flow in one direction but not the other. Diodes can emit light (e.g., laser diodes) or detect it (photodiodes).

Diode laser A semiconductor laser in which a current flowing through the device causes electrons and holes to recombine at the junction of *p*- and *n*-doped regions, where stimulated emission (laser action) takes place.

Direct modulation Control of diode laser output by modulating the drive current.

Divergence The angular spreading of a laser beam with distance.

Doppler broadening Spreading of absorption or emission lines caused by motion of gas atoms or molecules

Electrooptic The interaction of light and electric fields, typically changing the light wave. Used in some modulators, Q switches, and beam deflectors.

Electromagnetic radiation Waves moving at the speed of light and made up of oscillating electrical and magnetic fields perpendicular to one another. Also behaves as photons or quanta of electromagnetic energy. Electromagnetic radiation includes radio waves, microwaves, infrared, visible light, ultraviolet, X-rays, and gamma rays.

Electromagnetic spectrum The range of wavelengths or frequencies at which electromagnetic radiation is emitted.

Electronic transition Change in energy level of an electron.

Etalon An optical resonator, typically with a short distance between reflective surfaces, placed inside laser resonators to restrict the range of wavelengths.

Excimer laser A pulsed ultraviolet laser in which the active medium is a short-lived molecule containing a rare gas such as xenon and a halogen such as chlorine.

Exciton An electron–hole pair in a semiconductor; the two are bound to each other, but the electron has not dropped from the conduction band to fill the hole in the valence band.

Extreme ultraviolet Wavelengths from 1 to 100 nm.

Far-infrared laser One of a family of gas lasers emitting light at the far-infrared wavelengths of 30 to 1000 micrometers.

Forward bias Voltage applied across a diode so it carries current easily.

Free-electron laser A laser in which stimulated emission comes from electrons passing through a magnetic field that varies in space.

Frequency For light waves, the number of wave peaks per second passing a point. Measured in hertz, or cycles per second.

Fused silica Synthetic silica (SiO_2) formed from highly purified materials.

Gain bandwidth Range of wavelengths over which a laser medium has gain.

Gallium aluminum arsenide A semiconductor used in LEDs, diode lasers, and certain detectors. Chemically, $Ga_{1-x}Al_xAs$, where x is a number less than one.

Gallium arsenide A semiconductor used in LEDs, laser diodes, detectors, and electronic components; chemically, GaAs.

Gallium nitride (GaN) A semiconductor used in blue, violet, and ultraviolet lasers and LEDs; often contains indium.

Glass An amorphous solid, typically made mostly of silica (SiO_2) unless otherwise identified. Silica glasses transmit visible light.

Harmonic generation Multiplication of the frequency of a light wave.

Hertz Frequency in cycles per second.

Heterojunction A boundary between semiconductors that differ in composition, such as GaAs and GaAlAs.

Index of refraction The ratio of the speed of light in vacuum to the speed of light in a material, a crucial measure of a material's optical characteristics Usually abbreviated n.

Indium gallium arsenide phosphide A semiconductor used in lasers, LEDs, and detectors. The band gap and, hence, the wavelength emitted by light sources and detected by detectors, depends on the mixture of the four elements. Abbreviated InGaAsP.

Inertial confinement fusion Nuclear fusion produced by focusing powerful laser pulses to implode targets containing hydrogen fuel.

Infrared Invisible wavelengths longer than 700 nm and shorter than about one millimeter. Wavelengths from about 700 nm to 2000 nm are called the near infrared.

InGaAsP Indium gallium arsenide phosphide, a semiconductor compound used in light sources and detectors. The band gap depends on composition, often written $In_{1-x}Ga_xAs_{1-y}P_y$, where x and y are numbers less than one.

Injection laser Another name for the semiconductor laser, derived from the fact that current carriers are injected to produce light.

Integrated optics Optical elements analogous to integrated electronic circuits, with multiple devices on one substrate.

Intensity Power per unit solid angle.

Interference The addition of the amplitudes of light waves. In destructive interference the waves cancel; in constructive interference they combine to make more intense light.

Irradiance Power per unit area.

Joule Unit of energy equal to one watt of power delivered for one second.

Junction The boundary between *p*- and *n*-type materials in a semiconductor, where positive and negative carriers recombine.

Junction laser A semiconductor diode laser.

Laser Acronym of light amplification by stimulated emission of radiation; one of the wide range of devices that generate light in a resonant cavity by that principle. Laser light is directional, covers a narrow range of wavelengths, and is more coherent than ordinary light.

Lattice constant Atomic spacing in a semiconductor crystal.

LED Light-emitting diode; a semiconductor diode that emits incoherent light by spontaneous emission.

Light Strictly speaking, light is electromagnetic radiation visible to the human eye. Commonly, the term is applied to electromagnetic radiation close to the visible spectrum that acts similarly, including the near infrared and near ultraviolet.

Light-emitting diode Abbreviation: LED. A semiconductor diode that produces incoherent light by spontaneous emission.

Longitudinal modes Oscillation modes of a laser along the length of its cavity, so twice the length of the cavity equals an integral number of wavelengths. Distinct from transverse modes, which are across the width of the cavity.

Maser Microwave analog of a laser, an acronym for microwave amplification by stimulated emission of radiation.

Metastable An excited energy level that has an unusually long lifetime.

Micrometer One millionth of a meter, abbreviated μm. The μ is a Greek mu.

Mode A manner of oscillation in a laser. Modes can be longitudinal or transverse.

Mode locking The locking together of many longitudinal modes in a laser cavity, producing a series of short laser pulses. Mode locking can be visualized as clumping together a group of photons that bounce back and forth in the laser cavity.

Monochromatic Containing only a single wavelength or frequency.

Multimode Containing multiple modes of light. Typically refers to lasers that operate in two or more transverse modes.

***n* region** Part of a semiconductor doped so it has an excess of electrons as current carriers.

Nanometer A unit of length equal to 10^{-9} meter. Its commonest use is to measure visible wavelengths.

Nanosecond A billionth of a second, 10^{-9} second.

Near infrared The part of the infrared nearest the visible spectrum, typically 700 to 1500 or 2000 nm, but not rigidly defined.

Negative lens A lens that spreads out or diverges light.

Nonlinear effect An effect proportional to the second or higher power of the input, as distinct from linear effects proportional to the first power of the input.

Normal (angle) Perpendicular to a surface.

Optical amplifier A laser without a resonant cavity; a device that amplifies light by stimulated emission but does not resonate.

Optical density Attenuation of light passing through a material, defined as $-\log_{10}$(Output/Input)

Orthogonal Perpendicular.

Oscillator A laser cavity with mirrors so stimulated emission can oscillate within it. To some purists, only an oscillator can be a laser.

Output mirror The mirror through which a laser emits its beam.

***p* region** Part of a semiconductor doped with electron acceptors in which "holes" (vacancies in the valence electron level) are the dominant current carriers.

Peak power Highest instantaneous power level in a pulse.

Phase Position of a wave in its oscillation cycle.

Photodetector A light detector.

Photodiode Usually, a semiconductor diode that produces an electrical signal proportional to light falling upon it. There also are vacuum photodiodes that can detect light.

Photometer An instrument for measuring the amount of light visible to the human eye.

Photons Quanta of electromagnetic radiation. Light can be viewed as either a wave or a series of photons.

Polarization Alignment of the electric and magnetic fields that make up an electromagnetic wave. Normally, this refers to the electric field. If light waves all have a particular polarization pattern, they are called polarized.

Polarization vector A vector indicating the direction of the electric field in an electromagnetic wave.

Polarizer A device that transmits light of only one polarization.

Population Inversion The condition when more atoms are in an upper energy level than in a lower one. A population inversion is needed for laser action.

Positive lens A lens that focuses light to a point

Pumping The way a laser gets the energy to produce a population inversion.

Q factor Quality factor of a resonant laser cavity, a measure of loss with in cavity.

Q switch A device that changes the Q (quality factor) of a laser cavity to produce a short, powerful pulse.

Quantum Divided into discrete pieces or levels. A photon is a quantum of light energy. Quantum levels are discrete states with specific values of energy.

Quantum cascade laser A semiconductor laser in which energy is extracted in a series of steps as electrons pass through a series of quantum wells.

Quantum well A thin layer in a semiconductor diode with smaller bandgap than the layers above and below it, which traps electrons that lack the energy needed in the adjacent higher-bandgap layers.

Quartz A natural crystalline form of silica (SiO_2).

Quaternary A compound made of four elements, for example, InGaAsP.

Radian A unit of angular measure; 2π radians equals 360°, a circle.

Radiant flux Instantaneous power level in watts.

Radiometer An instrument to measure power (watts) in electromagnetic radiation. Distinct from a photometer, which measures light perceived by the human eye.

Raman scattering Scattering of light by an atom that changes its vibrational state during the scattering process, so the scattered wavelength is shifted from that of the input photon.

Rays Straight lines that represent the path taken by light.

Real image An image that can be projected onto a surface.

Recombination In a semiconductor laser, dropping of a conduction-band electron into a vacancy in the valence band of an atom. Free electrons also can recombine with gas atoms by dropping into a bound energy state.

Refraction The bending of light as it passes between materials of different refractive index.

Refractive index The ratio of the speed of light in vacuum to the speed of light in a material, a crucial measure of a material's optical characteristics. Abbreviated n.

Repetition rate Number of pulses per second.

Resonator A region with mirrors on the ends and a laser medium in the middle. Stimulated emission from the laser medium resonates between the mirrors, one of which lets some light emerge as a laser beam.

Retroreflector An optical device that reflects incident light back in precisely the direction from which it came. It typically is a prism or a cluster of three mirrors forming the corner of a cube.

Reverse bias Voltage applied across a diode so it does not carry current.

Semiconductor diode laser A laser in which recombination of current carriers at a *p–n* junction generates stimulated emission.

Silica Silicon dioxide, SiO_2, the major constituent of ordinary glass.

Silica glass Glass in which the main constituent is silica.

Single-frequency laser A laser that emits only a very narrow range of wavelengths, nominally a single frequency but actually a very narrow range small enough to be considered a single frequency.

Single mode Containing only a single mode. Beware of ambiguities because of the difference between transverse and longitudinal modes. A laser operating in a single transverse mode often does not operate in a single longitudinal mode.

Slope efficiency The fraction of each additional watt of drive power turned into laser output above a threshold.

Solid-state laser A laser made of a nonconductive solid that contains atoms that produce stimulated emission when excited by light from an external source. Different from semiconductor lasers.

Speckle Coherent noise produced by laser light. It gives a mottled appearance to holograms viewed in laser light.

Spectral reflection Reflection from a mirror-like smooth surface.

Spectroscopy Study of the wavelengths emitted and absorbed by materials.

Spontaneous emission Emission of a photon without outside stimulation when an atom or molecule drops from a high-energy state to a lower one.

Stimulated emission Emission of a photon that is stimulated by another photon of the same energy; the process that makes laser light.

Ternary Compound made of three elements, for example, GaAlAs.

III–V semiconductor A semiconductor compound made of one (or more) elements from the IIIA column of the periodic table (Al, Ga, and In) and one (or more) elements from the VA column (N, P, As, or Sb). Used in LEDs, diode lasers, and detectors.

Threshold The excitation level at which laser emission starts.

Threshold current The minimum current needed to sustain laser action, usually in a semiconductor laser.

Time response The time it takes to react to a change in signal level.

Total internal reflection Total reflection of light back into a material when it strikes the interface with a material having lower refractive index at a glancing angle.

Transition Shift between energy levels.

Transverse modes Modes across the width of laser. Distinct from longitudinal modes, which are along the length.

Tunable Adjustable in wavelength.

Ultraviolet Part of the electromagnetic spectrum at wavelengths shorter than 400 nm to about 10 nm, invisible to the human eye.

Valence band Energy levels of outer electrons in an atom, which form bonds to other atoms in a solid. Electrons are bound to the atom.

VCSEL Vertical-cavity surface-emitting laser, an important type of semiconductor laser.

Vibronic A electronic transition accompanied by a change in vibrational energy level.

Virtual image An image that can be seen by the eye but cannot be projected onto a surface.

Visible light Electromagnetic radiation visible to the human eye, at wavelengths of 400 to 700 nm.

Waveguide A structure that guides electromagnetic waves along its length. An optical fiber is an optical waveguide.

Wavelength The distance an electromagnetic wave travels during one cycle of oscillation. Wavelengths of light usually are measured in nanometers (10^{-9} meter) or micrometers (10^{-6} meter). The standard symbol is λ (Greek lambda).

X-ray A photon emitted when an electron drops into an inner shell of an atom. Typically, wavelengths are 0.03 to 10 nanometers, but the boundaries are not well defined.

YAG Yttrium aluminum garnet, a crystalline host for neodymium lasers.

YLF Yttrium lithium fluoride ($YLiF_4$), a solid-state laser host.

INDEX

Understanding Lasers: An Entry-Level Guide, Third Edition. By Jeff Hecht